Chemistry and Safety
of Acrylamide in Food

ADVANCES IN EXPERIMENTAL MEDICINE AND BIOLOGY

Editorial Board:
NATHAN BACK, *State University of New York at Buffalo*
IRUN R. COHEN, *The Weizmann Institute of Science*
DAVID KRITCHEVSKY, *Wistar Institute*
ABEL LAJTHA, *N. S. Kline Institute for Psychiatric Research*
RODOLFO PAOLETTI, *University of Milan*

Recent Volumes in this Series

Volume 553
BIOMATERIALS: From Molecules to Engineered Tissues
Edited by Nesrin Hasırcı and Vasıf Hasırcı

Volume 554
PROTECTING INFANTS THROUGH HUMAN MILK: Advancing the Scientific Evidence
Edited by Larry K. Pickering, Ardythe L. Morrow, Guillermo M. Ruiz-Palacios, and Richard J. Schanler

Volume 555
BREAST FEEEDING: Early Influences on Later Health
Edited by Gail Goldberg, Andrew Prentice, Ann Prentice, Suzanne Filteau, and Elsie Widdowson

Volume 556
IMMUNOINFORMATICS: Opportunities and Challenges of Bridging Immunology with Computer and Information Sciences
Edited by Christian Schoenbach, V. Brusic, and Akihiko Konagaya

Volume 557
BRAIN REPAIR
Edited by M. Bähr

Volume 558
DEFECTS OF SECRETION IN CYSTIC FIBROSIS
Edited by Carsten Schultz

Volume 559
CELL VOLUME AND SIGNALING
Edited by Peter K. Lauf and Norma C. Adragna

Volume 560
MECHANISMS OF LYMPHOCYTE ACTIVATION AND IMMUNE REGULATION X: INNATE IMMUNITY
Edited by Sudhir Gupta, William Paul, and Ralph Steinman

Volume 561
CHEMISTRY AND SAFETY OF ACRYLAMIDE IN FOOD
Edited by Mendel Friedman and Don Mottram

A Continuation Order Plan is available for this series. A continuation order will bring delivery of each new volume immediately upon publication. Volumes are billed only upon actual shipment. For further information please contact the publisher.

Chemistry and Safety of Acrylamide in Food

Edited by

Mendel Friedman
Agricultural Research Service, USDA
Albany, California

and

Don Mottram
University of Reading
Reading, United Kingdom

Library of Congress Cataloging-in-Publication Data

Chemistry and safety of acrylamide in food/edited by Mendel Friedman and Don Mottram.
    p.   cm. (Advances in experimental medicine and biology;. v. 561)
  Includes bibliographical references and index.
  ISBN 0-387-23920-0
    1. Acrylamide—Toxicology. 2. Food—Toxicology. I. Friedman, Mendel. II. Mottram D. S. (Donald S.). III. Series

RA1242.A44C447 2005
615.9′54—dc22

2005040244

Proceedings of ACS Symposium on "Chemistry and Safety of Acrylamide in Food" held at the American Chemical Society Meeting in Anaheim, California, March 28–April 1, 2004

ISSN: 0065 2598
ISBN-10: 0-387-23920-0 (Hardbound)   Printed on acid-free paper.
ISBN-13: 987-0387-23920-0

©2005 Springer Science+Business Media, Inc.
All rights reserved. This work may not be translated or copied in whole or in part without the written permission of the publisher (Springer Science+Business Media, Inc., 233 Spring Street, New York, NY 10013, USA), except for brief excerpts in connection with reviews or scholarly analysis. Use in connection with any form of information storage and retrieval, electronic adaptation, computer software, or by similar or dissimilar methodology now known or hereafter developed is forbidden.
The use in this publication of trade names, trademarks, service marks and similar terms, even if they are not identified as such, is not to be taken as an expression of opinion as to whether or not they are subject to proprietary rights.

Printed in the United States of America

9 8 7 6 5 4 3 2 1

springeronline.com

PREFACE

Acrylamide (CH_2=CH-$CONH_2$), an industrially produced conjugated reactive molecule, is used worldwide to synthesize polyacrylamide. The polymer has found numerous applications as a soil conditioner, in wastewater treatment, in the cosmetic, paper, and textile industries, and in the laboratory as a solid support for the separation of proteins by electrophoresis. Monomeric acrylamide is also widely used by researchers as an alkylating agent for the selective modification of protein SH groups and in fluorescence studies of tryptophan residues in proteins. Because of the potential of exposure by individuals to acrylamide, effects of acrylamide in cells, tissues, animals, and humans have been extensively studied. Neurotoxicity, reproductive toxicity, genotoxicity, clastogenicity (chromosome-damaging effects), and carcinogenicity have been demonstrated to be potential human health risks that may be associated with exposure to acrylamide.

In 2002, reports that acrylamide was found in plant-derived foods at levels up to 3 mg/kg, formed during their processing under conditions that also induce the formation of Maillard browning products, has resulted in heightened worldwide interest in the chemistry and safety of acrylamide. The recent realization that exposure of humans to acrylamide can come from the diet as well as from external sources has pointed to the need for developing a better understanding of its formation and distribution in food and how its presence in the diet may affect human health. A better understanding of the chemistry and biology of pure acrylamide in general and its impact in a food matrix in particular, can lead to the development of improved food processes to decrease the acrylamide content and thus the safety of the diet.

To contribute to this effort, we organized a Symposium on the "Chemistry and Safety of Acrylamide in Food". The three-day Symposium, which was the first international symposium on the subject, took place in Anaheim, California on March 29-31, 2004 and attracted 34 speakers from 8 countries. Because 'acrylamide' encompasses many disciplines, in organizing the Symposium, we sought participants with a wide range of interests, yet with a common concern for basic and applied aspects of acrylamide chemistry and safety.

The Proceedings of the Symposium are published as a volume in the series *Advances in Experimental Medicine and Biology*. We would like to emphasize the diversity of the subject matter and of the contributors' backgrounds and interests presented in this volume. The widest possible viewpoints and interactions of ideas are needed to transcend present limitation in our knowledge and to catalyze progress in all areas that are relevant to the human diet. These include mechanisms of formation of acrylamide in food, distribution of acrylamide and its precursors in foods, analysis, impact of other biologically active dietary ingredients on the safety of acrylamide, toxicology, pharmacology and metabolism, epidemiology, and risk assessment. This volume brings together all elements needed for such interactions. The content includes a great variety of specific and general topics listed in the following order: toxicology/epidemiology/risk assessment; mechanisms; kinetics; analysis; acrylamide reduction; formation in different foods. It is our hope that the reader will examine not only those articles of primary interest but others as well and so profit from a broad overview. The most important function of this volume, we believe, is dissemination of insights and exchange of ideas so as to permit synergistic interaction among related disciplines. This volume brings together elements needed for such interaction.

We are grateful to all contributors for their help in bringing this volume to fruition, to Carol E. Levin for assistance with formatting of manuscripts and to Wallace Yokoyama, Program Chair of the Division of Agricultural and Food Chemistry of the American Chemical Society for inviting us to organize the Symposium.

We hope that "Chemistry and Safety of Acrylamide in Food" will be a valuable record and resource for further progress in this very active interdisciplinary field. We are confident that the effort of all concerned will be most worthwhile and rewarding.

Mendel Friedman
Albany, California
January 10, 2005

Don Mottram
Reading, UK
January 10, 2005

CONTENTS

Acrylamide in Food: The Discovery and its Implications 1
    Margareta Törnqvist

Acrylamide Neurotoxicity: Neurological, Morphological and
    Molecular Endpoints in Animal Models 21
    Richard M. LoPachin

The Role of Epidemiology in Understanding the Relationship
    Between Dietary Acrylamide and Cancer Risk in
    Humans 39
    Lorelei A. Mucci and Hans-Olov Adami

Mechanisms of Acrylamide Induced Rodent Carcinogenesis 49
    James E. Klaunig and Lisa M. Kamendulis

Exposure to Acrylamide.. 63
    Barbara J. Petersen and Nga Tran

Acrylamide and Glycidamide: Approach Towards Risk
    Assessment Based on Biomarker Guided Dosimetry of
    Genotoxic/Mutagenic Effects in Human Blood............... 77
    Mathias Baum, Evelyne Fauth, Silke Fritzen, Armin
    Herrmann, Peter Mertes, Melanine Rudolphi, Thomas

Spormann, Heinrich Zankl, Gerhard Eisenbrand and Daniel Bertow

Pilot Study of the Impact of Potato Chips Consumption on Biomarkers of Acrylamide Exposure.............................. 89
Hubert W. Vesper, Hermes Licea-Perez, Tunde Myers, Maria Ospina, and Gary L. Mayers

LC/MS/MS Method for the Analysis of Acrylamide and Glycidamide Hemoglobin Adducts................................. 97
Maria Ospina, Hubert W. Vesper, Hermes Licea-Perez, Tunde Myers, Luchuan Mi, and Gary L. Mayers

Comparison of Acrylamide Metabolism in Humans and Rodents... 109
Timothy R. Fennell and Marvin A. Friedman

Kinetic and Mechanistic Data for a Human Physiologically Based Pharmacokinetic (PBPK) Model for Acrylamide . 117
Melvin E. Andersen, Joseph Scimeca, and Stephen S. Olin

In Vitro Studies of the Influence of Certain Enzymes on the Detoxification of Acrylamide and Glycidamide in Blood 127
Birgit Paulsson, Margareta Warholm, Agneta Rannug, and Margareta Törnqvist

Biological Effects of Maillard Browning Products that May Affect Acrylamide Safety in Food 135
Mendel Friedman

Acrylamide: Formation in Different Foods and Potential Strategies for Reduction.. 157
Richard H. Stadler

Mechanisms of Acrylamide Formation: Maillard-Induced Transformations of Asparagine.. 171
I. Blank, F. Robert, T. Goldmann, P. Pollien, N. Varga, S. Devaud, F. Saucy, T. Hyunh-Ba, and R. H. Stadler

CONTENTS

Mechanistic Pathways of Formation of Acrylamide from
    Different Amino Acids... 191
  Varoujan A. Yalayan, Carolina Perez Locas, Andrzej
    Wronowski and John O'Brien

New Aspects on the Formation and Analysis of Acrylamide...... 205
  Peter Schieberle, Peter Köhler and Michael Granvogl

Formation of Acrylamide from Lipids... 223
  Stefan Ehling, Matt Hengel and Kakyuki Shibamoto

Kinetic Models as a Route to Control Acrylamide Formation in
    Food 235
  Bronislaw L. Wedzicha, Donald S. Mottram, J. Stephen
    Elmore, Georgios Koutsidis and Andrew T. Dodson

The Effect of Cooking on Acrylamide and its Precursors in
    Potato, Wheat, and Rye.. 255
  J. Stephen Elmore, Georgios Koutsidis, Andrew T. Dodson
    and Donald S. Mottram

Determination of Acrylamide in Various Food Matrices Using a
    Single Extract; Evaluation of LC and GC Mass
    Spectrometric Method... 271
  Adam Becalski, Banjamin P.-Y. Lau, David Lewis, Stephen
    W. Seaman, and Wing F. Sun

Some Analytical Factors Affecting Measured Levels of
    Acrylamide in Food Products 285
  Sune Eriksson and Patrik Karlsson

Analysis of Acrylamide in Food... 293
  Reinhard Matissek and Marion Raters

On Line Monitoring of Acrylamide Formation 303
  David J. Cook, Guy A. Channell, and Andrew J. Taylor

Factors That Influence the Acrylamide Content of Heated
    Foods... 317

Per Rydberg, Sune Eriksson, Eden Tareke, Patrik Karlsson, Lar Ehrenberg, and Margareta Törnqvist

Model Systems for Evaluating Factors Affecting Acrylamide Formation in Deep Fried Foods 329
R. C. Lindsay and S. Jang

Controlling Acrylamide in French Fry and Potato Chip Models and a Mathematical Model of Acrylamide Formation 343
Yeonhwa Park, Heewon Yang, Jayne M. Storkson, Karen J. Albright, Wei Liu, Robert C. Lindsay, and Michael W. Pariza

Quality Related Minimization of Acrylamide Formation – An Integrated Approach.. 357
Knut Franke, Marco Sell, and Ernst H. Reimerdes

Genetic, Physiological, and Environmental Factors Affecting Acrylamide Concentration in Fried Potato Products 371
Erin M. Silva and Philipp W. Simon

Acrylamide Reduction in Processed Foods 387
A. B. Hanley, C. Offen, M. Clarke, B. Ing, M. Roberts and R. Burch

Chemical Intervention Strategies for Substantial Suppression of Acrylamide Formation in Fried Potato Products 393
Robert C. Lindsay and Sungjoon Jang

Acrylamide in Japanese Processed Foods and Factors Affecting Acrylamide Level in Potato Chips and Tea 405
Misuro Yoshida, Hiroshi Ono, Yoshimiro Chuda, Hiroshi Yada, Mayumi Ohnishi-Kameyama, Hidetaka Hobayashi, Akiko Ohara-Takada, Chie Matsura-Endo, Motoyuki Mori, Nobuyuki Hayashi, and Yuichi Yamaguchi

The Formation of Acrylamide in UK Cereal Products 415
Peter Sadd and Colin Hamlet

Factors Influencing Acrylamide Formation in Gingerbread 431
  Thomas M. Amrein, Barbara Schonbachler, Felix Escher,
  and Renato Amado

Effects of Consumer Food Preparation on Acrylamide
  Formation 447
  Lauren S. Jackson and Fadwa Al-Taher

Index 467

ACRYLAMIDE IN FOOD: THE DISCOVERY AND ITS IMPLICATIONS
A Historical Perspective

Margareta Törnqvist
Dept. of Environmental Chemistry, Stockholm University, SE-106 91 Stockholm, Sweden; e-mail: margareta.tornqvist@mk.su.se

Abstract: The unexpected finding that humans are regularly exposed to relatively high doses of acrylamide (AA) through normal consumption of cooked food was a result of systematic research and relevant developments in methodology over decades, as well as a chain of certain coincidences. The present paper describes the scientific approach, investigations and events leading to the discovery of the formation of AA during cooking of foods. In addition, related issues concerning assessment, communication and management of cancer risks and associated ethical questions raised by the finding of the presence of AA in foods will be discussed.

Key words: Acrylamide; glycidamide; cooking; hemoglobin adducts; health risk; occupational exposure; grouting

1. INTRODUCTION

1.1 Acrylamide, a potential genotoxic factor in vivo

Acrylamide (AA) is an electrophilically reactive unsaturated amide containing an activated double bond. In vivo, AA is metabolized through oxygenation of the double bond to the epoxide glycidamide, which also possesses electrophilic reactivity (Calleman et al., 1990; Bergmark et al., 1993). Both compounds are reactive towards nucleophilic sites such as sulfhydryl and amino groups forming adducts, e.g. in proteins (see Fig. 1) (reviewed by Friedman, 2003). Compared to glycidamide, AA has relatively higher reactivity towards sulfhydryl groups (Bergmark et al., 1993) and low reactivity towards DNA (Solomon et al., 1985). Glycidamide, however, has been shown to form adducts with DNA amino groups (Segerbäck et al., 1995; Gamboa da Costa et al., 2003).

For a long time it was known that chemical mutagens, i.e. genotoxic compounds, and most known carcinogens are metabolized to reactive electrophiles (Miller and Miller, 1966). The reverse of this observation, which also appears to be valid, means that AA has the ability of a metabolite to form DNA adducts. It may, therefore, constitute a potential genotoxic and cancer risk-increasing agent. In fact, AA has been shown to be an animal carcinogen. It is classified as a "probable human carcinogen" by IARC, and has also been shown to be a neurotoxic agent in animals and humans (IARC 1994; EC, 2000).

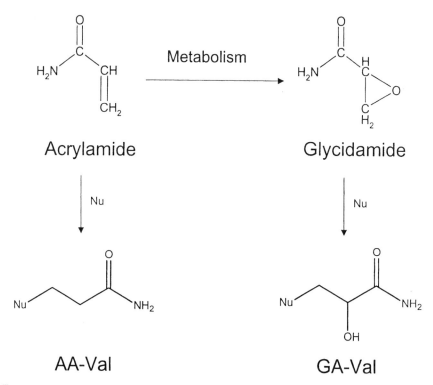

Figure 1. Acrylamide is metabolized to glycidamide. Both componds are electrophilic and thus able to form adducts with nucleophilic sites, in proteins.

1.2 Approach for risk assessment of genotoxic chemicals

Our cross-scientific research group is concerned with the development and application of methods and models for detection and quantification of cancer risks from electrophilically reactive compounds. The research line was initiated in the 1970-ies by Lars Ehrenberg at Stockholm University, and is to a large extent stimulated by experience from radiation biology (Ehrenberg et al., 1974). With the aim to transfer the experience from radiological protection, where measurement of dose is essential, a dose concept was defined for chemical carcinogens. With dose defined as the integral over time of the concentration of the electrophilic compound/metabolite, it became meaningful to explore the same models for cancer risk estimation of genotoxic chemicals as used for ionizing radiation, and to study dose-response relationships (Ehrenberg et al., 1983). The dose could also be denoted as the "Area Under the Curve"; *cf.* Törnqvist and Hindsø Landin (1995).

1.3 Measurement of electrophilic and genotoxic compounds in vivo through adducts

To prevent exposure to potential mutagens/carcinogens, it is desirable to be able to detect, identify and measure doses of genotoxic compounds/metabolites *in vivo*. Electrophiles undergo chemical and enzymatic reactions and are eliminated from tissues with short half-lives and are thus not accumulated. Furthermore, compounds like AA and glycidamide are hydrophilic and have no specific property that facilitates isolation and detection. An analytical approach was therefore developed, which shifted to analysis of stable reaction products of electrophiles with biomacromolecules (Ehrenberg et al., 1974).

The long life span renders hemoglobin (Hb) in red blood cells a useful monitor molecule for electrophiles (Osterman Golkar et al., 1976). Several methods for measuring Hb adducts have been developed (reviewed by Törnqvist et al., 2002). In addition to cysteines and histidines, N-termini, valines are major reactive nucleophilic sites in human Hb. In early work, adducts to histidines in the interior of the Hb chain were measured (Calleman et al., 1978). The requirements of a simpler, faster and more sensitive method initiated the development of a modified Edman protein degradation method for analysis of adducts to N-termini of the Hb chains (Jensen et al., 1984). The developed method (Törnqvist et al., 1986), with analysis by gas chromatography – mass spectrometry (GC-MS), was the result of overcoming many difficulties (Törnqvist, 1989). This so-called N-alkyl Edman method came into use for studies of a range of low-molecular weight electrophiles in exposed animals and humans. It was a major a break-through in this area (reviewed by Törnqvist et al., 2002). Using this method as a sensitive quantitative tool for in vivo studies of electrophiles was the basis for work leading to the detection of AA formation in heated foods. Fig. 2 outlines the N-alkyl Edman method.

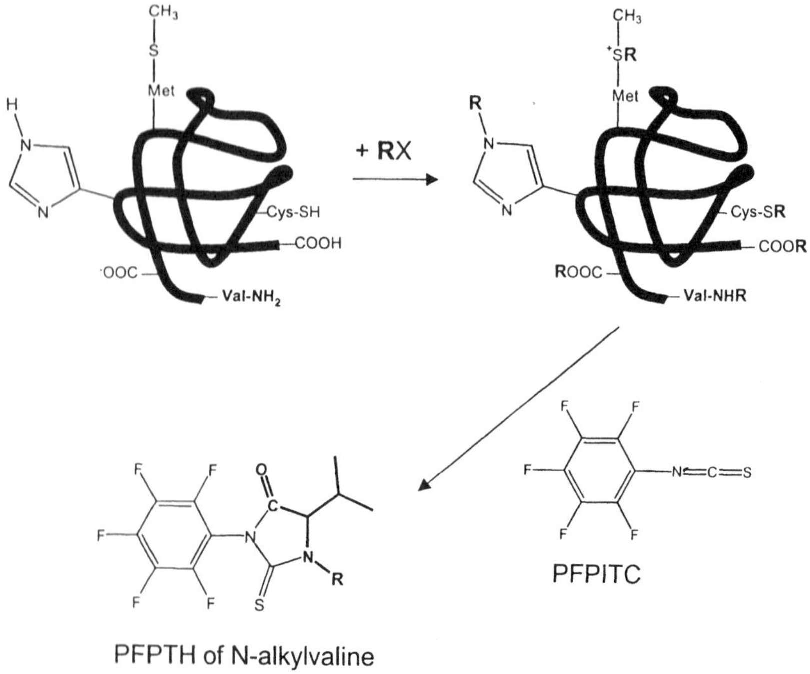

Figure 2. Outline of the N-alkyl Edman method for adducts to N-terminal valines in hemoglobin (Hb). An electrophile (RX) forms adducts to different nucleophilic sites in Hb, e.g. N-terminal valines. The reagent pentafluorophenyl isothiocyanate (PFPITC) detaches and derivatizes the N-terminal N-alkylvaline in one step. The pentafluorophenylthiohydantoin (PFPTH) of the N-alkylvaline is then isolated from the rest of the peptide and the unalkylated N-termini still bound to the peptide.

1.4 Background adducts - demonstration of exposure to electrophiles and genotoxic risks?

In investigations of individuals with exposure to chemical carcinogens (workers and smokers), background levels of Hb adducts from studied compounds have often been observed in control persons without known exposure (review: Törnqvist et al., 2002). The question arose whether the observation of the so-called background Hb adducts truly reflect exposures to electrophiles (Törnqvist, 1988, 1989). The relevance of the observations was investigated using as a model the background adduct to N-terminal valine in Hb from ethylene oxide. Analysis of adducts to N-termini by the N-alkyl Edman method was shown to be advantageous due to reduced risk of artefact formation. The occurrence of a true background of the adduct from ethylene oxide was demonstrated, which reflected exposure from endogenously produced ethylene metabolized to ethylene oxide (reviewed in Törnqvist and Kautiainen, 1993; Törnqvist 1996). Intestinal flora was shown to be one important determinant of adduct formation (Törnqvist et al., 1989; Kautiainen et al., 1993).

The demonstration of the occurrence of an electrophilic agent through adducts to proteins in vivo implies there is also a certain reactivity towards critical sites in DNA. The assumption that there is no threshold in dose-risk relationships of mutagens/genotoxic agents implies that the demonstration of an adduct reflects a genotoxic risk (Ehrenberg et al, 1996).

In the case of the background exposure from endogenously formed ethylene oxide, classified as a human carcinogen (IARC, 1994), comprehensive studies led to the estimation that the corresponding average lifetime doses imply a non-negligible cancer risk (Törnqvist, 1996).

This work, particularly the demonstration that there exists a true background Hb adduct from ethylene oxide, which reflects a true exposure, led to the formulation of two questions (Törnqvist, 1988, 1989). That is, could unknown general exposures to electrophiles (and cancer risk enhancing agents) in the population be detected and identified as background Hb adducts? Furthermore, could background Hb adducts be used to identify unknown contributors to today's cancer incidence? Thus far, a range of background adducts to N-terminal valines in Hb from simple epoxides and aldehydes have been identified with the N-alkyl Edman method (Törnqvist and Kautiainen, 1993). This research line also contributed to the finding of the general background exposure to AA through the consumption of cooked food.

2. BASIC WORK ON ACRYLAMIDE

2.1 Pioneering quantitative work based on Hb adduct measurements

Pioneering studies on AA concerning metabolism, pharmaco-kinetics and dose-response relationships for neurotoxic symptoms were done by the former members of the research group at Stockholm University: C.J. Calleman and E. Bergmark. The human studies concerned workers at a

Chinese factory producing monomeric AA and AA polymers. The dose of AA in exposed humans was measured by the above-mentioned modified Edman method as adducts from AA to N-terminal valine in Hb (at levels up to 34 nanomol/g Hb). A no-observed-adverse-effect level (NOAEL) for neurotoxic symptoms was estimated (Calleman et al., 1994; Calleman, 1996). Furthermore, they demonstrated, through Hb adducts, that glycidamide was formed as a metabolite of AA in exposed animals (Calleman et al., 1990) and humans (Bergmark et al., 1993). It was shown that glycidamide, but not AA, gives rise to detectable DNA adduct levels in rodents exposed to AA (Segerbäck et al., 1995). Fig. 1 shows adducts formed from AA and glycidamide, respectively.

Bergmark followed up her doctoral work with an investigation of AA exposure of laboratory workers using polyacrylamide gels (Bergmark, 1992; 1997). In the control group, a background level of adducts from AA was observed (about 30 picomol/g Hb), with an increased level in smokers (about 6 picomol/g Hb per cigarette per day). A higher level in smokers was not unexpected due to the occurrence of AA in tobacco smoke (Schumacher et al., 1977). However, no obvious explanation could be given for the relatively high background level found in non-smokers. The author of this paper and Bergmark carried out a collaborative study on background levels of Hb adducts from AA. In contrast to other studied adducts, e.g. from ethylene oxide, no clues to the background adduct levels were found in the case of AA (studies of twins, to be published).

Results on reaction-kinetic and pharmaco-kinetic parameters from the investigations on AA by Calleman (1996) and Bergmark et al. (1993) stimulated quantitative reasoning that led to the examination of heated food as a source of the general background exposure to AA.

2.2 Investigation of acrylamide exposure due to leakage from a tunnel construction work

In 1997, the competence within the research group was challenged with regard to measurement of in vivo doses of electrophiles and health risk assessment, as well as the above-mentioned knowledge collected concerning AA. Construction work of a railway tunnel through the mountain ridge Hallandsås in the south-west of Sweden initiated in 1992 encountered many problems, including heavy water inflow into the tunnel, causing delays in the construction work (described in Tunnel Commission, 1998). For sealing the tunnel walls, a chemical grout containing the monomers AA and N-methylolacrylamide was used in large quantitites from August 1997. Water from the tunnel was pumped into Vadbäcken, a rivulet close to one entrance of the tunnel construction. At the end of September there were signs that something was wrong; in down-stream Vadbäcken dead fish, in a fish culture, and paralysed cows, in a herd drinking water from the rivulet, were found. Could there be a leakage of the grouting agent into the environment?

We were contacted about AA analysis in water. We suggested that analysis of Hb adducts should be done to clarify whether AA was the cause of the observed effects in animals. Thanks to the earlier work of E. Bergmark (who had left the university) and the availability of methodology, it was possible to carry out adduct analysis relatively fast. The analysis of Hb adducts from AA in poisoned cattle and fish showed very high levels (Godin et al., 2002), consistent with the (unbelievably) high levels (about 100 mg/L) of AA and N-methylolacrylamide measured in the contaminated rivulet (Tunnel Commission, 1998). It was then obvious that the injected grouting agent had not polymerized properly and that the acrylamide monomers leaked out from the tunnel walls. The use of the chemical grout was immediately stopped. However, an acute and uncontrollable situation had already developed, with leakage of AA into the ground water. The local authority declared a state of emergency and risk areas were defined. A massive coverage by media was also took place. (Sequence of events described by the Tunnel Commission, 1998.)

In this situation, it appeared urgent to clarify the exposure situation for the more than 200 tunnel workers and also for residents in the area. We were contacted by Prof. L. Hagmar at University Hospital in Lund in this matter. Blood sampling for Hb adduct measurements was initiated in the beginning of October. The situation involved potential (and terminated) exposure and uptake of AA and N-methylolacrylamide both through the skin and from ingestion, and for tunnel workers also through inhalation. The only way of measuring the uptake of AA was to use analysis of Hb adducts. The first preliminary study disclosed exposure levels in tunnel workers that exceeded an estimated NOAEL for symptoms of impairment of the peripheral nervous system (PNS) (AA adduct level 0.3 – 1 nanomol/g Hb) (Törnqvist et al., 1998, with data from Calleman et al., 1994). In the complete investigation, it was shown that about 160 workers had increased adduct levels (0.07 – 17.7 nanomol/g Hb) from AA exposure (Hagmar et al., 2001). It should be noted that AA and N-methylolacrylamide were shown to give rise to the same Hb adduct (Tareke, 1998; Paulsson et al., 2002).

Fifty tunnel workers claimed neurotoxic symptoms and for 23 of them there was strong evidence for PNS impairment due to the occupational exposure to acrylamides (see Fig. 3). The anxiety among concerned persons made it necessary to present an assessment of cancer risks. The risk assessment showed that the probability was small that anyone of the workers would contract a cancer due to the occupational AA exposure (Törnqvist et al., 1998; and to be published). The communication of the results of the exposure investigations and health risk estimations was a challenging issue.

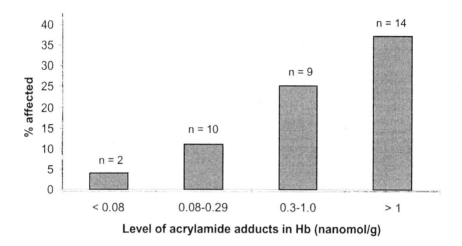

Figure 3. Neurotoxic effects and acrylamide exposure in tunnel workers (from Hagmar et al., 2001). Mild symptoms of impairment of peripheral nervous system (numbness and tingling in feet and legs) at different exposure levels. Percentage and number (n) of persons affected out of the total number of persons in the respective exposure group.

The leakage of AA from the tunnel construction and the contamination of the environment, the potential exposure of residents and the high exposure of workers led to discontinuation of the tunnel work and turmoil of events. Media reported on everything. Besides a questioning the justification of the tunnel construction, the leakage of the grouting agent had economical consequences, prosecution and legal consequences, and was followed by many investigations and debates of a range of different issues. The leakage of the grouting agent was denoted a scandal. The construction of the tunnel restarted in 2004 (using concrete lining instead of chemical grout) is expected to be complete in 2012.

It is interesting that during autumn 1997, food products from the risk zones at Hallandsås, due to possible contamination by AA, were not allowed to be offered for sale. They were destroyed as a precautionary measure. These included milk and butter from the area, vegetables and potato irrigated with possibly contaminated water, and cattle, which had used Vadbäcken as a drinking water supply. Furthermore, for a rather long time after abolishment of the risk zones, there was a buyers' resistance to food products, such as potatoes, from the area. One difficult problem was the leakage of AA and N-methylolacrylamide into ground water and wells. The WHO drinking water guideline (1996) value of 0.5 µg/L was applied and it was recommended that water with concentrations of AA above this level should not be used as drinking water. This problem lasted until 1998.

The consequences and the heavy media coverage of the Hallandsås "scandal" led to AA becoming well-known and perceived as a highly toxic chemical by the general public in Sweden.

3. TRACING A GENERAL BACKGROUND EXPOSURE TO ACRYLAMIDE

3.1 Significance of the background Hb-acrylamide adduct

The clarification of the exposure situation at Hallandsås also involved studies of residents and of control persons living outside the area (in collaboration with University Hospital in Lund). In agreement with Bergmark's earlier study, a general background (at about the same level) of Hb adducts from AA was observed in control persons without known exposure to AA, with higher levels in smokers. The assessment of uptake of AA and of cancer risks for the exposed workers also elucidated the possible cancer risks associated with the background exposure to AA, provided that the observed background Hb adducts really originated from AA. Based on the phamaco-kinetic data from Calleman (1996), it could be calculated that the background would correspond to a daily intake of about 100 µg of AA (cf. Törnqvist et al., 1998). It was suggested that a life-long uptake of AA at this rate could be associated with a considerable cancer risk. In the light of the cancer risk assessment for the exposed tunnel workers (involving evaluation with different cancer risk models, including a relative model developed within the group; Granath et al., 1999), and comparing with findings of other background Hb adducts, e.g. from ethylene oxide, the conclusions seemed reasonable.

Further clues concerning the background adduct, assumed to originate from AA, was obtained from the investigations of Hb adducts in cattle and wild animals potentially exposed to the leakage of AA at Hallandsås. (Studies of exposure to animals carried out in collaboration with Swedish University of Agricultural Science, and National Veterinary Institute of Sweden, Uppsala; studies of cows: Godin et al., 2002; studies of deer, moose, fish, etc.: to be published.) It was notable that very low AA adduct levels were observed in most animals (lower than the background in humans). This further strengthened that the background AA adduct observed in humans really corresponded to a true exposure. (In this context it should be mentioned that there is also species-differences in Hb accumulated adduct levels (see e.g. Törnqvist et al., 2002). Considering that AA is found in tobacco smoke, probably formed during combustion, the different levels of the background AA adduct in humans and in wild animals also were taken as an indication that heated food could be a source of AA. All known possible human exposure sources of AA (e.g. AA in cosmetics), were estimated to be insignificant contributions to a background exposure.

In this situation it appeared urgent a) to prove whether AA was really the origin to the observed background signal in Hb adducts from unexposed humans, and if so, b) find the source and, c) further improve estimations of the corresponding cancer risk.

3.2 Proof that background Hb adduct originates from acrylamide, with cooked food as the likely source

Strong indications led the hypothesis that AA formed in cooking was the source of the assumed, but still hypothetical, background exposure to AA. To test this hypothesis a feeding experiment with rats was carried out within the PhD work of E. Tareke in Spring 1998. After feeding rats (n = 3) for 2 months with fried animal standard feed, an increase by about a factor of 10 of the Hb adduct level from AA was observed compared to controls (Tareke et al., 2000). This was an unusually convincing result considering the small size of the experiment. The result was repeated in a second experiment (see Fig. 4). With the objective to prove the identity of the Hb adduct, a careful interpretation of a product ion mass spectrum obtained by gas chromatography/tandem-mass spectrometry (GC-MS/MS) was done with aid of isotope-substituted standards of the adduct (Tareke et al., 2000).

However, even though we had proven that the adduct was identical with the adduct from AA, we could not exclude that something else associated with the feed could give rise to the same adduct. Therefore, we realized that we had to analyse for AA in the feed given to rats, which led to contacts with S. Eriksson at AnalyCen in Lidköping. This laboratory had been involved in AA analysis, e.g. in vegetables, due to the leakage in the Hallandsås tunnel, and was also specialized in analysis of food (*cf.* Rydberg et al., 2004). A GC-MS method for analysis of AA in water (based on bromination) was adapted for the analysis of AA in the animal feed. The analysis showed that the content of AA in the feed was compatible with the measured increase of the level of the adduct in the rats fed the fried feed. This work was included as a manuscript in Tareke's licentiate thesis in 1998 (Tareke, 1998), and part of the results were made public as poster presentations (1998 and 2000). It was not published in a scientific journal until 2000 (Tareke et al., 2000).

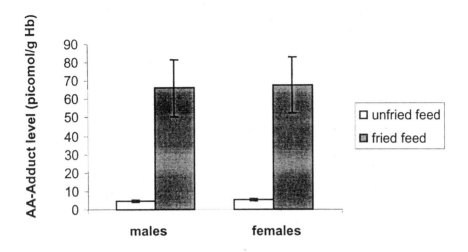

Figure 4. An animal feeding experiment with measurement of the level of adducts to N-terminal valine in Hb. One month feeding with fried and unfried feed of four rats in each group.

3.3 Demonstration of high contents of acylamide in cooked food

On the basis of the known formation of heterocyclic amines from proteins during frying (Sugimura, 2000), hamburger meat was a natural starting point for studies of AA formation during cooking of foodstuffs. An increased content of AA was observed in a preliminary experiment with fried hamburgers, and this preliminary result was in fact mentioned as a footnote in the publication concerning the animal feeding experiment (Tareke et al., 2000). However, we realized that the results were controversial and required incontestable proof, and we initiated laboratory controlled heating experiments.

We demonstrated that the AA formation in hamburgers was temperature-related and that it reached levels in the range < 5 – 50 µg/kg in different types of meat (Tareke et al., 2002). However, the content in meat products could account only for minor fractions of the estimated daily intake of AA. Also, the higher contents observed in the fried animal feed compared to the meat products indicated sources other than protein-rich food. Tareke started experiments with heating of potato as an example of staple food. A real surprise came with the first analysis of AA in fried potato in January 2001 (*cf.* Rydberg et al., 2004). Heating of potato and some other carbohydrate-rich foods, led to contents of AA in the range 100 – 900 µg/kg. During the autumn 2001, these findings were confirmed in restaurant-prepared foods (e.g. *ca.* 500 µg AA/kg in French fries), and finally very high values (up to 4 mg AA/kg) were found in potato crisps. Different cooking methods were also tested, including heating in a microwave oven. All methods, except boiling, led to the formation of AA. In the uncooked food studied AA was below detection level. Table 1 summarizes some of the results (from Tareke et al., 2002). The determined content of AA in food led to rough estimations of an average daily intake amounting to a few tenths of µg of AA per person.

With the experience from the investigations at Hallandsås, we realized that certain compounds, like N-methylolacrylamide, would give the same analyte as AA with the methods used. We therefore found it useful to confirm the analysis of the content of AA in food by a milder and independent analytical method. This was achieved by development of a liquid chromatography-tandem mass spectrometry (LC-MS/MS) method. The results obtained with these two methods (GC-MS and LC-MS/MS) were in full agreement (first results in January 2002). Analysis under different conditions with the two methods further supported the results. Method development concerning analysis of AA in food was done in collaborative work, primarily performed by S. Eriksson et al. at AnalyCen, Lidköping.

However, the finding of high levels of AA in cooked potato was something we understood would become controversial. Clues on the formation mechanism and prevention of formation would facilitate the communication of these findings.

Table 1. Levels of acrylamide found in cooked foods (Tareke et al., 2002). Analysis of foods heated under controlled laboratory conditions, in a frypan at 220°C, or in a microwave oven. Verification that acrylamide occurs at high levels in restaurant-prepared foods.

Type of food (number of samples/analysis)	Acrylamide content;median (range) (µg/kg heated food)
Laboratory-fried protein-rich food	
Beef, minced (5)	17 (15 – 22)
Chicken, minced (2)	28 (16 – 41)
Cod, minced (3)	< 5 (< 5 – 11)
Laboratory-fried carbohydrate-rich food	
Potato, grated (5)	447 (310 – 780)
Beetroot, grated (2)	850 (810 – 890)
Boiled or raw food	
Potato, beef, cod (12)	< 5
Restaurant-prepared etc. foods	
Hamburger (4)	18 (14 – 23)
French fries (6)	424 (314 – 732)
Potato crisps (8)	1740 (1300 – 3900)
Crisp bread (different types) (3)	208 (37 – 1730)
Beer (3)	< 5

4. COMMUNICATIONS AND PUBLISHING OF THE FINDING OF ACRYLAMIDE IN HEATED FOOD

The finding of high contents of AA in a staple food such as potato could be expected to be a "hot potato", particularly in Sweden because of the bad reputation of AA as the "Hallandsås poison". One could anticipate large requirements for information from media and the general public to be initiated by the publishing of the findings, which might be difficult to manage for a small research group. The Swedish National Food Administration (Sw NFA) had, as a responsible authority, been involved in the Hallandås case, due to the contamination of AA in water and food. Considering the responsibility of Swedish Universities to inform the general public, we therefore informed the Sw NFA about our studies and the findings on high AA contents in potato (in Spring 2001). This contact led to an article in the Swedish journal "Vår Föda" (Tareke and Törnqvist, 2001), where the finding on AA formation in hamburgers was mentioned. This article received limited attention (*cf.* Löfstedt, 2003). However, the finding on high contents of AA in heated potato was a fact about which only those involved in the research and some persons at Sw NFA were informed. After requests for information and discussion with us in Autumn 2001, Sw NFA initiated follow-up and verifying studies regarding AA in heated food. In the beginning of 2002, with the aid of AnalyCen, Sw NFA then verified our findings of AA in heated foodstuffs.

When the Sw NFA realized the significance and scope of the problem and had verified our analyses, this authority considered it as its duty to inform the public about the findings. On the other hand, I had the responsibility to protect priorities in collaborators contributions, including a Ph.D. work, as well as the finding as a result of our research. However, it was realized that a scientific publication of the findings, would strengthen a public communication of the difficult issue, which otherwise could be jeopardized as unbelievable and unreliable. The pressure to publish our results had led to a conflict with time and other duties. A communication sent to Nature was rejected, and instead a full manuscript was submitted to Journal of Agricultural and Food Chemistry (JAFC). It was agreed with Sw NFA that a public communication should not occur until a scientific paper from our group at Stockholm University was published.

Now, however, too many knew about the findings and this resulted in information leaks. One acute leak was stopped but a wide leak to policy makers, industrialists and other stakeholders was imminent (*cf.* Löfstedt, 2003). We had to face a situation which might become uncontrollable, with spreading of data in media which in the worst case might result in biased information and unnecessary anxiety. It was also considered that communication concerning cancer risks is particularly difficult, because of the fact that many people find it hard to envision a risk. In a stage when the referee comments on the manuscript sent to JAFC had been dealt with, it was decided that a press conference about the findings should be held when the paper was accepted. The press conference April 24th 2002 initiated an avalanche of reactions (*cf.* Löfstedt, 2003).

Our scientific publication of the results on AA in food encountered difficulties and was somewhat delayed, until the Summer 2002 (Tareke et al., 2002). However, it could be included as an original finding within a PhD thesis (Tareke, 2003). The follow-up and verifying studies on AA in food by Sw NFA were published about the same time as our publication in 2002 (Rosén and Hellenäs, 2002). The release about our findings caused extraordinary pressure on the research group, with the consequence that the publication on our follow-up studies on determinants for the formation was delayed (Rydberg et al., 2003, and this volume). Studies on AA in coffee (and snuff) were taken up early by colleagues at Stockholm University (Licea-Pérez, 2000; Licea-Pérez and Osterman-Golkar, 2003), after our initial findings (Tareke, 1998; Tareke et al., 2000).

5. ESTIMATIONS OF INTAKE OF ACRYLAMIDE IN THE GENERAL POPULATION

From data on the average background Hb adducts from AA, the intake of AA have been calculated. The level of about 30 picomol/g Hb corresponds to a daily intake of about 1.2 µg/kg, i.e. ca. 85 µg in a 70 kg person (*cf.* Bergmark, 1997; Törnqvist et al., 1998). Sw NFA obtained a figure on the average daily intake of AA of about 35 µg per person on the basis of analytic data on a large number of foods in Sweden (Svensson et al., 2003). A similar figure has been obtained in Norway (Dybing and Sanner, 2003). Considering the uncertainties in the pharmaco-kinetic parameter values (Calleman, 1996) behind the calculations from adduct data and that there are also uncertainties in the estimations from food analysis, the agreement is surprisingly good. A further improvement of the calculations from adduct data will be obtained through re-determination of involved parameters and scrutinizing other possible sources of exposure. Also, the analytical data on AA in food has to be further evaluated (*cf.* Eriksson, this volume).

6. WORK ON IMPROVED CANCER RISK ASSESSMENT OF ACRYLAMIDE

Besides the studies within our research group on background exposure to AA, work aimed at improved cancer risk assessment of AA has been performed as a follow-up of the Hallandsås case, partly as PhD work (Paulsson, 2003). The method for analysis of Hb adducts from glycidamide has been improved for studies of metabolism of AA and glycidamide (Paulsson et al., 2003a). For instance, the influence of enzymatic polymorphism is being studied (*cf.* Paulsson et al., this volume). The work aimed at improved cancer risk estimation concerns evaluation of glycidamide as the genotoxic agent in AA exposure (Paulsson et al., 2003b), and evaluation of cancer risk models applicable to AA (Törnqvist et al., 1998; Granath et al., 1999).

Epidemiological methods have been too insensitive to demonstrate or exclude a cancer risk from AA (Granath et al., 2001; Hagmar and Törnqvist, 2003; Granath and Törnqvist, 2003). A risk in the magnitude of 1 % or less of the background cancer risk is hardly detectable by epidemiological methods, even if the corresponding annual cases expected is unacceptable (cf. Törnqvist et al., 1998; Tareke et al., 2000; Dybing and Sanner, 2003). Therefore, it becomes important to fill in data gaps by continued experimental studies at low doses.

In this context, it should be mentioned that the earlier estimated NOAEL for neurotoxic symptoms from AA (Calleman, 1996; Törnqvist et al., 1998 with data from Calleman et al., 1994) was supported by the data from the tunnel workers. An AA adduct level in the interval 0.3 – 1 nmol/g Hb was estimated for mild symptoms of PNS impairment (Hagmar et al., 2001). This would correspond to an uptake of a few mg of AA per day (cf. Tareke et al., 2002).

7. ISSUES AND REFLECTIONS

The research results and the consequences of publishing brought up a number of issues in addition to the scientific questions. For instance, concerning risk management: How to deal with the fact that consumption of a genotoxic substance formed during cooking borders or greatly exceeds the threshold for the same compound in other exposures? Questions concerning risk communication, as well as ethical issues were raised: How should health risks be communicated to the general public (*cf.* Löfstedt, 2003)? To whom and at which stage in their work must researchers communicate health risks? What role do researchers have, and what role do the authorities have? How are "hot" research results with great public to be managed interest and in relation to intellectual property and scientific publishing ? To these remarks must be added to the questions about nuance, authenticity and objectivity in the reporting by media.

The public communication of the findings in Sweden in April 2002 was questioned and criticized by several parties, possibly with an exception of those acquainted with the research field. One thing was apparent, the intention to avoid alarm in media by giving a balanced information totally failed (*cf.* Löfstedt, 2003). However, the alarm in media, followed by the criticism seemed to be counter productive. In the general public in Sweden, AA in food now seems to be perceived to constitute an insignificant health risk, because of the low levels. In contrast, it is perceived as AA constituted a health risk when used in the tunnel work at Hallandsås, because of the high exposure levels (but relatively short times of exposure). This reasoning presumes a threshold dose for the health effects, as is expected for neurotoxic symptoms. However, in case where a genotoxic mechanism is operating, a threshold dose-response relationship is not expected.

One task is to clarify mechanisms for toxicity and dose-response relationships at low doses of AA. Even though there is now available good information about formation mechanisms of AA in cooking and its prevention, it will probably not be possible to reduce AA in food to negligible levels. On the other hand, it should be considered that AA is only one among many electrophilic compounds observed to occur as a natural background exposure, e.g. through background Hb adducts.

One thing is certain, the finding of high contents of the toxic compound AA in food opened areas for new research and calls for new thinking in several fields and it demonstrated the strength of the applied scientific approach.

ACKNOWLEDGEMENTS

Dr. U. Olsson and Prof. L. Ehrenberg are acknowledged for critical reading of the manuscript. The Swedish Research Council for Environment, Agricultural Science and Spatial Planning (FORMAS), the Swedish Cancer Society, and the Swedish Cancer and Allergy Fund, are acknowledged for economical support.

REFERENCES

Bergmark, E., 1992, *Hemoglobin Dosimetry and Comparative Toxicity of Acrylamide and its Metabolite Glycidamide*, Doctoral Thesis, Dept. of Radiobiology, Stockholm University, Sweden.

Bergmark, E., 1997, Hemoglobin adducts of acrylamide and acetonitrile in laboratory workers, smokers, and nonsmokers, *Chem. Res. Toxicol.* **10**: 78–84.

Bergmark, E., Calleman, C.J., He, F., and Costa, L.G., 1993, Determination of hemoglobin adducts in humans occupationally exposed to acrylamide, *Toxicol. Appl. Pharmacol.* **120**: 45-54.

Calleman, C.-J., 1996, The metabolism and pharmacokinetics of acrylamide: Implications for mechanisms of toxicity and human risk, *Drug Metab. Rev.* **28**: 527-590.

Calleman, C.J., Ehrenberg, L., Jansson, B., Osterman-Golkar, S., Segerbäck, D., Svensson, K., and Wachtmeister, C.A., 1978, Monitoring and risk assessment by means of alkyl groups in hemoglobin in persons occupationally exposed to ethylene oxide, *J. Environ. Pathol. Toxicol.* **2**: 427-442.

Calleman, C.-J., Bergmark, E., and Costa, L.G., 1990, Acrylamide is metabolized to glycidamide in the rat: evidence from hemoglobin adduct formation, *Chem. Res. Toxicol.* **3**: 406-412.

Calleman, C.-J., Wu, Y., He, F., Tian, G., Bergmark, E., Zhang, S., Deng, H., Wang, Y., Crofton, K. M., et al., 1994, Relationships between biomarkers of exposure and neurological effects in a group of workers exposed to acrylamide, *Toxicol. Appl. Pharmacol.* **126**: 361-71.

Dybing, E., and Sanner, T., 2003, Risk assessment of acrylamide in foods. *Toxicol. Sci.* **75**(1): 7-15.

Ehrenberg, L., 1974, Genetic toxicology of environmental chemicals, *Acta Biol. Iugosl. Ser. F. Genetica* **6**: 367-398.

Ehrenberg, L., Moustacchi, E., and Osterman-Golkar, S., 1983, Dosimetry of genotoxic agents and dose–response relationships of their effects, *Mutat. Res.* **123**: 121-182.

Ehrenberg, L., Granath, F., and Törnqvist, M., 1996, Macromolecule adducts as biomarkers of exposure to environmental mutagens in human populations, *Environ. Health Perspect.* **104**: Suppl. 3, 423-428.

EC, 2000, (June 18, 2001) Risk assessment of acrylamide. Draft Risk Assessment Report. October 2000, http://ecb.jrc.it/existing-chemicals/.

Friedman, M., 2003, Chemistry, biochemistry, and safety of acrylamide. A review, *J. Agric. Food Chem.* **51**: 4504-4526.

Gamboa da Costa, G., Churchwell, M.I., Hamilton, L.P., Von Tungeln, L.S., Beland, F.A., Marques, M.M. and Doerge, D.R., 2003, DNA Adduct Formation from acrylamide via conversion to glycidamide in adult and neonatal mice, *Chem. Res. Toxicol.* **16**: 1328 – 1337.

Godin, A.C., Bengtsson, B., Niskanen, R., Tareke, E., Törnqvist, M., and Forslund, K., 2002, Acrylamide and N-methylolacrylamide poisoning in a herd of Charolais crossbreed cattle, *Vet. Rec.* **151**: 724-728.

Granath, F., and Törnqvist, M. 2003 Who knows whether acrylamide in food is hazardous to humans? *J. Nat. Cancer Inst.* **95**, 842-843.

Granath, F., Vaca, C., Ehrenberg, L., Törnqvist, M., 1999, Cancer risk estimation of genotoxic chemicals based on target dose and a multiplicative model, *Risk Analysis*, **19**, 309-320.

Granath, F., Ehrenberg, L., Paulsson, B., Törnqvist, M., 2001, Cancer risk from exposure to occupational acrylamide, *Occup. Environ. Med.* **58**: 608.

Hagmar, L., and Törnqvist, M. 2003 Inconclusive results from an epidemiological study on dietary acrylamide and cancer. *Br. J. Cancer* **89**: 774-776.

Hagmar, L., Törnqvist, M., Nordander, C., Rosén, I., Bruze, M., Kautiainen, A. et al., 2001, Health effects of occupational exposure to acrylamide using hemoglobin adducts as biomarkers of internal dose, *Scand. J. Work Environ. Health* **27**(4): 219-226.

IARC. 1994, *IARC Monographs on the evaluation of carcinogen risk to humans: some industrial chemicals.* No 60. 1994. Lyon, International Agency for Research on Cancer.

Jensen, S., Törnqvist, M, and Ehrenberg, L., 1984, Hemoglobin as a dose monitor of alkylating agents: Determination of alkylation products of N-terminal valine, in: *Individual Susceptibility to Genotoxic Agents in the Human Population. Environmental Science Research,* F.J. de Serres and R.W. Pero, eds, Vol. 30, Plenum Press, New York, pp. 315-320.

Kautiainen, A., Midtvedt, T., and Törnqvist, M., 1993, Intestinal bacteria and endogenous production of malonaldehyde and alkylators in mice, *Carcinogenesis* **14**: 2633-2636.

Licea-Pérez, H., 2000, *In Vivo Dosimetry of Some Important Industrial Chemicals by Measurement of Their Reaction Products with Hemoglobin,* Doctoral Thesis, Dept. of Molecular Genome Research, Stockholm University, Sweden.

Licea Pérez, H., and Osterman-Golkar, S., 2003, A sensitive gas chromatographic–tandem mass spectrometric method for detection of alkylating agents in water: Application to acrylamide in drinking water, coffee and snuff, *The Analyst* **128** (8): 1033-1036.

Löfstedt, R.E., 2003, Science communication and the Swedish acrylamide "alarm", *J. Health Comm.* **8**: 407-432.

Miller, E.C., and Miller, J.A., 1966, Mechanism of chemical carcinogenesis: nature of proximate carcinogens and their interactions with macromolecules. *Pharmacol. Rev.* **18**: 805-838.

Osterman-Golkar, S., Ehrenberg, L., Segerbäck, D., and Hällström, I., 1976, Evaluation of genetic risks of alkylating agents. II. Haemoglobin as a dose monitor. *Mutat Res.* **34**: 1-10.

Paulsson, B., 2003, *Dose Monitoring for Health Risk Assessment of Exposure to Acrylamides,* Doctoral Thesis, Department of Environmental Chemistry, Stockholm University.

Paulsson, B., Athanassiadis, I., Rydberg, P., and Törnqvist, M, 2003a, Hemoglobin adducts from glycidamide: acetonisering of hydrophilic groups for reproducible gas chromatography/tandem mass spectrometric analysis. *Rapid Commun. Mass Spectrom.* **17**: 1859-1865.

Paulsson, B., Kotova, N., Grawé, J., Granath, F., Henderson, A., Golding, B., and Törnqvist, M., 2003b, Induction of micronuclei in mouse and rat by glycidamide, the genotoxic metabolite of acrylamide, *Mutat. Res.* **535**: 15–24.

Rosén, J., and Hellenäs, K.-E., 2002, Analysis of acrylamide in cooked foods by liquid chromatography tandem mass spectrometry, *Analyst* **127**: 880-882.

Rydberg, P., Eriksson, S., Tareke, E., Karlsson, P., Ehrenberg, L., and Törnquist, M, 2005, Factors that influence the acrylamide content of heated foods, in: *Chemistry and Safety of Acrylamide in Food,* M. Friedman and D.S. Mottram, eds, Springer, New York, pp. 317-328.

Rydberg, P., Eriksson, S., Tareke, E., Karlsson, P., Ehrenberg, L., and Törnquist, M, 2004, Factors that influence the acrylamide content of heated foods, in: *Chemistry and Safety of Acrylamide in Food,* Kluwer Academic, New York. (this volume)

Schumacher, J.N., Green, C.R., Best, F.W., and Newell, M.P., 1977, Smoke composition. An extensive investigation of the water-soluble portion of cigarette smoke, *J. Agric. Food Chem,* **25**: 310-320.

Segerbäck, D., Calleman, C.J., Schroeder, J.L., Costa, L.G., and Faustman, E.M., 1995, Formation of N-7-(2-carbamoyl-2hydroxyethyl)guanine in DNA of the mouse and the rat following intraperitoneal administration of [^{14}C]acrylamide. *Carcinogenesis* **16**: 1161-1165.

Solomon, J. J., Fedyk, J., Mukai, F., and Segal, A., 1985, Direct alkylation of 2'-deoxynucleosides and DNA following in vitro reaction with acrylamide. *Cancer Research* **45**(8): 3465-3470.

Svensson, K., Abramsson, L., Becker, W., Glynn, A., Hellenäs, K.-E., Lind, Y., and Rosén, J., 2003, Dietary intake of acrylamide in Sweden. *Food Chem Toxicol* **41**(11): 1581-1586.

Sugimura, T., 2000, Nutrition and dietary carcinogens. *Carcinogenesis* **21**: 387-395.

Tareke, E., 1998, *Studies on Background Carcinogens*, Ph. Lic. Thesis, Dept. of Environmental Chemistry, Stockholm University, Sweden.

Tareke, E., 2003, *Identification and Origin of Potential Background Carcinogens: Endogenous Isoprene and Oxiranes, Dietary Acrylamide*, Doctoral Thesis, Dept. of Environmental Chemistry, Stockholm University, Sweden.

Tareke, E., Rydberg, P., Karlsson, P., Eriksson, S., and Törnqvist, M., 2000, Acrylamide: A cooking carcinogen? *Chem. Res. Toxicol.* **13**: 517-522.

Tareke, E., Rydberg, P., Karlsson, P., Eriksson, S., and Törnqvist, M., 2002, Analysis of acrylamide, a carcinogen formed in heated foodstuffs, *J. Agric. Food Chem.* **50**: 4998-5006.

Tareke, E., and Törnqvist, M., 2001, Akrylamid – inte bara i Hallandsåsen utan även i stekta hamburgare, *Vår Föda* **2**: 28-29.

The Tunnel Commission, 1998, *Kring Hallandsåsen* [About Hallandsåsen]. SOU 1998:60, Stockholm, Miljödepartementet.

Törnqvist, M., 1988, Search for unknown adducts: Increase of sensitivity through preselection by biochemical parameters, in: *Methods for Detecting DNA Damaging Agents in Humans: Applications in Cancer Epidemiology and Prevention*, H. Bartsch, K. Hemminki, and I.K. O'Neill, eds, IARC Sci. Publ. **89**, International Agency for Research on Cancer, Lyon, pp. 378-383.

Törnqvist, M., 1989, *Monitoring and Cancer Risk Assessment of Carcinogens, Particularly Alkenes in Urban Air*, Doctoral Thesis, Dept. of Radiobiology, Stockholm University, Sweden.

Törnqvist, M., 1996, Ethylene oxide as a biological reactive intermediate of endogenous origin, in: *Biological Reactive Intermediates V*, R. Snyder et al. eds, Plenum Press, New York, pp. 275-283.

Törnqvist, M., and Hindsø Landin, H., 1995, Hemoglobin adducts for in vivo dose monitoring and cancer risk estimation, *J. Occup. Environ. Med.* **37**: 1077-1085.

Törnqvist, M., and Kautiainen, A., 1993, Adducted proteins for identification of endogenous electrophiles. *Environ. Health Perspect.* **99**: 39-44.

Törnqvist, M., Mowrer, J., Jensen, S., and Ehrenberg, L., 1986, Monitoring of environmental cancer initiators through hemoglobin adducts by a modified Edman degradation method, *Anal. Biochem.* **154**: 255-266.

Törnqvist, M., Gustafsson, B., Kautiainen, A., Harms-Ringdahl, M., Granath, F., and Ehrenberg, L. 1989, Unsaturated lipids and intestinal bacteria as sources of endogenous production of ethene and ethylene oxide, *Carcinogenesis* **10**: 9-41.

Törnqvist, M., Bergmark, E., Ehrenberg, L., and Granath, F., 1998, [*Risk Assessment of Acrylamide*] (in Swedish), National Chemicals Inspectorate, Sweden, PM 7/98.

Törnqvist, M., Fred, C., Haglund, J., Helleberg, H., Paulsson, B., and Rydberg, P., 2002, Protein adducts: Quantitative and qualitative aspects of their formation, analysis and applications. *J. Chromatogr. B* **778**(1-2): 279-308.

WHO, 1996, *Guidelines for Drinking-Water Quality*, 2nd ed., World Health Organisation, Geneva, Vol **2**, pp. 940-949.

ACRYLAMIDE NEUROTOXICITY: NEUROLOGICAL, MORHOLOGICAL AND MOLECULAR ENDPOINTS IN ANIMAL MODELS

Richard M. LoPachin
Department of Anesthesiology, Albert Einstein College of Medicine, 111 E. 210th st., Bronx NY 10467; e-mail: lopachin@aecom.yu.edu

Abstract: Acrylamide (AA) monomer is used in numerous chemical industries and is a contaminant in potato- and grain-based foods prepared at high temperatures. Although experimental animal studies have implicated carcinogenicity and reproductive toxicity as possible consequences of exposure, neurotoxicity is the only outcome identified by epidemiological studies of occupationally exposed human populations. Neurotoxicity in both humans and laboratory animals is characterized by ataxia and distal skeletal muscle weakness. Early neuropathological studies suggested that AA neurotoxicity was mediated by distal axon degeneration. However, more recent electrophysiological and quantitative morphometric analyses have identified nerve terminals as primary sites of AA action. A resulting defect in neurotransmitter release appears to be the pathophysiological basis of the developing neurotoxicity. Corresponding mechanistic research suggests that AA impairs release by adducting cysteine residues on functionally important presynaptic proteins. In this publication we provide an overview of recent advances in AA research. This includes a discussion of the cumulative nature of AA neurotoxicity and the putative sites and molecular mechanisms of action.

Key words: toxic neuropathy, distal axonopathy, protein adducts, nerve terminals, neurotoxicity

1. INTRODUCTION

Acrylamide (AA) is a water-soluble, vinyl monomer that has multiple industrial applications: e.g., waste water management, ore processing, and is used extensively in molecular laboratories for gel chromatography.

Neurotoxicity, characterized by ataxia, distal skeletal muscle weakness, and numbness of the hands and feet, is currently the only documented outcome in occupationally exposed human populations (Deng et al., 1993; Garland and Paterson, 1967; He et al., 1989; Spencer and Schaumburg, 1974a). In addition, experimental data from rodent studies suggest that AA produces reproductive toxicity (e.g., reduced litter size, DNA strand breaks, dominant lethal mutations; Sega et al., 1990; Tyl et al., 2000a,b; Working et al., 1987). Studies of AA-exposed laboratory animals (primarily rodents) have also revealed an increased incidence of tumors in certain tissues (e.g., mammary gland fibroadenomas in female rats, tunica vaginalis mesotheliomas in male rats; e.g., Bull et al., 1984a,b; Friedman et al., 1995; Johnson et al., 1986). Given this toxic potential, significant concern was expressed when a Swedish research group recently (April 2002) announced preliminary findings of significant AA concentrations in certain potato or grain-based foods (e.g., 3500 µg AA /kg potato chips) that had been prepared at high temperatures (>160°C). These early data were later confirmed by the Swedish group (Tareke et al., 2002) and by other researchers in Europe and America (e.g., Rosen and Hellenas, 2002; Sanders et al., 2002). Recent evidence suggests that AA in food is generated from pyrolytic fragments of asparagine and that this reaction is facilitated by concomitant pyrolysis of Maillard-active dicarbonyl and hydroxycarbonyl precursors (Becalski et al., 2002; Sanders et al., 2002). Based on the AA content of various foods, it has been estimated that the average consumer is exposed to approximately 0.8 – 3 µg AA/kg BW/day (FAO/WHO report 2002). However, it is important to emphasize that the toxicological risk of this daily AA intake has not been established in humans. Nonetheless, the obvious health implications of food-borne AA has initiated substantial public and scientific concern (World Health Organization meeting June, 2002; US FDA meeting September, 2002) and has significantly increased interest in the toxic effects of AA. Since neurotoxicity has demonstrated relevance to human exposure, we thought it was important to discuss recent advances in related neurological and morphological research. In particular, new evidence will be reviewed, which suggests that the nerve terminal is a primary site of AA action and that inhibition of neurotransmission contributes significantly to the development of corresponding neurological deficits. Our work in this area suggests that AA disrupts membrane fusion processes that mediate neurotransmission and membrane turnover in nerve terminals. A final goal of this review is to identify data gaps concerning the pathophysiological processes of AA neurotoxicity.

2. ACRYLAMIDE NEUROTOXICITY

AA is a neurotoxicant in both humans and laboratory animals. As indicated above, occupational exposure of humans to AA produces neurotoxicity characterized by ataxia, skeletal muscle weakness and numbness of the hands and feet. In laboratory animal models, AA intoxication (10-50 mg/kg/d) produces neurological signs that, in many respects, resemble the neurotoxicity occurring in humans; i.e., ataxia (open field gait abnormalities), skeletal muscle weakness (decreased fore- and hindlimb grip strength) and hindlimb foot splay (Burek et al., 1980; Crofton et al., 1996; Edwards and Parker, 1977; LoPachin et al., 2002b; Moser et al., 1992; Shell et al., 1992; Tilson and Cabe, 1979). Assessment of neurological function over a 90-day exposure period at lower AA dose-rates (0.05-6.7 mg/kg/d) did not reveal evidence of neurotoxicity (Burek et al., 1980; Crofton et al., 1996) and it was concluded that these exposure conditions were non-neurotoxic.

2.1 CUMMULATIVE NEUROTOXICITY

Recent studies (LoPachin et al., 2002b) measuring multiple neurological parameters (gait, foot splay, grip strength and extensor thrust) across two intoxication schedules (21 mg/kg/d in drinking water vs. 50 mg/kg/d i.p.) have indicated that AA produced cumulative neurotoxicity. Specifically, dose-rate did not determine the final magnitude of neurological deficit, but rather determined the time of onset and development of neurotoxicity; i.e., quantitatively similar maximum neurological deficits were observed on day 11 of the 50 mg/kg dose-rate, whereas a similar level of neurotoxicity was achieved on day 40 of the 21 mg/kg exposure rate (Fig. 1). The time-dependent nature of AA neurotoxicity (i.e., dose-rate determines onset of neurological effect) implies cumulative intoxication and a cumulative threshold neurotoxic dose (Kuperman, 1958). Therefore, the failure of previous studies to find neurotoxicity at dose-rates less than 10 mg/kg/d (see Section 2. above), might be a function of an inappropriately short experimental window (i.e., 90 days observation period). The cumulative nature of AA toxicity has important implications for human exposure; i.e., it cannot be assumed that the low exposure rate associated with food-borne AA does not carry risk, since neurotoxicity could result from daily long-term intoxication. Furthermore, the majority of animal studies to date have involved subchronic AA administration ($\leq$90 days exposure) at relatively high daily dose rates (mg/kg/d) via the i.p. or oral (gavage, drinking water) route. In contrast, the daily rate of human AA intoxication is much less (ng - µg/kg/d) with variations in duration (near whole life?) and route (inhalation, dermal, oral) of exposure. Clearly, future studies are needed to characterize neurotoxic potential of AA in lower dose-rate models and to discern the potential long-term cumulative effects of AA.

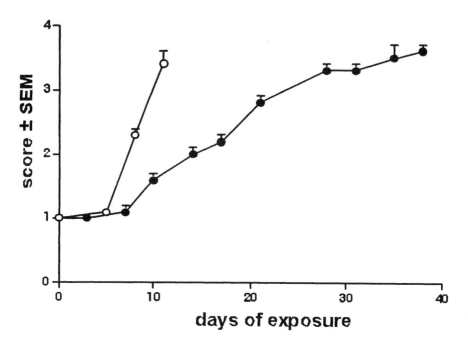

Figure 1. Effects of AA on mean gait scores. AA was administered to groups of rats (n = 8-10 per group) at a daily dose rate of either 50 or 21 mg/kg per day. Age-matched control rats for the 50mg/kg per day dose-rate group were administered saline vehicle by daily i.p. injections. To assess gait scores rats were placed in a clear Plexiglas box and observed for three minutes. Observations were converted into numerical values ranging from 1-4. The assigned scores were 1= a normal gait; 2 = a slightly abnormal gait(slight ataxia, hopping gait and foot splay); 3= a moderately abnormal gait (obvious ataxia and foot splay with limb abduction during ambulation); 4 = a severely abnormal gait (inability to support body weight and foot splay; See LoPachin et al, 2002b for methodological details).

2.2 Parent chemical vs. active metabolite

Previous toxicokinetic studies have shown that AA is rapidly absorbed from most sites of exposure and then evenly distributed among tissues (Barber et al., 2001; Edwards, 1975; Miller et al., 1982). Following uptake, AA can be oxidized to an epoxide metabolite, glycidamide, presumably by the activity of cytochrome P450 2E1 (Sumner et al., 1999). A study by Abou-Donia et al. (1993) suggested that glycidamide played a causal role in producing the neurological deficits and axonal degeneration induced by AA intoxication of rats. In contrast, other research has indicated that the parent compound (AA) and not glycidamide is primarily responsible for induction of neurotoxicity (Barber et al., 2001; Brat and Brimijion, 1993; Costa et al., 1992, 1995). Additional studies are needed to confirm that glycidamide is not involved in the production of neurological or behavioral toxicity and to discern the role of glycidamide in the reproductive toxicity or carcinogenic actions of AA.

3. ACRYLAMIDE NEUROPATHY: A CENTRAL-PERIPHERAL DISTAL AXONOPATHY?

Early morphological studies of AA neuropathy revealed that low-dose subchronic induction of neurological toxicity was associated with nerve damage in both the central and peripheral nervous systems (reviewed in LoPachin et al., 2002a, 2003). The morphological hallmark of this toxic neuropathy was considered to be distal nerve terminal and preterminal axon swellings of the longest myelinated fibers. These swellings contained an abundance of neurofilaments, tubulovesicular profiles and effete, probably degenerating, mitochondria (Prineas, 1969; Schaumburg et al., 1974; Suzuki and Pfaff, 1973). As exposure continued, progressive retrograde degeneration of these distal axon regions ensued with preservation of more proximal segments (reviewed in Spencer and Schaumburg, 1974b, 1976). This pattern of neuropathological expression (i.e., initial nerve terminal damage and subsequent retrograde axon degeneration) was consistent with the theory of toxic "dying-back" neuropathies proposed by Cavanagh (1964; 1979). According to this hypothesis, direct neurotoxicant actions at cell body sites caused deficient manufacture and transport of axon-directed materials. The resulting decrease in distal axon delivery preferentially damaged nerve terminals of long axons and prompted centripetal fiber degeneration. However, Spencer and Schaumburg (1976, 1977a,b) reported that degeneration did not start at the nerve terminal and move rostrally in a seriatim fashion as stipulated by the dying-back hypothesis. Instead, degeneration "bloomed" simultaneously at multifocal sites along distal preterminal axons. Work by Jennekins et al. (1979) indicated that nerve terminals of long axons were not preferentially damaged by AA intoxication as was predicted by the dying-back theory. Based on accumulating evidence that the dying back theory might not accurately explain AA neuropathy, Spencer and Schaumburg (1976) formulated a hypothesis that emphasized direct axonal injury. They proposed that large diameter axons in the central nervous system (CNS) and peripheral nervous sysem (PNS) were most sensitive to development of simultaneous, multifocal paranodal axon swellings in distal regions and that these swellings served as initiation points for subsequent degeneration. In the PNS, AA preferentially affected axons in tibial nerve branches supplying calf muscles, plantar sensory nerves innervating the digits and plantar nerve branches supplying the flexor digitorum brevis muscle (Spencer and Schaumburg, 1977b). Axon swelling and degeneration were noted in certain CNS regions; e.g., dorsal spinocerebellar tract, gracile fasciculus, cerebellar white matter (Ghetti et al., 1973; Prineas, 1969; Spencer and Schaumburg, 1977b). The characteristic spatiotemporal pattern of axon damage in the central and peripheral nervous systems, lead Spencer and Schaumburg (1976) to classify AA neuropathy as a "central-peripheral distal axonopathy".

4. ACRYLAMIDE NEUROPATHY: A TERMINALOPATHY?

Other morphological evidence generated during the past thirty years has indicated that early nerve terminal damage might be importantly involved in the pathophysiological process leading to AA neurotoxicity (Cavanagh, 1982; DeGrandchamp and Lowndes, 1990; DeGrandchamp et al., 1990; Prineas, 1969; Tsujihata et al., 1974). Electrophysiological studies showed that neurotransmission was impaired at spinal cord primary afferent nerve terminals, peripheral neuromuscular junctions and autonomic synapses of AA-intoxicated laboratory animals (Abelli et al., 1991; DeRojas and Goldstein, 1987; Goldstein and Lowndes, 1979, 1981, Goldstein, 1985; Lowndes and Baker, 1976; Lowndes et al., 1978a,b; Munch et al., 1994; Tsujihata et al, 1974). Based on evidence of early structural and functional damage, LoPachin et al. (2002a) suggested that nerve terminals were the primary site of AA action and that synaptic dysfunction and subsequent degeneration were necessary and sufficient steps for production of AA neurotoxicity. Corroborative research using the de Olmos silver stain method to detect neurodegeneration showed that AA intoxication (50 or 21 mg/kg/d) caused early, generalized nerve terminal damage in the PNS and CNS. The early appearance of this effect indicated that the nerve terminal was a primary site of direct action. That axonopathy might not be pathophysiologically relevant was suggested by observations that axon degeneration in PNS and CNS was not linked to the expression of neurological deficits, but rather appeared to be an exclusive product of lower AA dose-rates (≤ 21 mg/kg/d); i.e., intoxication at a higher dose-rate (50 mg/kg/d) did not produce peripheral axon degeneration (Lehning et al., 1998, 2002a,b; 2003). The axon orientation of formative morphological investigations (e.g., Spencer and Schaumburg, 1977a,b) is likely due to a focus on lower dose-rate, subchronic AA dosing schedules. In addition, a growing awareness of dose-rate impact on neurotoxicological expression (LoPachin et al., 2000) and advances in computer-assisted quantitative morphometric and histochemical techniques (de Olmos silver stain) have contributed to the changing view of axon and nerve terminal damage in AA neurotoxicity. Thus, AA intoxication is associated with a terminalopathy characterized by primary nerve terminal damage in the PNS and CNS. Consequently, AA neuropathy cannot be classified as a distal axonopathy and; instead, nosological schemes should consider nerve terminal sites and mechanisms of action. The majority of axon/nerve terminal studies discussed above were conducted at dose-rates between 20 and 50 mg/kg/d. Therefore, future work related to structural and functional damage at distal nerve sites should include lower dose-rates (e.g., 0.05 – 2 mg/kg/d) that are, in accordance with the cumulative neurotoxicity of AA, conducted over appropriately long experimental durations. Such studies would accurately determine the lowest observed and no observed adverse effect levels (LOAEL and NOAEL, respectively), which are important parameters for risk characterization.

Table 1. Density (mean ±SEM) of nerve terminal degeneration in spinal cord gray matter of AA intoxicated rats.

mid thoracic	21 mg/kg/d	50 mg/kg/d
L1-4	1.7±0.6	2.0±0.9
L5-7	3.7±0.5	3.3±0.6
L8-9	2.7±0.6	1.3±0.5
area 10	3.0±0.9	2.3±0.6
IML Nucleus	2.0±0.0	3.7±1.0
Clarke's Nucleus	3.7±0.6	2.7±0.6
lumbar		
L1-4	2.0±0.0	2.3±0.7
L5-7	3.3±0.6	3.7±0.9
L8-9	3.0±0.0	3.3±0.6
area 10	3.0±0.0	4.3±0.4

The density of degeneration was rated according to the following scale: 0=None, 1=Rare, 2=Occasional, 3=Slight, 4=Moderate or 5=Heavy. Rats were exposed to AA at 50 mg/kg/d x 5 or 11 days or 21 mg/kg/d x 14 or 28 days. No argyrophilic changes were evident at day 5 of the higher dose-rate or day 14 of the lower dose-rate. Degenerating neurons or their processes were not found in silver stained spinal cord sections from age-matched control rats.
Abbreviations: L = laminae; IML = intermediolateral nucleus.

5. THE DIRECT EFFECT OF ACRYLAMIDE ON NERVE TERMINAL STRUCTURE AND FUNCTION

The preceding discussion suggests that the nerve terminal is a neurotoxicologically relevant site of AA action and that corresponding dysfunction and eventual degeneration play a significant role in mediating the characteristic neurological defects. How AA produces nerve terminal damage is not known. It is possible that deficient cell body synthesis and/or delivery of presynaptic components causes secondary nerve terminal damage (Cavanagh, 1979; LoPachin and Lehning, 1994; Sickles et al., 2002). However, in previous studies of AA neuropathy, we reasoned that if either cell body synthesis or subsequent transport of vital materials were significantly affected, then disbursement of these materials to distal axon/nerve terminal sites should be impaired and distal dependent processes (e.g., axonal ion regulation) should be rendered defective. Accordingly, Na^+/K^+-ATPase is delivered to axon and nerve terminal sites by kinesin-based rapid anterograde transport (Lombet et al., 1986; Mata et al., 1993). Yet, in distal tibial nerve axons of severely affected AA-intoxicated rats (50 mg/kg/d x 11 d), we found that axolemmal Na^+/K^+-ATPase was normal with respect to corresponding protein content, enzyme activity, and Rb^+ (K^+) transport rate (Lehning et al., 1994, 1997, 1998). In addition, these tibial axons exhibited normal structural indices (e.g., g ratio, axon area, axon perimeter) and subaxonal ion distribution and regulation (i.e., mitochondrial, axoplasmic). These structural and functional parameters are highly dependent upon synthesis and anterograde delivery of integral components; e.g., channel proteins, axolemmal ion transporters, membrane constituents (Amaratunga et al., 1995; Lasek et al., 1984). Therefore, the observation that these distal axon processes are not affected in AA-exposed animals suggests that neither defective cell body synthesis nor deficient anterograde transport are importantly involved in AA-induced nerve terminal damage. Finally, nerve terminal toxicity induced by AA could develop secondary to an energy deficit (Erecinska and Nelson, 1994; Sabri and Spencer, 1980). However, AA exposure does not alter either anaerobic or aerobic energy production in central and peripheral nervous tissues (Brimijion and Hammond, 1985; LoPachin et al., 1984; Medrano and LoPachin, 1989; Sickles et al., 1990). Together, these data suggest that, rather than being a secondary phenomenon, nerve terminal damage is due to a direct effect of AA. This possibility is supported by results from recent studies showing that direct, in vitro exposure of isolated brain synaptosomes to AA decreased transmitter release (LoPachin et al., 2004). The neurotoxicological significance of this direct presynaptic effect is suggested by our observation that brain synaptosomes prepared from AA-intoxicated rats were also release incompetent (LoPachin et al., 2004).

6. THE MECHANISM OF ACRYLAMIDE NERVE TERMINAL DAMAGE: IMPAIRMENT OF MEMBRANE FUSION PROCESSES

Several lines of evidence now indicate that AA acts directly at nerve terminals to cause structural and functional damage. If this is the case, what sites and molecular mechanism of action might be responsible for this damage? The ultrastructural (e.g., accumulation of tubulovesicular profiles) and functional (e.g., reduced neurotransmitter release) defects suggest that AA interferes with the fusion of vesicles (synaptic or transport) and their cognate membrane targets (Chretein et al., 1981; DeGrandchamp et al., 1990; DeGrandchamp and Lowndes, 1990; Goldstein and Lowndes, 1979, 1981; Jennekens et al., 1979; LoPachin et al., 2004; Prineas, 1969; Tsujihata et al., 1974). The fusion between presynaptic and vesicular (i.e., synaptic vesicles, Golgi-derived transport vesicles) membranes is involved in neurophysiological events that are critical to nerve terminal structure (membrane turnover and integrity) and function (neurotransmitter release). AA disruption of fusion events can, therefore, provide a rational explanation for the presynaptic toxic effects of AA; i.e., build-up of tubulovesicular profiles, decreased docked synaptic vesicle, reduced neurotransmitter release, and nerve terminal degeneration. If inhibition of membrane fusion is responsible for presynaptic dysfunction in AA neurotoxicity, what is the mechanism of this effect? The fusion of cognate membranes requires a highly conserved, common set of proteins known as the SNAREs (soluble NSF attachment protein receptor) and their accessory proteins (Jahn and Sudhof, 1999; Lin and Scheller, 2000; Nichols and Pelham, 1998; Rothman, 1994). The proper targeting and docking of, for example, synaptic vesicles at presynaptic release sites involves the association of a specific vesicle (v)-SNARE protein (synaptobrevin) with corresponding presynaptic membrane target (t)-SNAREs (SNAP-25 and syntaxin 1). Protein-protein interaction of the v- and t-SNAREs forms the 7S core complex, which mediates vesicle pore formation and neurotransmitter release. Membrane fusion is terminated by disassembly of the SNARE core, a process involving the sequential binding of α-SNAP (soluble NSF attachment proteins) and NSF (N-ethylmaleimide (NEM)-sensitive fusion protein) to the 7S complex. NSF-mediated disassembly of SNARE complexes into their monomeric components is an absolute requirement for continuation of membrane fusion processes (Littleton et al., 1998; Malhotra et al., 1988; Tolar and Pallanack,

1998). Therefore, AA might impair fusion by disrupting the formation or disassembly of the 7S complex.

We think the molecular effect of AA on SNARE core kinetics is mediated by adduction of NSF and/or the SNARE proteins. AA is a soft electrophile that forms adducts with nucleophilic sulfhydryl groups on protein cysteine residues (Bergmark et al., 1991, 1993; Calleman, 1996; Cavins and Friedman, 1968; Dixit et al., 1986; Hashimoto and Aldridge, 1970; Kempley and Cavanagh, 1984a,b; LoPachin et al., 2004; Sega et al., 1989). This indicates that AA will adduct numerous proteins and, consequently, it is difficult to understand how widespread adduction might translate into a specific effect on membrane fusion. However, whether a protein adduct is disruptive to a physiological process (e.g., membrane fusion) depends upon certain toxicodynamic conditions (Hinson and Roberts, 1992). Specifically, unless the cysteine target is involved in the tertiary structure or function of a protein, adduction does not necessarily cause dysfunction of that protein. Furthermore, the toxic impact of an adducted, dysfunctional protein on a physiological process (e.g., membrane fusion) is dependent upon the corresponding role (e.g., regulatory or rate-limiting) of that protein. This means that much of AA adduct formation is non-specific with regard to neurotoxic mechanisms. Instead, it is now recognized that the redox-sensing properties of certain protein sulfhydryl sites constitute a physiological regulatory mechanism and that disruption (e.g., chemical alkylation) of these groups can initiate pathophysiological cascades (Dschida et al., 1995; Gilbert, 1982; Lipton et al., 1993; Pan et al., 1995). Among thiol-directed systems (e.g., glycolysis, anterograde transport), membrane fusion processes are most sensitive to inhibition by general sulfhydryl reagents such as *N*-ethylmaleimide (NEM; reviewed in Lin and Scheller, 2000; Nichols and Pelham, 1998). Moreover, NSF, SNAP-25 and other SNAREs are cysteine-containing proteins that play a critical role in membrane fusion. These proteins have been shown to be exquisitely susceptible to adduction and subsequent inhibition by sulfhydryl alkylating chemicals like NEM (Beckers et al., 1989; Chapman et al., 1994; Mastrogiacomo and Gundersen, 1995; Rothman, 1994). The susceptibility of these proteins to inhibition is likely related to the nucleophilic reactivity of specific cysteine groups (Friedman, 1973) and to the role these residues play in determining protein function; e.g., the ATPase activity of NSF is dependent upon a cysteine residue (254) in the D1 domain (Tagaya et al., 1993; Whiteheart et al., 1994). Therefore, we have hypothesized that, although AA forms adducts with many proteins, nerve terminal membrane fusion processes fail first due to adduction of functionally important cysteine residues on key regulatory proteins; i.e., NSF and/or the SNARE proteins (reviewed in LoPachin et al., 2002a, 2003).

7. SUMMARY

Exposure of both laboratory animals and humans to AA produces neurotoxicity characterized by ataxia and skeletal muscle weakness. Reproductive toxicity and carcinogenicity are also potential human health concerns based on substantial data from animal studies. These different toxic potentials have recently drawn substantial attention following the announcement that AA is present in certain types of foods prepared at high temperatures; e.g., French fries, potato chips, and some breads. In this review we have discussed evidence that experimental AA neurotoxicity is cumulative; i.e., different dose-rates produce equivalent neurotoxic effects that differ with respect to onset and development. We also discussed the growing body of evidence, which suggests that AA acts directly at nerve terminal sites to cause primary presynaptic dysfunction and eventual degeneration. Nerve terminal damage in the PNS and CNS could account for the sensory, motor and autonomic deficits that characterize AA neurotoxicity. Also according to this database, the expression of axon degeneration is exclusively related to dose-rate; i.e., degeneration in the PNS and CNS is evident only during intoxication at lower AA dose-rates. This indicates that axon degeneration during AA exposure is not a critical event in the pathophysiological process that mediates neurotoxicity. AA intoxication is also associated with cerebellar Purkinje cell injury, and in conjunction with nerve terminal damage, these neuropathological lesions and their dysfunctional consequences likely mediate neurotoxicity. At present the respective mechanisms of toxic injury are unknown, although we have suggested that both might involve an inhibition of membrane fusion processes. Future research concerning AA neurotoxicity should focus on molecular sites and mechanisms of nerve cell damage and explore poorly defined issues such as neurodevelopmental toxicity and the dose-rate relationship between the carcinogenic, reproductive and neurotoxic actions of AA.

ACKNOWLEGDEMENT

Research supported by NIEHS grant to RML (ES3830-17)

REFERENCES

Abelli L., Ferri G.-L., Astolfi M., Conte, B., Geppetti P., Parlani M., Dahl, D., Polak, J.M. and Maggi, C.A. 1991, Acrylamide-induced visceral neuropathy: evidence for the involvement of capsaicin-sensitive nerve to the rat urinary bladder. *Neuroscience* **41**: 311-321.

Abou-Donia, M.B., Ibrahim S.M., Corcoran J.J., Lack L., Friedman, M.A. and Lapadula, D.M. 1993, Neurotoxicity of glycidamide, an acrylamide metabolite, following intraperitoneal injections in rats. *J. Toxicol. Environ. Health* 39: 447-464.

Amaratunga, A., Leeman, S.E., Kosik, K.S. and Fine, R.E. 1995, Inhibition of kinesin synthesis in vivo inhibits the rapid transport of representative proteins for three transport vesicle classes into the axon. *J. Neurochem.* **64**: 2374-2376.

Barber, D., Hunt, J.R., Ehrich, M., Lehning, E.J. and LoPachin, RM. 2001, Metabolism, toxicokinetics and hemoglobin adduct formation in rats following subacute and subchronic acrylamide dosing. *NeuroToxicology*, **22**:341-353.

Becalski, A., Lau, B., Lewis, D. and Seaman, S. 2002, Acrylamide in food: occurrence and sources. Presented at the Annual Meeting of AOAC International, September 26.

Beckers, C.J.M., Block, M.R., Glick, B.S., Rothman, J.E and Balch, W.E. 1989, Vesicular transport between the endoplasmic reticulum and the Golgi stack requires the NEM-sensitive fusion protein. *Nature* **339**: 397-398.

Bergmark, E., Calleman, C.J. and Costa, L.G. 1991, Formation of hemoglobin adducts of acrylamide and its epoxide metabolite glycidamide in the rat. *Toxicol. Appl. Pharmacol.* **111**: 352-363.

Bergmark, E., Calleman, C.J., He, F. and Costa, L.G. 1993, Determination of hemoglobin adducts in humans occupationally exposed to acrylamide. *Toxicol. Appl. Pharmacol.* **120**: 45-54.

Brat, D.J. and Brimijion, S. 1993, Acrylamide and glycidamide impair neurite outgrowth in differentiating N1E.115 neuroblastoma without disturbing rapid bi-directional transport of organelles observed by video microscopy. *J. Neurochem.* **60**: 2145-2152.

Brimijion, W.S. and Hammond, P.I. 1985, Acrylamide neuropathy in the rat: effects on energy metabolism in sciatic nerve. *Mayo Clin. Proc.* **60**: 3-8.

Bull, R.J., Robinson, M., Laurie, R.D., Stoner, G.D., Greisiger, E., Meier, J.R. and Stober, J. 1984a, Carcinogenic effects of acrylamide in sencar and A/J mice. *Cancer Res* **44**: 107-111.

Bull, R.J., Robinson, M. and Stoner, G.D. 1984b, Carcinogenic activity of acrylamide in the skin and lung of Swiss-ICR mice. *Cancer Lett* **24**: 209-212.

Burek, J.D., Albee, R.R., Beyer, J.E., Bell, T.J., Carreon, R.M., Morden, D.C., Wade, C.E., Hermann, E.A. and Gorzinski, S.J. 1980, Subchronic toxicity of acrylamide administered to rats in drinking water followed by up to 144 days of recovery. *J. Environ. Pathol. Toxicol.* **4**: 157-182.

Calleman, C.J. The metabolism and pharmacokinetics of acrylamide: implications for mechanisms of toxicity and human risk estimation. *Drug Met. Rev.* **28**: 527-590.

Cavanagh, J.B. 1964, The significance of the "dying-back" process in experimental and human neurological disease. *Int. Rev. Exp. Pathol.* **3**: 219-267.

Cavanagh, J.B. 1979, The dying back process. *Arch. Pathol. Lab. Med.* **103**: 659-664.

Cavanagh, J.B. 1982, The pathokinetics of acrylamide intoxication: a reassessment of the problem. *Neuropath Appl Neurobiol* **8**: 315-336.

Cavins, J.F. and Friedman, M. 1968, Specific modification of protein sulfhydryl groups with α,β-unsaturated compounds. *J. Biol. Chem.* **243**: 3357-3360.

Chapman, E.R., An, S., Barton, N. and Jahn, R. 1994, SNAP-25, a t-SNARE which binds to both syntaxin and synaptobrevin via domains that may form coiled coils. *J. Biol. Chem.* **269**: 27427-27432.

Chretien, M., Patey, G., Souyri, F. and Droz, B. 1981, Acrylamide-induced neuropathy and impairment of axonal transport of proteins. II. Abnormal accumulations of smooth endoplasmic reticulum as sites of focal retention of fast transported proteins. Electron microscope radioautographic study. *Brain Res.* **205**: 15-28.

Costa, L.G., Deng, H., Gregotti, C., Manzo, L., Faustman, E.M., Bergmark, E. and Calleman C.J. 1992, Comparative studies on the neuro- and reproductive toxicity of acrylamide and its epoxide metabolite glycidamide in the rat. *NeuroToxicology* **13**: 219-224.

Costa, L.G., Deng, H., Calleman, C.J. and Bergmark, E. 1995, Evaluation of the neurotoxicity of glycidamide, an epoxide metabolite of acrylamide: behavioral, neurochemical and morphological studies. *Toxicology* **98**: 151-161.

Crofton, K.M., Padilla, S., Tilson, H.A., Anthony, D.C., Raymer, J.H. and MacPhail, R.C. 1996, The impact of dose rate on the neurotoxicity of acrylamide: The interaction of administered dose, target tissue concentrations, tissue damage, and functional effects. *Tox. Appl. Pharmacol.* **139**: 163-176.

DeGrandchamp, R.L. and Lowndes, H.E. 1990, Early degeneration and sprouting at the rat neuromuscular junction following acrylamide administration. *Neuropath. Appl. Neurobiol.* **16**: 239-254.

DeGrandchamp, R.L., Reuhl, K.R. and Lowndes, H.E. 1990, Synaptic terminal degeneration and remodeling at the rat neuromuscular junction resulting from a single exposure to acrylamide. *Tox Appl Pharmacol* **105**: 422-433.

De Rojas, T.C, and Goldstein, B.D. 1987, Primary afferent terminal function following acrylamide: alterations in the dorsal root potential and reflex. *Tox. Appl. Pharmacol.* **88**: 175-182.

Deng, H., He, S. and Zhang, S. 1993, Quantitative measurements of vibration threshold in healthy adults and acrylamide workers. *Int. Arch. Occup. Environ. Health* **65**: 53-56.

Dixit, R., Sas, M., Seth, P.K. and Mukhtar, H. 1986, Interaction of acrylamide with bovine serum albumin. *Environ. Res.* **40**: 365-371.

Dschida, W.J.A, and Bowman, B.J. 1995, The vacuolar ATPase: sulfite stabilization and the mechanism of nitrate inactivation. *J. Biol. Chem.* **270**: 1557-1563.

Edwards, P.M. 1975, The distribution and metabolism of acrylamide and its neurotoxic analogues in rats. *Biochem. Pharmacol..* **24**: 1277-1282.

Edwards, P.M. and Parker, V.H. 1977, A simple, sensitive and objective method for early assessment of acrylamide neuropathy in rats. *Tox. Appl. Pharmacol.* **40**: 589-591.

Erecinska, M. and Nelson, D. 1994, Effects of 3-nitropropionic acid on synaptosomal energy and transmitter metabolism: relevance to neurodegenerative brain diseases. *J. Neurochem.* **63**: 1033-1041.

Friedman, M.A., Dulak, L.H. and Stedham, M.A. 1995, A lifetime oncogenicity study in rats with acrylamide. *Fund. Appl. Toxicol.* **27**: 95-105.

Friedman, M., 1973, Nucleophilic additions. In: *The Chemistry and Biochemistry of the Sulfhydryl Group in Amino Acids, Peptides, and Proteins*, Chapter 4. New York: Pergamon Press; 1973. pp. 88-134.

Garland, T.O. and Patterson M. 1967, Six cases of acrylamide poisoning. *Brit Med J* **4**: 134-138.

Gilbert, H.F. 1982, Biological disulfides: the third messenger? *J. Biol. Chem.* **257**: 12086-12091.

Ghetti, B., Wisneiwski, H.M., Cook, R.D. and Schaumburg, H.H. 1973, Changes in the CNS after acute and chronic acrylamide intoxication. *Am. J. Pathol.* **70**: 78A.

Goldstein, B.D. and Lowndes, H.E. 1979, Spinal cord defect in the peripheral neuropathy resulting from acrylamide. *NeuroToxicology* **1**: 75-87.

Goldstein, B.D. and Lowndes, H.E. 1981, Group Ia primary afferent terminal defect in cats with acrylamide neuropathy. *NeuroToxicology* **2**: 297-312.

Hashimoto, K. and Aldridge, W.N. 1970, Biochemical studies on acrylamide, a neurotoxic agent. *Biochem Pharmacol* **19**, 2591-2604.

He, F., Zhang,, S. and Wang, H. 1989, Neurological and electroneuromyographic assessment of the adverse effects of acrylamide on occupationally exposed workers. *Scand. J. Work Environ. Health* **15**: 125-129.

Hinson, J.A. and Roberts, D.W. 1992, Role of covalent and noncovalent interactions in cell toxicity: effects on proteins. *Annu. Rev. Pharmacol. Toxicol.* **32**: 471-510.

Jahn, R. and Sudhof, T.C. 1999, Membrane fusion and exocytosis. *Ann. Rev. Biochem.* **68**: 863-911.

Jennekens, F.G.I., Veldman, H., Schotman, P. and Gispen, W.H. 1979, Sequence of motor nerve terminal involvement in acrylamide neuropathy. *Acta Neuropath* **46**: 57-63.

Johnson, K.A., Gorzinski, S.J., Bodner, K.M., Campbell, R.A., Wolf, C.H., Friedman, M.A. and Mast, R.W. 1986, Chronic toxicity and oncogenicity study on acrylamide incorporated in the drinking water of Fischer 344 rats. *Toxicol. Appl. Pharmacol.* **85**: 154-168.

Kemplay, S, and Cavanagh, J.B. 1984a, Effects of acrylamide and other sulfhydryl compound *in vivo* and *in vitro* on staining of motor nerve terminals by the zinc iodide-osmium technique. *Musc Nerve* **7**: 94-100.

Kemplay, S. and Cavanagh, J.B. 1984b, Effects of acrylamide and some other sulfhydryl reagents on spontaneous and pathologically induced terminal sprouting from motor end-plates. *Musc Nerve* **7**: 101-109.

Kuperman, A.S. 1958, Effects of acrylamide on the central nervous system of the cat. *J Pharmacol Exp Ther* **123**: 180-192.

Lasek, R.J., Garner, J.A. and Brady, S.T. 1984, Axonal transport of the cytoplasmic matrix. *J Cell Biol* **99**: 212-221.

Lehning, E.J., LoPachin, R.M., Matthew, J. and Eichberg, J. 1994, Changes in Na-K ATPase and protein kinase C activities in peripheral nerve of acrylamide-treated rats. *J. Tox. Environ. Health* **42**: 331-342.

Lehning, E.J., Gaughan, C.L. and LoPachin, R.M. 1997, Acrylamide intoxication modifies in vitro responses of peripheral nerve axonal to anoxia. *J Periph Nerv Sys* **2**: 165-174.

Lehning, E.J., Persaud, A., Dyer, K.R., Jortner, B.S. and LoPachin, R.M. 1998, Biochemical and Morphologic characterization of acrylamide peripheral neuropathy. *Toxicol. Appl. Pharmacol.* **151**: 211-221.

Lehning, E.J., Balaban, C.D., Ross, J.F., Reid, M.L and LoPachin, R.M. 2002a, Acrylamide neuropathy. I. Spatiotemporal characteristics of nerve cell damage in rat cerebellum. *NeuroToxicology* **23**: 397-414.

Lehning, E.J., Balaban, C.D., Ross, J.F. and LoPachin, R.M. 2002b, Acrylamide neuropathy. II. Spatiotemporal characteristics of nerve cell damage in rat brainstem and spinal cord. *NeuroToxicology* **23**: 415-429.

Lehning, E.J., Balaban, C.D., Ross, J.F. and LoPachin, R.M. 2003, Acrylamide neuropathy. III. Spatiotemporal characteristics of nerve cell damage in rat forebrain. *NeuroToxicology* **24**: 124-136.

Lin, R.C. and Scheller, R.H. 2000, Mechanisms of synaptic vesicle exocytosis. *Annu Rev Cell Dev Biol* **16**: 19-49.

Lipton, S.A, Choi, Y.B., Pan, Z.H., Lei, S.Z., Chen, H.S.V., Sucher, N.J., Loscalzo, J., Singel, D.J. and Stamier, J.S. 1993, A redox-based mechanism for the neuroprotective and neurodestructive effects of nitric oxide and related nitroso-compounds. *Nature* **364**: 626-631.

Littleton, J.T., Chapman, E.R., Kreber, R., Garment, M.B., Carlson, S.D. and Ganetzky, B. 1998, Temperature-sensitive paralytic mutations demonstrate that synaptic exocytosis requires SNARE complex assembly and disassembly. *Neuron* **21**: 401-413.

Lombet, A., Laduron, P., Mourre, C., Jacomet, Y. and Lazdunski, M. 1986, Axonal transport of Na, K-ATPase identified as an ouabain binding site in rat sciatic nerve. *Neurosci Letts* **64**: 177-183.

LoPachin, R.M., Moore, R.W., Menahan, L.A. and Peterson, R.E. 1984, Glucose-dependent lactate production by homogenates of neuronal tissues prepared from rats treated with 2,4-dithiobiuret, acrylamide, *p*-bromophenylacetylurea and 2,5-hexanedione, *Neuro Toxicology* **5**: 25-36.

LoPachin, R.M. and Lehning, E.J. 1994, Acrylamide-induced distal axon degeneration: A proposed mechanism of action. *NeuroToxicology* **15**: 247-260.

LoPachin, R.M., Lehning, E.J., Opanashuk, L.A. and Jortner, B.S. 2000, Rate of neurotoxicant exposure determines morphologic manifestations of distal axonopathy. *Tox Appl Pharmacol* **167**: 75-86.

LoPachin, R.M., Ross, J.F. and Lehning, E.J. 2002a, Nerve terminals as the primary site of acrylamide action. *NeuroToxicology* **23**: 43-59.

LoPachin, R.M., Ross, J.F., Reid, M.L., Dasgupta, S., Mansukhani, S. and Lehning, E.J. 2002b, Neurological evaluation of toxic axonopathies in rats: acrylamide and 2,5-hexanedione. *NeuroToxicology* **23**: 95-110.

LoPachin, R.M., Balaban, C.D. and Ross, J.F. 2003, Acrylamide axonopathy revisited. *Tox. Appl. Pharmacol.* **188**:135-153.

LoPachin, R.M., Schwarcz, A.I., Gaughan, C.L., Mansukhani, S. and Das, S. 2004, In vivo and in vitro effects of acrylamide on synaptosomal neurotransmitter uptake and release. *NeuroToxicology* **25**: 349-363.

Lowndes, H.E. and Baker, T. 1976, Studies on drug-induced neuropathies. III. Motor nerve deficit in cats with experimental acrylamide neuropathy. *Europ. J. Pharmacol.* **35**: 177-184.

Lowndes, H.E., Baker, T., Michelson, L.P. and Vincent-Ablazey, M. 1978a, Attenuated dynamic responses of primary endings of muscle spindles: A basis for depressed tendon responses in acrylamide neuropathy. *Ann. Neurol.* **3**: 433-437.

Lowndes, H.E., Baker, T., Cho, E.-S. and Jortner, B.S. 1978b, Position sensitivity of de-efferented muscle spindles in experimental acrylamide neuropathy. *J. Pharmacol. Exp. Ther.* **205**: 40-48.

Malhotra, V., Orci, L., Glick, B.S., Block, M.R. and Rothman, J.E. 1988, Role of an *N*-ethylmaleimide-sensitive transport component in promoting fusion of transport vesicles with cisternae of the Golgi stack. *Cell* **54**: 221-227.

Mastrogiacomo, A. and Gundersen, C.B. 1995, The nucleotide and deduced amino acid sequence of a rat cysteine string protein. *Mole. Brain Res.* **28**: 12-18.

Mata, M., Datta, S., Jin, C.F. and Fink, D.J. 1993, Differential axonal transport of individual Na,K-ATPase catalytic (α) subunit isoforms in rat sciatic nerve. *Brain Res.* **618**: 295-298.

Medrano, C.J. and LoPachin, R.M. 1989, Effects of acrylamide and 2,5-hexanedione on brain mitochondrial respiration. *NeuroToxicology* **10**: 249-256.

Miller, M.S., Carter, D.E. and Sipes, I.G. 1982, Pharmacokinetics of acrylamide in Fisher-334 rats. *Tox Appl Pharmacol* **63**: 36-44.

Moser, V.C., Anthony, D.C., Sette, W.F. and MacPhail, R.C. 1992, Comparison of subchronic neurotoxicity of 2-hydroxyethyl acrylate and acrylamide in rats. *Fundam. Appl. Toxicol.* **18**: 343-352.

Munch, G., Lincoln, J., Maynard, K.I., Belai, A. and Burnstock, G. 1994, Effects of acrylamide on cotransmission in perivascular sympathetic and sensory nerves. *J. Auton. Nerv. Sys.* **49**: 197-205.

Nichols, B.J. and Pelham, H.R.B. 1998, SNAREs and membrane fusion in the Golgi apparatus. *Biochen. Biophys. Acta* **1404**: 9-31.

Pan, Z.H., Bahring, R., Grantyn, R. and Lipton, S.A. 1995, Differential modulation by sulfhydryl redox agents and glutathione of GABA- and glycine-evoked currents in rat retinal ganglion cells. *J. Neurosci.* **15**: 1384-1391.

Prineas, J. 1969, The pathogenesis of dying-back polyneuropathies. Part II. An ultrastructural study of experimental acrylamide intoxication in the cat. *J. Neuropath. Exp. Neurol.* **28**: 598-621.

Rosen, J. and Hellenas, K.E. 2002, Analysis of acrylamide in cooked foods by liquid chromatography and tandem mass spectrometry. *The Analyst* **127**: 880-882.

Rothman, J.E. 1994, Mechanisms of intracellular protein transport. *Nature* **372**: 55-63.

Sabri, M.I. and Spencer, P.S. 1980, Toxic distal axonopathy: biochemical studies and hypothetical mechanisms. In: Spencer PS, Schaumburg HH, editors. Experimental and clinical neurotoxicology. Baltimore, MD: Williams & Wilkins, p. 206-219.

Sanders, R.A., Zyzak, D.V., Stojanovic, M., Tallmadge, D.H., Eberhart, B.L. and Ewald, D.K. 2002, An LC/MS acrylamide method and it's use in investigating the role of asparagine. Presented at the Annual Meeting of AOAC International, September 26.

Schaumburg, H.H., Wisniewski, H.M. and Spencer, P.S. 1974, Ultrastructural studies of the dying-back process. I. Peripheral nerve terminal and axon degeneration in systemic acrylamide intoxication. *J. Neuropath. Exp. Neurol.* **33**: 260-284.

Sega, G.A., Valdivia, Alcota, R.P., Tancongco, C.P. and Brimer, P. 1989, Acrylamide binding to the DNA and protamine of spermiogenic stages in the mouse and its relationship to genetic damage. *Mutat. Res.* **216**: 221-230.

Shell, L., Rozum, M., Jortner, B.S. and Ehrich, M. 1992, Neurotoxicity of acrylamide and 2,5-hexanedione in rats evaluated using a functional observational battery and pathological examination. *Neurotox. Teratol.* **14**: 273-283.

Sickles, D.W., Fowler, S.R. and Testino, A.R. 1990, Effects of neurofilamentous axonopathy-producing neurotoxicants on *in vitro* production of ATP by brain mitochondria. *Brain Res.* **528**: 25-31.

Sickles, D.W., Stone, J.D. and Friedman, M.A. 2002, Fast axonal transport: a site of acrylamide neurotoxicity. *NeuroToxicology* **23**: 223-251.

Spencer, P.S. and Schaumburg, H.H. 1974a, A review of acrylamide neurotoxicity. Part I. Properties, uses and human exposure. *Can. J. Neurol. Sci.* **1**: 151-169.

Spencer, P.S. and Schaumburg, H.H. 1974b, A review of acrylamide neurotoxicity. Part II. Experimental animal neurotoxicity and pathologic mechanisms. *Can J Neurol Sci* **1**: 170-192.

Spencer, P.S. and Schaumburg, H.H. 1976, Central-peripheral distal axonopathy- The pathology of dying-back polyneuropathies. In: *Progress in Neuropathology*. Zimmerman H., ed., New York, Grune & Stratton, **3**: 253-276.

Spencer, P.S. and Schaumburg, H.H. 1977a, Ultrastructural studies of the dying-back process. III. The evolution of experimental peripheral giant axonal degeneration. *J Neuropath Exp Neurol* **36**: 276-299.

Spencer, P.S. and Schaumburg, H.H. 1977b, Ultrastructural studies of the dying-back process. IV. Differential vulnerability of PNS and CNS fibers in experimental central-peripheral distal axonopathy. *J. Neuropath. Exp. Neurol.* **36**: 300-320.

Sumner, S., Fennell, T., Moore, T.A., Chanas, B., Gonzalez, F. and Ghanayem, B.I. 1999, Role of cytochrome P450 2E1 in the metabolism of acrylamide and acrylonitrile in mice. *Chem Res Toxicol* **12**: 1110-1116.

Suzuki, K. and Pfaff, L. 1973, Acrylamide neuropathy in rats. An electron microscopic study of degeneration and regeneration. *Acta Neuropathol.* **24**: 197-203.

Tagaya, M., Wilson, D.W., Brunner, M., Arango, N. and Rothman, J.E. 1993, Domain structure of an *N*-ethylmaleimide-sensitive fusion protein involved in vesicular transport. *J. Biol. Chem.* **268**: 2662-2666.

Tareke, E., Rydberg, P., Karlsson, P., Eriksson, S. and Tornqvist, M. 2000, Acrylamide: A cooking carcinogen? *Chem. Res. Toxicol.* **13**: 517-522.

Tareke, E., Rydberg, P., Karlsson, P., Eriksson, S. and Tornqvist, M. 2002, Analysis of acrylamide, a carcinogen formed in heated foodstuffs. *J. Agric. Food Chem.* **50**: 4998-5006.

Tilson, H.A. and Cabe, P.A. 1979, The effects of acrylamide given acutely or in repeated doses on fore- and hindlimb function of rats. *Tox. Appl. Pharmacol.* **47**; 253-260.

Tolar, L.A. and Pallanck, L. 1998, NSF function in neurotransmitter release involves rearrangement of the SNARE complex downstream of synaptic vesicle docking. *J. Neurosci.* **18**: 10250-10256.

Tsujihata, M., Engel, A.G. and Lambert, E.H. 1974, Motor end-plate fine structure in acrylamide dying-back neuropathy: A sequential morphometric study. *Neurology* **24**: 849-856.

Tyl, R.W., Marr, M.C., Myers, C.B., Ross, W.P. and Friedman, M.A. 2000a, Relationship between acrylamide reproductive and neurotoxicity in male rats. *Reprod Toxicol* **14**: 147-157.

Whiteheart, S.W., Rossnagel, K., Buhrow, S.A., Brunner, M., Jaenicke, R. and Rothman, J.E. 1994, *N*-Ethylmaleimide-sensitive fusion protein: a trimeric ATPase whose hydrolysis of ATP is required for membrane fusion. *J. Cell Biol.* **126**: 945-954.

Working, P., Bentley, K., Hurtt, M. and Mohr, K. 1987, Comparison of the dominant lethal effects of acrylonitrile and acrylamide in male Fischer 344 rats. *Mutagenesis* **2**: 215-220.

THE ROLE OF EPIDEMIOLOGY IN UNDERSTANDING THE RELATIONSHIP BETWEEN DIETARY ACRYLAMIDE AND CANCER RISK IN HUMANS

Lorelei A. Mucci[1,2] and Hans-Olov Adami[1,3]
[1]*Department of Epidemiology, Harvard School of Public Health, Boston MA, USA;* [2] *Channing Laboratory, Harvard Medical School/Brigham and Women's Hospital, Boston, MA USA;* [3]*Department of Medical Epidemiology and Biostatistics, Karolinska Institutet, Stockholm, Sweden; e-mail: lmucci@hsph.harvard.edu*

Abstract: Since April 2002, when the Swedish National Food Administration first reported its finding of elevated levels of the substance acrylamide in commonly consumed foods (Swedish National Food Administration, 2002), there has been considerable debate about the health effects of dietary exposure to acrylamide. In particular, researchers have speculated on whether the amount of acrylamide consumed through the typical diet could increase the risk of cancer in humans. In this paper, we review the epidemiological data to date examining dietary acrylamide in relation to cancer risk. We highlight the strengths and limitations of using epidemiology to address this public health question. Finally, we provide an overview of future directions of epidemiological research on the health effects of dietary acrylamide.

Key words: Acrylamide, diet, colorectal cancer, kidney cancer, bladder cancer, epidemiology, case-control study, cohort study

1. BACKGROUND

The finding of acrylamide in commonly consumed foods, such as coffee, fried potato products, and breads, generated substantial public health alarm that intake of these foods could increase the risk of human cancer. Indeed, some researchers speculated that dietary sources of acrylamide could be a causal factor in 30% of human cancer cases. This concern stems from the classification of acrylamide as a "probable human carcinogen" by the

International Agency for Cancer Research(International Agency for Research on Cancer, 1994). However, the scientific evidence on the potential carcinogenicity of acrylamide relies primarily on data from experimental models. In vivo and in vitro experiments have shown that cells exposed to high levels of acrylamide undergo genetic mutations and cellular transformation (Park et al., 2002). In addition, animals exposed to very high levels of acrylamide develop several tumors, including those of the mammary gland, lung, and intestinal and reproductive tract (Friedman et al., 1995; Johnson et al., 1986).

In assessing the carcinogenicity of acrylamide, there have been only limited studies among humans. However, the classification of acrylamide as a probable human carcinogen was sufficient to generate concern that intake of foods high in acrylamide could increase risk of cancer. In fact, under California's proposition 65, public officials are now considering whether to impose warning labels on those food items containing acrylamide, including breads, cereals, potato products, and coffee, that "these products contain chemicals known by the State of California to cause cancer" (OEHHA, 2003). Given the lack of empirical data in humans to address the concerns, epidemiological studies addressing the association between dietary exposure to acrylamide and cancer risk are warranted.

2. ACRYLAMIDE FORMATION IN FOODS

Since the initial discovery by the Swedish National Food Administration, scientists around the world have confirmed the detection of acrylamide in foods, and have also quantified acrylamide levels in several additional food items. Elevated levels of acrylamide have been detected in potato chips, French fries, crisp and soft breads, cereals, chocolate, and coffee (Center for Food Safety and Applied Nutrition, 2002; WHO/FAO, 2003). Table 1 presents acrylamide concentration for several food groups from the Swedish National Food Administration.

During the months following the initial report, researchers were able to elucidated the potential mechanism of acrylamide formation in foods, which occurs as a result of a reaction between amino acids, particularly asparagine, and reducing sugars during the heating of starch-rich foods to high temperatures(Mottram et al., 2002; Stadler et al., 2002). In this way, acrylamide occurs as a natural process of cooking, rather than as a food contaminant. Acrylamide concentrations appear to be a function of asparagine levels in the foods, the temperature at which foods are cooked, and the duration of cooking (Friedman, 2003). Thus, there is variability in acrylamide levels between and within food items.

Table 1. Acrylamide concentration for several food groups: Swedish National Food Administration, 2002

Food group	Acrylamide concentration (microgram/kilogram)	
	Median	min-max
Potato crisps	1200	330-2300
French fries	450	300-1100
Pan fried potatoes	300	
Biscuits and crackers	410	<30-650
Crisp breads	140	<30-1900
Breakfast cereals	160	<30-1400
Corn chips	150	120-180
Soft breads	50	<30-160
Various fried foods (pizza, pancakes, waffles, fish fingers, meatballs, chickenbits, deep fried fish, vegetarian schnitzel and cauliflower gratin)	40	<30-60
Coffee	25	8-40

3. OCCUPATIONAL EXPOSURE TO ACRYLAMIDE AND CANCER RISK

Workers in certain settings, such as biomedical laboratories and chemical plants, can be exposed to acrylamide occupationally (Marsh et al., 1999; Schettgen et al., 2002). The most common routes of exposure in the workplace are through inhalation or dermal absorption. Although the intensity varies across settings, an average workplace exposure is 30 micrograms/kg body weight per day.

There is some epidemiological evidence available from cohorts of workers exposed to acrylamide occupationally assessing cancer risk. In a small study of 371 workers with potential exposure to acrylamide, there was no statistically significant excess mortality observed in the cohort (Sobel et al., 1986). Although the number of cancer deaths in the cohort was greater than expected, this excess was attributed to increased respiratory cancers in the group exposed to organic dyes. In a larger cohort of workers, there was no evidence of a statistically significant increase in cancer mortality (Marsh et al., 1999).

Among two hundred and ten tunnel workers exposed to short-term but intensive doses of acrylamide, researchers examined the health effects of occupational acrylamide exposure using hemoglobin (Hb) adducts as biomarkers of internal dose (Hagmar et al., 2001). Exposed workers had elevated levels of Hb adduct. Moreover, there was a dose-response relationship between adduct levels on the one hand and demyelinating and axonal changes in peripheral nerves(Kjuus et al., 2004) as well as neurological symptoms(Hagmar et al., 2001). Short-term follow-up of these works demonstrated that these changes were mild and generally reversible; long-term follow-up will be required in order to assess potential carcinogenic effects from the exposure.

In summary, the results of these occupational studies suggest that there is little evidence of an increased risk of cancer associated with occupational acrylamide exposure. Very high exposure doses are correlated with HB adduct formation, although the potential impact of short-term exposure is unclear. The routes of acrylamide exposure occupationally differ substantially from dietary exposure, and the exposure dose is several times higher.

4. DIETARY ACRYLAMIDE AND CANCER RISK

Acrylamide formation in foods was first reported by the Swedish National Food Administration in April, 2002. Thus, there were no epidemiological studies assessing the health effects of acrylamide exposure through diet prior, since this source of exposure was unknown before this time. Thus far, the results of three case-control studies examining the relationship between dietary acrylamide and cancer risk have been published (Mucci et al., 2003a; Mucci et al., 2003b; Mucci et al., 2004; Pelucchi et al., 2003).

The first published epidemiological study of dietary exposure to acrylamide, undertaken in response to the Swedish report, examined acrylamide in relation to risk of cancer of the large bowel, bladder and kidney (Mucci et al., 2003a). An updated analysis including additional data on coffee consumption was published soon afterwards (Mucci et al., 2003b). This investigation was undertaken within an existing population-based case-control study in Sweden, and included incident cases of cancer of the large bowel (N=591), bladder (N=263) and kidney (N=133) and healthy controls (N=538) frequency matched on age and gender. Information on dietary habits in the 5 years prior to study was assessed through semi-quantitative food frequency questionnaire of 88 food items. The majority of the foods found to contain high levels of acrylamide were assessed in the food frequency questionnaire including; french fries, potato chips, fried potatoes, crisp and soft bread, breakfast cereals, biscuits, cakes and coffee. Information on acrylamide levels within foods was obtained through the Swedish National Food Administration database on acrylamide (Swedish National Food Administration, 2002). A summary measure of acrylamide intake was calculated by ranking the food items of acrylamide dose, and multiplying an individual's intake of each food item per day by the ranking, and summing across all foods (Mucci et al., 2003a).

Among the controls in the study, the major dietary source of acrylamide was crisp breads (28%), fried potato products (22%), and coffee (20%) (Mucci et al., 2003b). The estimated daily mean (standard error) dietary acrylamide dose (μg) was: 34.0 (0.6) for controls, 34.8 (0.6) for colorectal, 36.8 (1.0) for bladder, and 34.5 (1.4) for kidney cancers. These estimates of daily acrylamide intake are in line with data from cohorts in the US, UK, Holland, Sweden and Norway (Dybing and Sanner, 2003; Konings et al., 2003; Svensson et al., 2003; WHO/FAO, 2003). Adjusting for potential confounders, there was no evidence that intake of food items with elevated levels of acrylamide was associated with cancer of the large bowel, kidney or bladder. Likewise, there was no positive association between total dietary acrylamide intake and risk of the studied cancers. Indeed, there was evidence of an inverse association for large bowel cancer, with a 40% reduced risk in the extreme quartiles of acrylamide. Smoking is an important source of non-occupational acrylamide exposure, with intake of 1-2 micrograms of acrylamide per cigarette (Smith et al., 2000). The authors examined whether the effect of dietary acrylamide intake was similar for smokers and nonsmokers, and found no increased risk for either group.

The same authors undertook a reanalysis of a larger case-control study of renal cell cancer using data from the Swedish component of an international collaborative population-based study (Mucci et al., 2004). Incident cases of renal cell cancer were identified through regional cancer registries, and controls were randomly selected from the study base through the register of total population and frequency matched on age and sex. The study included 379 cases and 353 controls. Data were obtained through structured interviews. A food frequency questionnaire asked about intake of selected food items prior to cancer diagnosis. Acrylamide content in foods was derived from national databases, and a summary measure of total acrylamide intake was calculated for each individual.

In this study, estimated mean (standard deviation) daily acrylamide exposure through dietary sources was 27.6 (0.6) μg among controls and 27.6 (0.7) μg among renal cell cancer cases (Mucci et al., 2004). In line with the earlier Swedish study, intake of coffee, crisp bread and fried potatoes were the major contributors to acrylamide through diet. Adjusting for potential confounders, there was no association between intake of foods with elevated acrylamide, including coffee, crisp bread or fried potatoes, and risk of renal cell cancer. In addition, comparing the highest (>31.9 micrograms/day) and lowest (<20.1 micrograms/day) quartiles of estimated dietary acrylamide intake, the relative risk was 1.1 (95% Confidence Interval = 0.7-1.8). There was no difference in the effect of acrylamide among smokers or nonsmokers.

In the final published study to date, researchers examined intake of fried potato products and risk of several cancers (Pelucchi et al., 2003). Fried potato consumption is an important contributor to dietary acrylamide exposure in the studied populations. Data were analyzed from several large, hospital-based case-control studies conducted in Italy and Switzerland between 1991 and 2000. The cancer sites included oral cavity and pharynx (749 cases, 1772 controls), esophagus (395 cases, 1066 controls), larynx (527 cases, 1297 controls), large bowel (1225 colon and 728 rectum cases, 4154 controls), breast (2569 cases, 2588 controls) and ovary (1031 cases, 2411 controls). In these studies, controls were sampled from patients admitted to the same network of hospitals of cases for acute, non-neoplastic conditions. For all of the studied cancers, there was no evidence of a positive association between intake of fried potato products and risk. All the odds ratios (OR) for the highest vs. the lowest tertile of intake ranged between 0.8-1.1. Similar to the above studies, there was no difference among smokers versus nonsmokers.

In summary, intake of acrylamide through diet in the Swedish populations was about 25 to 35 micrograms per day, substantially lower than that through smoking or occupational settings. In addition, data from these three case-control studies provide converging evidence that the intake of acrylamide in the amounts generally consumed through diet is not associated with a higher risk of several cancers in humans. The lack of positive effect was similar for smokers and nonsmokers. To date, there have been no prospective studies examining acrylamide and cancer. Moreover, studies of other cancers and in additional populations are needed to confirm the findings.

5. STRENGTHS AND LIMITATIONS OF EPIDEMIOLOGICAL STUDIES

Understanding whether dietary sources of acrylamide increase cancer risk in humans will require the culmination of scientific evaluation across research disciplines. Epidemiological studies play and important and prerequisite role in assessing the potential risk. One of the major strengths of epidemiological studies is that one can directly assess the effect of dietary acrylamide within the relevant study population. Such studies avoid the uncertainties of extrapolating results from experimental models (Ruden, 2004). For example, in animal models, rats were exposed to acrylamide levels 3 to 5 orders of magnitude greater than what humans generally intake through diet (Friedman et al., 1995; Johnson et al., 1986). Moreover, the routes of exposure are different, with animals exposed to acrylamide orally in aqueous solutions or through IP injection. Currently, it is unclear to what extent acrylamide in foods is bioavailable. Initial evidence suggests that update of acrylamide is reduced in the presence of dietary proteins (Schabacker et al., 2004).

There are some limitations to consider when evaluating the epidemiological literature. The published studies thus far have relied on self-report of food items, and may be subject to reporting error. There is also variability in acrylamide dose across brands of a given food. Thus, estimated daily intake of dietary acrylamide probably constitutes an underestimate. However, ranking of individuals with respect to exposure, rather than the absolute intake, determines the calculated relative risk in case-control studies. In fact, when comparing an abbreviated versus extensive food frequency questionnaire, increasing the number of food items improves the ranking, and thereby the relative risk, only to a small degree (Voskuil et al., 1999). In the Swedish studies, the relative risk estimates comparing quartiles of total acrylamide dose were insensitive to the concentration of acrylamide used to rank individual food items, and thus it appears one can effectively rank individuals which is valid for drawing conclusions.

Another possible limitations of these study is statistical power. Using data extrapolated from the animal models, risk assessment models have determined an expected relative risk of cancer of 1.006 to 1.05 for the highest dose of acrylamide > 70 micrograms per day (Dybing and Sanner, 2003; Hagmar and Tornqvist, 2003). In the two Swedish studies, less than 1.5% of the population consumed acrylamide in these levels. An epidemiological study would require a cohort of more than two million individuals to detect such a small relative risk given such a small proportion of the population exposed. In epidemiology, we lack the scientific means to document such a small effect, nor can we ever prove the negative that acrylamide does not cause human cancer. Beyond the limitations of risk assessment models, we should consider the relevance of such a small effect in terms of public health.

6. FUTURE DIRECTIONS

A well-conducted case-control study is an efficient design to examine the association between dietary exposure to acrylamide and risk of colorectal cancer. However, this study design is vulnerable to potential biases, including selection and recall biases. Thus, data from prospective cohort studies examining the association between dietary acrylamide and cancer risk are eagerly awaited. Our own group is now examining the risk of breast cancer and colorectal cancer associated with dietary acrylamide in large (>50,000) cohorts of Swedish women. The study findings must also be further replicated in additional study populations, which might have different quantity of acrylamide intake, and for additional cancers, which have different biological relevance for acrylamide.

No single study can provide conclusive evidence on the health effects of acrylamide in diet. However, an accumulation of evidence through well-conducted epidemiological studies can shed light on this important public health concern.

REFERENCES

Center for Food Safety and Applied Nutrition, 2002, Exploratory Data on Acrylamide in Foods, US Food and Drug Administration, http://www.cfsan.fda.gov/~dms/acrydata.html

Dybing, E., and Sanner, T., 2003, Risk assessment of acrylamide in foods, *Toxicol. Sci.* **75**:7-15.

Friedman, M., 2003, Chemistry, biochemistry and safety of acrylamide. A review, *J. Agric. Food Chem.* **51**:4504-4526.

Friedman, M. A., Dulak, L. H., and Stedham, M. A., 1995, A lifetime oncogenicity study in rats with acrylamide, *Fundam. Appl. Toxicol.* **27**:95-105.

Hagmar, L., and Tornqvist, M., 2003, Inconclusive results from an epidemiological study on dietary acrylamide and cancer, *Br. J. Cancer* **89**:774-775; author reply 775-776.

Hagmar, L., Tornqvist, M., Nordander, C., Rosen, I., Bruze, M., Kautiainen, A., Magnusson, A. L., Malmberg, B., Aprea, P., Granath, F., and Axmon, A., 2001, Health effects of occupational exposure to acrylamide using hemoglobin adducts as biomarkers of internal dose, *Scand. J. Work Environ. Health* **27**:219-226.

International Agency for Research on Cancer, 1994, IARC Monographs on the Evaluation of Carcinogen Risk to Humans: Some Industrial Chemicals, International Agency for Research on Cancer, Lyon.

Johnson, K. A., Gorzinski, S. J., Bodner, K. M., Campbell, R. A., Wolf, C. H., Friedman, M. A., and Mast, R. W., 1986, Chronic toxicity and oncogenicity study on acrylamide incorporated in the drinking water of Fischer 344 rats, *Toxicol. Appl. Pharmacol.* **85**:154-168.

Kjuus, H., Goffeng, L. O., Heier, M. S., Sjoholm, H., Ovrebo, S., Skaug, V., Paulsson, B., Tornqvist, M., and Brudal, S., 2004, Effects on the peripheral nervous system of tunnel workers exposed to acrylamide and N-methylolacrylamide, *Scand. J. Work Environ. Health* **30**:21-29.

Konings, E. J., Baars, A. J., van Klaveren, J. D., Spanjer, M. C., Rensen, P. M., Hiemstra, M., Kooij, J. A. v., and Peters, P. W., 2003, Acrylamide exposure from foods of the Dutch population and an assessment of the consequent risks, *Food Chem. Toxicol.* **41**:1569-1579.

Marsh, G. M., Lucas, L. J., Youk, A. O., and Schall, L. C., 1999, Mortality patterns among workers exposed to acrylamide: 1994 follow up, *Occup. Environ. Med.* **56**:181-190.

Mottram, D. S., Wedzicha, B. L., and Dodson, A. T., 2002, Acrylamide is formed in the Maillard reaction, *Nature* **419**:448-449.

Mucci, L. A., Dickman, P. W., Steineck, G., Adami, H. O., and Augustsson, K., 2003a, Dietary acrylamide and cancer of the large bowel, kidney, and bladder: absence of an association in a population-based study in Sweden, *Br. J. Cancer* **88**:84-89.

Mucci, L. A., Dickman, P. W., Steineck, G., Adami, H. O., and Augustsson, K., 2003b, Dietary acrylamide and cancer risk: additional data on coffee [Letter], *Br. J. Cancer* **89**:775-776.

Mucci, L. A., Lindblad, P., Steineck, G., and Adami, H. O., 2004, Dietary acrylamide and risk of renal cell cancer, *Int. J. Cancer* **109**:774-776.

OEHHA, 2003, Acrylamide Workplan, Office of Environmental Health Hazard Assessment, http://www.oehha.ca.gov/prop65/docs_state/arcyl2.html

Park, J., Kamendulis, L. M., Friedman, M. A., and Klaunig, J. E., 2002, Acrylamide-induced cellular transformation, *Toxicol. Sci.* **65**:177-183.

Pelucchi, C., Franceschi, S., Levi, F., Trichopoulos, D., Bosetti, C., Negri, E., and La Vecchia, C., 2003, Fried potatoes and human cancer, *Int. J. Cancer* **105**:558-560.

Ruden, C., 2004, Acrylamide and cancer risk--expert risk assessments and the public debate, *Food Chem. Toxicol.* **42**:335-349.

Schabacker, J., Schwend, T., and Wink, M., 2004, Reduction of acrylamide uptake by dietary proteins in a caco-2 gut model, *J. Agric. Food Chem.* **52**:4021-4025.

Schettgen, T., Broding, H. C., Angerer, J., and Drexler, H., 2002, Hemoglobin adducts of ethylene oxide, propylene oxide, acrylonitrile and acrylamide-biomarkers in occupational and environmental medicine, *Toxicol. Lett.* **134**:65-70.

Smith, C. J., Perfetti, T. A., Rumple, M. A., Rodgman, A., and Doolittle, D. J., 2000, "IARC group 2A Carcinogens" reported in cigarette mainstream smoke, *Food Chem. Toxicol.* **38**:371-383.

Sobel, W., Bond, G. G., Parsons, T. W., and Brenner, F. E., 1986, Acrylamide cohort mortality study, *Br. J. Ind. Med.* **43**:785-788.

Stadler, R. H., Blank, I., Varga, N., Robert, F., Hau, J., Guy, P. A., Robert, M. C., and Riediker, S., 2002, Acrylamide from Maillard reaction products, *Nature* **419**:449-450.

Svensson, K., Abramsson, L., Becker, W., Glynn, A., Hellenas, K. E., Lind, Y., and Rosen, J., 2003, Dietary intake of acrylamide in Sweden, *Food Chem. Toxicol.* **41**:1581-1586.

Swedish National Food Administration, 2002, Acrylamide in food, URL: http://www.slv.se/engdefault.asp

Voskuil, D. W., Augustsson, K., Dickman, P. W., van't Veer, P., and Steineck, G., 1999, Assessing the human intake of heterocyclic amines: limited loss of information using reduced sets of questions, *Cancer Epidemiol. Biomarkers Prev.* **8**:809-814.

WHO/FAO, 2003, Acrylamide, http://www.who.int/fsf/Acrylamide/Acrylamide_index.htm

MECHANISMS OF ACRYLAMIDE INDUCED RODENT CARCINOGENESIS

James E. Klaunig and Lisa M. Kamendulis
Division of Toxicology, Department of Pharmacology and Toxicology, Indiana University School of Medicine, 635 Barnhill Dr., MS 551, Indianapolis, Indiana 46202; e-mail jklauni@iupui.edu

Abstract: Acrylamide is a monomer of polyacrylamide, used in biochemistry, in paper manufacture, in water treatment, and as a soil stabilizer. The monomer can cause several toxic effects and has the potential for human exposure either through the environment or from occupational exposure. Recently, additional concern for the potential toxicity of acrylamide in humans has arisen with the finding of acrylamide formation in some processed foods. It has been established that following chronic exposure, rats exhibited an increase in the incidence of adrenal pheochromocytomas, testicular mesotheliomas, thyroid adenomas and mammary neoplasms in F344 rats. This has raised increased concerns regarding the carcinogenic risk to humans from acrylamide exposure. Studies examining the DNA reactivity of acrylamide have been performed and have had differing results. The tissue and organ pattern of neoplastic development seen in the rat following acrylamide exposure is not consistent with that seen with other strictly DNA reactive carcinogens. Based on the pattern of neoplastic development, it appears that acrylamide is targeting endocrine sensitive tissues. In the current monograph, studies on the effect of acrylamide on DNA reactivity and on altered cell growth in the target tissues in the rat are reported. DNA synthesis was examined in F344 rats treated with acrylamide (0, 2, or 15 mg/kg/day) for 7, 14, or 28 days. Acrylamide increased DNA synthesis in the target tissues (thyroid, testicular mesothelium, adrenal medulla) at all doses and time points examined. In contrast, in a non-target tissue (liver), no increase in DNA synthesis was seen. Examination of DNA damage using single cell gel electrophoresis (the Comet assay) showed an increase in DNA damage in the target tissues, but not in non-target tissue (liver). In addition, a cellular transformation model, (the Syrian Hamster Embryo (SHE) cell morphological transformation model), was used to examine potential mechanisms for the observed carcinogenicity of acrylamide. SHE cell studies showed that glutathione (GSH) modulation by acrylamide was important in the cell transformation process. Treatment with a sulfhydryl donor compound (NAC) reduced acrylamide transformation while depletion of

GSH (BSO) resulted in an enhancement of transformation. In summary, acrylamide caused both an increase in DNA synthesis and DNA damage in mammalian tissues and cells suggesting that DNA reactivity and cell proliferation, in concert, may contribute to the observed acrylamide-induced carcinogenicity in the rat and has implication on the possible risk for human neoplasm development.

Key words: Acrylamide, transformation, Comet assay, carcinogenicity, DNA synthesis, DNA damage

1. INTRODUCTION

Acrylamide (2-propenamide) is the monomer of polyacrylamide, and is used in a number of commercial and industrial applications (NIOSH, 1976; WHO, 1985; and U.S. EPA, 1994). In addition, recent studies have demonstrated that acrylamide is produced from the processing of selective foods (Weiss, 2002; Rosen et al., 2002; Tareke et al., 2000). As such, human exposure to acrylamide could occur during the production and handling of acrylamide and polyacrylamide, via environmental exposure and possibly through the ingestion of processed foods.

Chronic exposure to acrylamide via drinking water in the F344 rat resulted in a tissue-specific increase in tunica vaginalis mesotheliomas in male rats, mammary gland fibroadenomas in female rats, as well as thyroid follicular cell adenomas and adrenal pheochromocytomas in both male and female F344 rats (Johnson et al. 1986, and Friedman et al. 1995). Based on the observed v rodent carcinogenicity, acrylamide has been classified as a probable carcinogen in humans (U.S. EPA, 1994).

Acrylamide tested negative in bacterial gene mutation assays, both in the presence and absence of metabolic activation (Knaap et al., 1988; Tsuda et al., 1993; Dearfield et al., 1995). However, chromosomal aberrations, sister chromatid exchanges, unscheduled DNA synthesis, DNA fragments (breaks and deletions), cell transformation, and cellular mitotic disruptions have been observed in mammalian cell lines (Barfknecht et al., 1988; Knaap et al., 1988; Sega et al., 1990; Tsuda et al., 1993; U.S. EPA, 1994; Dearfield et al., 1995; Park et al., 2002). Acrylamide induced cellular transformation in a Syrian hamster embryo transformation system (SHE cell) was modulated by glutathione depletion (increased transformation) and addition (decreased transformation) (Park et al., 2002). Morphological transformation of Syrian Hamster Embryo (SHE) cells mimics the early stage of carcinogenesis and has been used to understand the mechanisms of chemical carcinogenesis (Barrett et al., 1984; Isfort et al., 1994). Acrylamide was previously found to induce morphological transformation in C3H/10T1/2, BALB/c 3T3 and NIH/3T3 cells (Banerjee and Segal, 1986; Tsuda et al., 1993).

Acrylamide is metabolized into glycidamide, a reactive epoxide, via cytochrome P450 2E1 (Sumner et al. 1999). Both acrylamide and glycidamide have been shown to interact with DNA *in vitro*, which may be suggestive of a potential genotoxic mechanism of action for the production of acrylamide-induced cancers. However, the tissue selective tumor response observed following acrylamide exposure is not characteristic of a strictly genotoxic mode of action and suggests that acrylamide (and/or glycidamide) may function through an epigenetic mechanism(s) in the production of neoplasia.

The induction of DNA synthesis and subsequent cell growth by chemical carcinogen exposure is an important component of the carcinogenesis process both for genotoxic and nongenotoxic carcinogens (Cohen and Ellwein, 1990; Ames et al., 1990; Butterworth et al., 1991). Cell proliferation may result in an increase in spontaneous mutations and/or a selective clonal expansion of initiated cells (Cohen and Ellwein, 1991; Schulte-Hermann, 1987; Klaunig et al., 1993). Additionally, since acrylamide and glycidamide have been shown to be DNA reactive, it is possible that exposure to acrylamide results in the induction of DNA strand breaks. If the damage to DNA is severe, and/or occurs persistently during exposure to acrylamide, cellular repair capabilities may be surpassed and DNA damage may accumulate, and if the cell undergoes DNA replication, a mutational event may occur. Therefore, examining the effect of acrylamide on the induction of DNA damage and cell proliferation is important to further our understanding of the mechanism for the tissue-specific carcinogenicity of acrylamide.

2. MATERIALS AND METHODS

2.1 Chemicals

Monoclonal primary antibody to 5-bromo-2'-deoxyuridine (BrdU), mouse IgG1, biotynilated anti-mouse link, peroxidase-conjugated streptavidin label, and 3-amino-9-ethylcarbazole (AEC) were purchased from Biogenix (San Ramon, CA.). Lysis buffer, LMAgarose, and 200 mM EDTA were purchased from Trevigen (Gaithersburg, MD). Acrylamide (99.9% purity), and all other reagents were from Sigma Chemical Co. (St. Louis, MO.) and were the highest analytical grade available.

2.2 Animals and Treatment

Seven week-old male F344 rats were purchased (Harlan Sprague Dawley Co, Indianapolis, IN.), and acclimated for 1 week prior to study initiation. Animals were maintained in an AALAC-certified facility, and in accordance with the NIH guide for the Care and Use of Laboratory Animals. Following the acclimation period, animals were randomly placed into groups of either 0 mg/kg/ or 15 mg/kg acrylamide/day. For the analysis of DNA synthesis, treatment was for 7, 14 or 28 days, and for the analysis of DNA damage, treatment was for 7 days. Acrylamide was administered in drinking water and was changed every other day.

2.3 Cell culture

Syrian Hamster Embryo (SHE) cells were isolated and cultured as described previously (Kerckaert et. al. 1996). The primary embryo cells were grown in LeBoeuf's Dulbecco's Modified Eagles Medium (pH 6.7), supplemented with 20% fetal bovine serum and 4 mM L-glutamate at 37°C in 10% CO_2 and 90% relative humidity. A feeder layer was prepared by plating 2×10^6 SHE cells into 30 mL complete media in a T-150 tissue culture flask. Cells were exposed to acrylamide with and without BSO and/or NAC continuously. Cells were treated with BSO or NAC for 24 hours in advance and then co-incubated with acrylamide for 6 days. At sampling, the cultures were then rinsed, fixed with methanol, stained with Giemsa (Sigma, St. Lois, MO), and evaluated for the presence of a morphologically transformed phenotype as described previously (Kerckaert et al., 1996). For each treatment, total colony number, morphological transformation frequency [(the number of transformed colonies/total number of colonies scored) × 100], and relative plating efficiency [RPE, (test group plating efficiency/solvent control plating efficiency) × 100] were determined.

2.4 Analysis of DNA Synthesis

Osmotic mini pumps containing BrdU (20 mg/mL in 0.9% saline, Alzet Co, Palo Alto, CA.), were implanted into animals seven days prior to sacrifice. At sampling, animals were killed, and the thyroid, adrenals, testes and liver were harvested. Tissues were fixed in formalin (10%, 48 hrs), processed for paraffin embedding, sectioned and stained for the determination of DNA synthesis. DNA synthesis was examined using immunohistochemical detection of BrdU (Eldridge et al., 1990). Approximately 2000 cells were counted for each animal. The labeling index (the percentage of BrdU labeled cells divided by the total number of cells scored) was calculated for thyroid, adrenal, and liver tissues. In the testicular mesothelium, the number of labeled cells/centimeter of tissue was determined.

2.5 Alkaline Comet Assay

To detect DNA strand breaks, alkaline single cell gel electrophoresis (the Comet assay) was performed using a modified procedure as described (Shimoi et al., 2001; Rojas et al., 1999; Tice, 1995). Briefly, tissues (liver, adrenal, and thyroid) obtained at sacrifice, were homogenized in buffer (0.075 M NaCl, 0.024 M Na_2EDTA, pH 7.5, 0°C, 1 mL/g tissue). The homogenate was then centrifuged (700 x g, 10 min.), and the supernatant discarded. The precipitate was then resuspended in homogenizing buffer (1 mL). 10 μL of the homogenate was mixed with 70 μL of LMAgarose (Trevigen; Gaithersburg, MD.) at 37°C, and samples (50 μL) were applied to sample wells of pre-coated CometSlide ™ (Trevigen, Gaithersburg, MD.) . After solidification, the slides were immersed in lysis solution (1% Triton X-100, 10 % DMSO, 2.5 M NaCl, 100 mM Na_2EDTA, 10 mM Tris, pH 10.0, 1hour, 4°C), followed by incubation in an alkali solution (0.3 M NaOH, 200 mM EDTA; final pH > 13, 40 min). The slides were then electrophoresed in buffer (1 mM Na_2EDTA and 0.3 M NaOH, pH >13.5) at 25 V, 300 mA for 30 min at 4°C. After electrophoresis, the slides were rinsed (3x, 0.4 M Tris, pH 7.5), stained with ethidium bromide (60 μl; 20 μg/mL), and covered with coverslips. A total of 100 nuclei/data point (50 cells chosen randomly per slide), were visualized with a Nikon inverted fluorescent microscope and Comet Tail Moment was quantified using the Komet 4.0 analysis software (Kinetic Imaging Ltd.).

2.6 Statistical Analysis

Statistical analysis was performed for the Comet assay using ANOVA followed by a Dunnett's post-hoc test (Gad, 1986). Statistical analysis for DNA damage and the SHE cell assay was performed using ANOVA followed by Fisher's Exact post hoc test (Gad, 1986). Values were considered significant at $p<0.05$.

3. RESULTS

No significant differences in animal or tissue weights of acrylamide treated F344 rats were observed at any treatment or time point examined (data not shown). DNA synthesis was measured in three acrylamide target tissues for neoplastic development (thyroid, testes, and adrenal medulla) and two non-target tissues (liver and adrenal cortex) in F344 rats. Acrylamide induced an increase in DNA synthesis in thyroid follicular cells (~2-fold over control) following treatment with 15 mg/kg acrylamide at all time points examined (Fig. 1). In the adrenal medulla, DNA synthesis was increased by acrylamide, and was highest after 7 days of treatment (Fig. 1). The third target tissue examined, the testicular mesothelium, also showed increased DNA synthesis following acrylamide treatment (Fig. 1). Similar to the other tissues examined, the 15 mg/kg dose of acrylamide increased DNA synthesis at all time points in the testicular mesothelium. In the two non-target tissues examined, the liver and the adrenal cortex, acrylamide failed to increase DNA synthesis at any time point examined (Fig. 1).

DNA damage in two target tissues for acrylamide neoplasia (thyroid and adrenal) and a non-neoplastic tissue (liver) was measured using the Comet assay, and is represented by comet tail moment (Fig. 2). The testicular mesothelium was logistically difficult to isolate and was not examined in this study. Treatment with 15 mg/kg acrylamide produced significant increases in comet tail moment in the thyroid (Fig. 2). Similarly, acrylamide induced DNA damage (~2-fold over control) in the adrenal of F344 rats (Fig. 2). In the non-target tissue examined, the liver, no increase in DNA damage was seen following acrylamide exposure (Fig. 2).

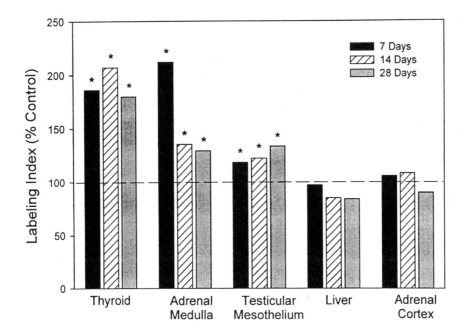

Figure 1. Effect of Acrylamide on DNA synthesis in the F344 rat. 8 week old, male F344 rats were treated with 15 mg/kg acrylamide for 7, 14, or 28 days. At sampling, DNA synthesis was measured in cells of the follicular thyroid, adrenal medulla, testicular mesothelium, liver and adrenal cortex. Labeled cells were scored for each animal (approximately 2000 cells/animal). Labeling Index = ((number of labeled cells/total number of cells counted)*100, except for the testis, in which labeling index was calculated by determining the number of labeled cells/cm of tissue. Data is expressed as % of untreated control. * Statistically different from controls, $p < 0.05$.

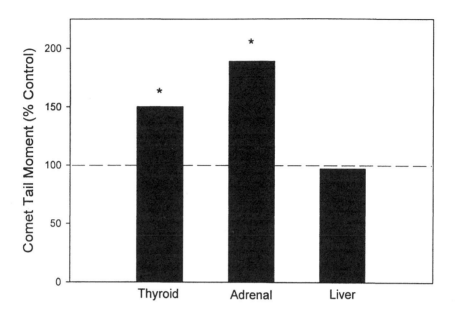

Figure 2. Effect of Acrylamide on DNA damage in the F344 rat. 8 week old, male F344 rats were treated with 15 mg/kg acrylamide for 7 days. DNA damage was assessed using the Comet assay in the thyroid, adrenal gland and liver. DNA damage was quantified as comet tail moment and is expressed as % of control. * Statistically different from controls, $p < 0.05$.

Cell transformation by acrylamide was examined in SHE cells treated with concentrations ranging from 0.1 to 0.7 mM acrylamide for 7 days. A concentration-dependent increase in morphological transformation relative to untreated control cells was observed at 0.5 and 0.7 mM acrylamide (Fig. 3). Treatment with either DL-buthionine-[S,R]-sulfoxamine (BSO), a selective inhibitor of γ-glutamylcysteine synthetase, the rate-limiting step in GSH synthesis or N-acetyl-L-cysteine (NAC), a thiol donor, were utilized to examine the effect of GSH depletion or supplementation, respectively. BSO enhanced acrylamide-induced transformation, while GSH supplementation (NAC treatment) reduced the transformation (Fig. 4).

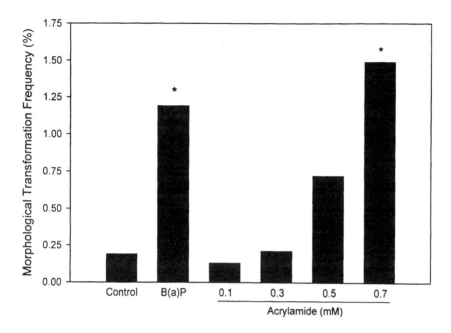

Figure 3. Acrylamide-induced morphological transformation in SHE cells. Morphological transformation was assessed in SHE cells treated with 0.1 to 0.7 mM acrylamide for 7 days of treatment. * Statistically different from control, $p < 0.05$.

The results of these experiments showed that maintenance of GSH levels appear to be involved in acrylamide-induced morphological transformation. Additional studies were performed to examine the effects of acrylamide, BSO, and NAC on SHE levels in SHE cells. Treatment with 5μM BSO reduced GSH to 35% of control levels. Co-treatment of 0.3 mM acrylamide with BSO reduced GSH levels to 20% of control, whereas co-treatment of 0.5 mM acrylamide with 2.5 μM NAC restored GSH to control levels (data not shown)

4. DISCUSSION

Chronic exposure to acrylamide resulted in neoplasms of the thyroid, adrenal medulla, testicular mesothelium and mammary gland in the rat (Johnson et al., 1986; Friedman et al., 1995). No clear mechanism for the observed neoplasms has been established. The present study examined the effect of acrylamide on the induction of DNA damage and DNA synthesis in tumor target tissues and non-target tissues in male F344 rats. The results showed that the in vivo induction of DNA damage and DNA synthesis correlated with the reported tissue site neoplasia.

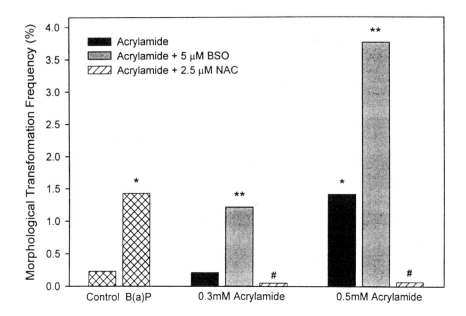

Figure 4. Effect of Glutathione Depletion and Supplementation on Acrylamide-Induced Morphological Transformation. Morphological transformation was assessed in SHE cells treated with 0.3 and 0.5 mM acrylamide in the presence or absence of 5 mM BSO or 2.5 mM NAC for 7 days of treatment. * Statistically different from control; ** Statistically increased from acrylamide treatment; # Statistically decreased from acrylamide treatment, $p < 0.05$.

Additionally, our group using co-exposure of acrylamide with 1-aminobenzotriazole (ABT), a non-specific suicide inhibitor of cytochrome P450 enzymes, showed that in the presence of ABT, acrylamide-induced DNA synthesis was blocked only in the adrenal medulla (Lafferty et al., 2004). This finding suggests the involvement of glycidamide in the induction of DNA synthesis and presumably adrenal medullary pheochromocytomas, but not in the induction of thyroid or testicular tumors. Thus oxidative metabolism and/or oxidative metabolites of acrylamide (i.e. glycidamide) do not appear to exclusively account for the induction of DNA synthesis and presumably the observed neoplasia seen following acrylamide treatment.

Following the observation that acrylamide increased in DNA synthesis in the 3 target tissues examined, and that previous investigations have reported that acrylamide produces mitotic spindle disturbances (Lapadula et al., 1989) as well as DNA damage in vitro (Barfknecht et al., 1988; Knapp et al.,

1988), the question of whether acrylamide had the ability to produce DNA damage and followed the tissue specific pattern observed for acrylamide tumor sites was of interest to study. Using the Comet assay (single cell gel electrophoresis), DNA damage was assessed in the thyroid and adrenal gland (tumor target tissues) and in the liver (non tumor target tissue). In both target tissues examined, acrylamide induced DNA damage whereas in the liver, no increase in DNA damage was seen. This observation showing an induction of DNA damage by acrylamide in the previously reported tumor target sites suggests that this event may contribute to acrylamide-induced carcinogenicity in the rat.

Other potential mechanisms have been suggested for acrylamide tumorigenicity, such as the activation of biochemical pathways involving thyroid hormone regulation (Dumont et al., 1992), and the induction of oxidative stress (Park et al., 2002). For the latter potential mechanism, both acrylamide and glycidamide have been shown to conjugate with glutathione (Sumner et al., 1999), which may lead to depletion of glutathione and a resulting oxidative stress. Furthermore, using the Syrian hamster embryo (SHE) cell transformation assay, a model system that has been used as a surrogate for in vivo carcinogenicity, it was shown that co-incubation of acrylamide with N-acetyl-L-cysteine, a sulfhydryl group donor, resulted in the reduction of acrylamide-induced morphological transformation in SHE cells (Park et al., 2002).

SHE cell transformation mimics the multistage properties of the carcinogenesis process and has been used to understand the mechanisms of genotoxic and nongenotoxic carcinogens (Barrett et al., 1984; Isfort et al., 1994). Acrylamide treatment at concentrations of 0.5 mM and higher induced morphological transformation in SHE cells. Acrylamide induced cellular transformation in these models at similar concentrations to that seen in the present study with SHE cells (0.2 mM to 1.0 mM, dependent on cell type).

Acrylamide has been reported to be clastogenic and to induce aneuploidy, chromosomal aberrations and sister chromatid exchange in Chinese hamster V79 cells (Tsuda et al., 1993; Shiraishi Y., 1978; Moore et al., 1987). Other chemicals that produce aneuploidy and have clastogenic activity in cultured cells have been shown to induce SHE cell transformation (Gibson et al., 1995; Tsutsui et al., 1997; Tsutsui and Barrett, 1997). The reactivity of acrylamide with proteins may also contribute to the carcinogenesis process by modifying cellular functions and signal pathways. Acrylamide treatment alters dopamine receptor affinity and changed hormone levels (Agrawal et al. 1981; Srivastava et al., 1986). Disruption of the cytoskeletal proteins by acrylamide may also contribute to its carcinogenicity. Microtubule disruption has been shown to change cellular response and to stimulate DNA synthesis of bovine endothelial cells (Liaw and Schwartz, 1993). Cytoskeletal disruption has been shown to modify growth factor responsiveness to mitogens, cell differentiation, altered shape and altered motility, and cytoskeleton in SHE cells (Pienta and Coffey, 1992; Isfort et al., 1994).

Depletion of GSH has been reported in vitro and in vivo following acrylamide exposure (Srivastava et al., 1986). Acrylamide and glycidamide GSH conjugates also have been detected in the urine of treated rats. Our laboratory has shown that BSO depletion of GSH enhanced acrylamide-induced SHE cell transformation. A reduction of cellular GSH level by acrylamide treatment was observed in our study. However, the reduction of GSH levels by BSO treatment only did not result in transformation (data not shown) indicating that transformation required additional events besides simple GSH depletion. Further studies are warranted to define the entire mechanism for acrylamide carcinogenicity.

In summary, acrylamide exposure in F344 rats produced increases in DNA synthesis and DNA damage in tumor target tissues of acrylamide while no increases in these endpoints were observed in non-tumor target tissues that were examined. Our studies with cellular transformation models suggest a role for GSH reactivity in at least the DNA reactivity of acrylamide. However, depletion of GSH has also been attributed to modification of second messengers important in cell proliferation and apoptosis. These studies also suggest a role for oxidative stress and oxidative damage in acrylamide-induced carcinogenesis. Since the induction of cancer requires both mutation and cell proliferation, the observations that acrylamide induced both DNA damage and DNA synthesis previously reported tumor target sites suggests that these events may act in concert and contribute to the observed acrylamide-induced carcinogenicity in the rat.

REFERENCES

Ames, B. N., and Gold, L. S., 1990, Chemical carcinogenesis: Too many rodent carcinogens, *Proc. Natl. Acad. Sci. USA* **87**:7772-7776.

Agrawal. A.K., Seth, P. K., Squibb, R.E., Tilson, H. A., Uphouse, L. L., and Bondy, S. C., 1981, Neurotransmitter receptors in brain regions of acrylamide-treated rats. I: Effects of a single exposure to acrylamide, *Pharmacol. Biochem. Behav.* **14**:527-531.

Banerjee, S, and Segal, A., 1986, In vitro transformation of C3H/10T1/2 and NIH/3T3 cells by acrylonitrile and acrylamide, *Cancer Lett.* **32**:293-304.

Barfknecht, T., Mecca, D., and Naismith, R., 1988, The genotoxic activity of acrylamide, *Environ. Mol. Mutagen.* **11** (Suppl. 11): 9.

Barrett, J. C., Hesterberg, T. W., and Thomassen, D. G., 1984, Use of cell transformation systems for carcinogenicity testing and mechanistic studies of carcinogenesis, *Pharmacol. Rev.* **36**:53s-7

Butterworth, B. E., and Goldsworthy, T. L., 1991, The role of cell proliferation in multi-stage carcinogenesis, *Proc. Soc. Exp. Biol. Med.* **198**:683-687.

Cohen, S. M., and Ellwein, L .B., 1990, Cell proliferation in carcinogenesis, *Science* **249**: 1007-1011.

Cohen, S. M., and Ellwein, L. B., 1991, Genetic errors, cell proliferation, and carcinogenesis, *Cancer Res.* **51**:6393-6505.

Dearfield, K., Douglas, G., Ehling, U., Moore, M., Sega, G., and Brusick, D., 1995, Acrylamide: a review of its genotoxicity and an assessment of heritable genetic risk, *Mutation Res.* **330**:71-99.

Dumont, J. E., Lamy, F., Roger, P., and Maenhaut, C., 1992, Physiological and pathological regulation of thyroid cell proliferation and differentiation by thyrotropin and other factors, *Physiol. Rev.* **72**:667-697.

Eldridge, S. R., Tilbury, L. F., Goldsworthy, T. L., and Butterworth, B. E., 1990, Measurement of chemically induced cell proliferation in rodent liver and kidney: a comparison of 5-bromo-2'-deoxyuridine and [^{3}H]-thymidine administered by injection or osmotic pump, *Carcinogenesis* **11**:2245-2251.

Friedman, M. A., Dulak, L., and Stedham, M., 1995, A lifetime oncogenicity study in rats with acrylamide, *Fundam. Appl. Toxicol.* **27**:95-105.

Gad, S., and Weil, C. S., 1986, *Statistics and Experimental Design for Toxicologists,* Telford Press: New Jersey.

Gibson, D.P., Aardema, M. J., Kerckaert, G. A., Carr, G. J. Brauniger, R. M. and LeBoeuf R. A., 1995, Detection of aneuploidy-inducing carcinogens in the Syrian hamster embryo (SHE) cell transformation assay, *Mutat. Res.* **343**:7-24.

Isfort, R. J., Cody, D. B., Doersen, C. J., Kerckaert, G. A., and LeBoeuf, R. F., 1994, Alterations in cellular differentiation, mitogenesis, cytoskeleton and growth characteristics during Syrian hamster embryo cell multistep in vitro transformation, *Int. J. Cancer* **59**:114-125.

Johnson, K. A., Gorzinski, S. J., Bodner, K. M., Campbell, R. A., Wolf, C. H., Friedman, M. A., and Mast, R. W., 1986, Chronic toxicity and oncogenicity study on acrylamide incorporated in the drinking water of Fischer 344 rats, *Toxicol. Appl. Pharmacol.* **85**:154-168.

Kerckaert, G. A., Isfort, R. J., Carr, G. J., Aardema, M. .J., and LeBoeuf, R. A., 1996, A comprehensive protocol for conducting the Syrian hamster embryo cell transformation assay at pH 6.70, *Mutat. Res.* **356**:65-84.

Klaunig, J. E., 1993, Selective induction of DNA synthesis in mouse preneoplastic and neoplastic hepatic lesions after exposure to phenobarbital, *Environ. Health Perspect.* **101** 235-249.

Knaap, A., Kramers, P., Voogd, C., Bergkamp, W., Groot, M., Langerbroek, P., Mout, H., van der Stel, J., and Verharen, H., 1988, Mutagenic activity of acrylamide in eukaryotic systems but not in bacteria, *Mutagenesis* **3**:263-268.

Lafferty, J. S., Kamendulis, L. M., Kaster, J, Jiang, J., and Klaunig, J. E., 2004, Subchronic acrylamide treatment induces a tissue-specific increase in DNA synthesis in the rat, *Toxicol. Letts*, in press.

Lapadula. D., Bowe, M., Carrington, C., Dulak, L., Friedman, M., and Abou-Donia, M., 1989, In Vitro binding of [^{14}C] acrylamide to neurofilament and microtubule proteins of rats, *Brain Res.* **481**:157-161.

Liaw, L., and Schwartz, S. M., 1993, Microtubule disruption stimulates DNA synthesis in bovine endothelial cells and potentiates cellular response to basic fibroblast growth factor, *Am. J. Pathol.* **143**:937-948.

Moore, M., Amtower, A., Doerr, C., Brock, K., and Dearfield, K., 1987, Mutagenicity and clastogenicity of acrylamide in L5178Y mouse lymphoma cells, *Environ. Mutagenesis* **9**: 261-267.

National Institute for Occupational Safety and Health (NIOSH), 1976, *Criteria for a recommended standard- Occupational exposure to acrylamide*, Publication No. 77-112, DHHS (NIOSH): Washington, DC.

Park, J., Kamendulis, L. M., and Klaunig, J. E., 2002, Acrylamide-induced cellular transformation, *Toxicol. Sci.* **65**:177-183.

Pienta, K. J., and Coffey, D. S., 1992, Nuclear-cytoskeletal interaction: evidence for physical connections between the nucleus and cell periphery and their alteration by transformation, *J. Cell Biochem.* **49**:357-365.

Rojas, E., Lopez, M. C. and Valverde, M., 1999, Single cell gel electrophoresis assay: methodology and applications, *J Chromatogr B Biomed Sci Appl*, **722**:225-54

Rosen J., and Hellenas K. E., 2002, Analysis of acrylamide in cooked foods by liquid chromatography tandem mass spectrometry, *Analyst* **127**:880-882.

Schulte-Hermann, R., 1987, Tumor promotion in the liver, *Arch Toxicol.* **57**:147-158.

Sega, G., Generoso, E., and Brimer, P., 1990, Acrylamide exposure induces a delayed unscheduled DNA synthesis in germ cells of male mice that is correlated with the temporal pattern of adduct formation in testis DNA, *Environ. Mol. Mutagen.* **16**:137-142.

Shimoi, K., Okitsu, A., Green, M. H. L., Lowe, J. E., Ohta, T., Kaji, K., Terato, K., Ide, H., and Kinae, N., 2001, Oxidative DNA damage induced by high glucose and its suppression in human umbilical vein endothelial cells, *Mutat. Res.* **480**:371-378.

Shiraishi, Y., 1978, Chromosome aberrations induced by monomeric actylamide in bone marrow and germ cells of mice, *Mutat. Res.* **57**:313-324.

Srivastava, S. P., Sabri, M. I., Agrawal, A K., and Seth, P. K., 1986, Effect of single and repeated doses of acrylamide and bis-acrylamide on glutathione-S-transferase and dopamine receptors in rat brain, *Brain Res.* **371**:319-323.

Sumner, S. C. J., Fennell, T. R., Moore, T. A., Chanas, B, Gonzalez, F., and Ghanayem, B.I., 1999, Role of cytochrome P450 2E1 in the metabolism of acrylamide and acrylonitrile in mice, *Chem. Res. Toxicol.* **12**:1110-1116.

Tareke E., Rydberg P., Karlsson P., Eriksson S., and Tornqvist M., 2000, Acrylamide: a cooking carcinogen? *Chem. Res. Toxicol.* **13**(6):517-22.

Tice, R.R. and Strauss, G. H., 1995, The single cell gel electrophoresis/comet assay: a potential tool for detecting radiation-induced DNA damage in humans, *Stem Cells* **13** (Suppl 1): 207-14.

Tsuda, H., Shimizu, C., Taketomi, M., Hasegawa, M., Hamada, A., Kawata, K., and Inui, N., 1993, Acrylamide; induction of DNA damage, chromosomal aberrations, and cell transformation without gene mutations, *Mutagenesis* **8**:23-29.

Tsutsui, T., and Barrett, J. C., 1997, Neoplastic transformation of cultured mammalian cells by estrogens and estrogenlike chemicals, *Environ. Health Perspect.* **105** (Suppl 3): 619-624.

Tsutsui, T., Hayashi, N., Maizumi, H., Huff, J., and Barrett, J. C., 1997, Benzene-, catechol-, hydroquinone-, and phenol-induced cell transformation, gene mutations, chromosome aberrations, aneuploidy, sister chromatid exchanges and unscheduled DNA synthesis in Syrian hamster embryo cells, *Mutat. Res.* **373**:113-123.

United States Environmental Protection Agency, 1994, *Chemicals in the Environment: Acrylamide* (CAS No. 79-06-01).

Weiss, G., 2002, Acrylamide in Food: Uncharted territory, *Science* **297**:27.

World Health Organization, 1985, *Acrylamide*, Environmental Health Criteria 49.

EXPOSURE TO ACRYLAMIDE
Placing exposure in context

Barbara J. Petersen, PhD and Nga Tran, Dr. PH
Exponent, Inc. 1730 Rhode Island Ave., NW suite 1100, Washington, DC 20036; e-mail bpetersen@exponent.com

Abstract: This paper attempts to assess possible risks that may result from human exposure to dietary intake of acrylamide.

Key words: Acrylamide, dietary intake; risk assessment

1. INTRODUCTION

In April 24, 2002 Swedish scientists reported finding acrylamide in food, sparking international concern (SNFA, 2002). Tornqvist et al identified food as a source of acrylamide as the result of investigations related to industrial use of acrylamide (Hagmar, et al, 2001) and a local contamination event (Tareke, 2002).

Since acrylamide has a long history of use for industrial purposes, it was initially assumed to be in the food as a result of exogenous contamination. Swedish scientists reported its presence in foods as a result of cooking. This finding was quickly confirmed by authorities in Europe, the United States and elsewhere (JIFSAN, 2004). Potential mechanisms of formation were investigated and it is now established that the most common formation mechanisms require the presence of asparagine, reducing sugars, and sufficient temperatures to produce a browning (Maillard) reaction (JIFSAN, 2004). These conditions occur widely in food as confirmed by the presence of acrylamide in many different types of foods (JIFSAN, 2004).

Acrylamide has been subjected to extensive toxicity testing. Depending upon the doses tested and the lengths of exposure acrylamide has been shown to have neurotoxic, reproductive and carcinogenic effects. The

World Health Organization and the Food and Agriculture Organization (WHO/FAO) convened an expert committee to review the available data in June 2002 (WHO/FAO, 2002). Although only preliminary data were available regarding the levels of acrylamide in the diets of the world's population, the experts concluded that at dietary levels neurotoxic and reproductive effects would not be of concern but recommended further study to assess the potential for dietary sources of acrylamide to have carcinogenic potential (WHO/FAO, 2002). Since that consultation, additional data have been generated that provide estimates of the levels of acrylamide in individual foods stuffs and in the overall diet. Studies are now available allowing the estimation of variability of acrylamide levels within foods and have identified the processes responsible for the formation of acrylamide in many if not all of the food types. Depending upon how the foods are categorized at least 50 different food categories have been reported to contain acrylamide by one or more authorities. JIFSAN has organized two workshops to allow scientists to review the data, identify critical gaps in our knowledge about acrylamide and to develop approaches to obtain additional data.

Research to address many key questions is now underway. In the meantime, the significance of the presence of acrylamide in so many foods can best be understood by placing that exposure in the context of other components of food and by evaluating the impact of changes to the characteristics of foods in the diet.

Food contains many bioactive compounds – some are "naturally" present in the food – either as a result of formation by the plant or by uptake from the soil during growth. Others may be xenobiotics that are introduced either intentionally or accidentally during cultivation or during subsequent processing and food preparation. As in all exposure assessments, the significance of any bioactive compound will depend upon the toxicity of the compounds at the doses encountered. Bioavailability of xenobiotics is also a key consideration. Acrylamide has been shown to be bioavailable as measured by the formation of adducts with hemoglobin (Tareke, 2002). The risk profiles of other endogenous substances that are present in food should also be considered.

Acrylamide is formed during many types of food processing and preparation. These processing and preparation steps are taken to improve palatability and nutrient availability as well as to create specific products such as ready-to-eat breakfast cereals, fried foods and potato chips. In this paper we summarize the estimates of exposure and place it in context of other substances that have been identified as potentially carcinogenic. We have estimates the contribution of different food groups to exposure as well as to nutritional status.

2. METHODS

2.1 Method 1 -- Indirect Method of Exposure Estimation

Exposure can be estimated by combining the levels of acrylamide in individual foods and the amounts of those foods that are eaten. The total acrylamide exposure is the sum of the products of exposures from individual foods.

Method 1 Formula

$$E_t = \sum_i (C_f)_i (L)_I \tag{1}$$

i = number of different food types consumed.
C_f = Concentration of acrylamide in foods(mg/Kg)
L = Amount of food consumed (kg/day)

Method 1 can estimate exposures using deterministic methods that select a single value for residues in each food and a single consumption value. This approach is most useful for estimating mean exposures to acrylamide. Attempts to estimate extremes of exposure (either high or low) require consideration of the full range of residues as well as the range of consumption amounts of each food. Selecting high residue levels in selected foods or high consumption levels of those foods (or both) will result in over-estimation of exposure. More realistic estimates of high (and low) ends of the exposure distribution can be obtained by using probabilistic models that allow inclusion of the distribution of residues and the distribution of consumption of those same foods. The most widely used technique is called Monte Carlo modelling (Petersen, 2000).

Our diets are extremely complex and variable. Of the potential number of foods that could be consumed, only a few are eaten by a single individual on any given day. No one eats every food on every day. This complexity and variability in food consumption pattern complicates our attempts to estimate the ranges of exposure that can be encountered by consumers. Further, the reliability of estimates generated using Method 1 is limited by the availability of data. For examples, identification of all foods that may contain acrylamide is incomplete and uncertainty about levels in the specific foods consumed by each individual remains unknown.

2.2 Method 2 – Direct Exposure Estimation using Duplicate Diet/Total Diet Type Studies

Exposure can also be estimated by conducting a study in which diets for a full day are collected and analyzed for acrylamide (method 2).

A common method of data collection for assessing actual intake of a particular group is the duplicate method. Participants are asked to provide a duplicate amount of every food they consumed over the duration of the study. The foods contributed by each participant are composited and then analysed for the substances of interest (Petersen and Barraj, 1996). Intake estimates derived by the duplicate method combine the actual amount of foods consumed with the actual amount of the substance detected in "identical" foods. Unlike most other methods, no surrogate data are used in the analysis. The duplicate method places a high burden on the respondent and does not allow the identification of the foods contributing most to the exposures.

2.3 Identification of Dietary Sources of Exposure and Risks

Both methods 1 and 2 have been used by various researchers to estimate exposure to acrylamide. These results are presented in this paper. Identification of individual food contribution to total exposure and risks, however, can be best identified by method 1 since method 2 typically combines many or all of the foods for a single analysis. As such, to identify dietary sources of acrylamide exposure, we used method 1 in our own analysis.

Two major data sources were used to conduct a 2-day average intake analyses for acrylamide: USDA's Continuing Survey of Food Intake by Individuals (CSFII) (1994-1996, 1998), and FDA's 2002/2003 combined acrylamide sampling data (DiNovi and Howard 2003) Probabilistic approach (i.e. Monte Carlo modeling) was carried out using Exponent's proprietary software, FARE™. Assuming the US population has concurrent exposure to all types of foods that tested positive for acrylamide, the total dietary exposure to acrylamide and corresponding excess lifetime cancer risks were estimated by summing intake and risks across all foods. Cancer risk estimate is calculated using the EPA's cancer slope factor for acrylamide (q^*=4.5 per mg/kg/day). Percent contribution to the total acrylamide exposures and cancer risks by each food type was also estimated.

2.4 Risk Reduction and Potential Impact to the Food Supply and Nutrition

Although an obvious risk mitigation strategy is to reduce or eliminate acrylamide in foods, how much modification to the food supply would be necessary to reduce risks to "safe" levels and what would be the impact to nutritional needs require further elucidation. To examine the level of risk reduction that could be achieved by eliminating acrylamide levels in certain foods we conduct a "what if" analysis by sequentially setting acrylamide levels in a chosen food to zero level (0 µg/kg) and rerun the model to develop revised estimates of exposure and risk. Alternatively, to examine impact on nutrition should the food supply is disrupted by efforts to reduce acrylamide exposures, we also estimated the contribution of acrylamide containing foods to the total dietary calories, micronutrients and macronutrients using USDA's calories and nutrients database.

3. RESULTS

3.1 Existing Acrylamide Exposure Estimates Based on Method 1

Experts at the WHO/FAO consultation (WHO/FAO, June 2002) reviewed preliminary estimates from the United States, the Netherlands, Norway, Australia, Sweden, and from the IARC EPIC Study (Slimani et al, 2002). These estimates were based on method 1. From the review, it was concluded that while the data were extremely limited the long term exposures would be in the range of 0.3 - 0.8 µg/kg bw/day.

Since that time many additional foodstuffs have been analysed and the estimates have been refined using method 1. Svenson at al reported that the mean, median and maximum intakes by Swedish consumers are 31,37 and 138 µg//kg bw/person, respectively (Svenson, 2003). Konings et al (2003) reported mean intakes of .48 µg/kg bw/day for the Dutch total population. The upper 99.99th percentile of the population intake among the Dutch population was 2.7 µg/kg bw/day. Exposures for children were also estimated: Children 1-6 yr of age was estimated to have mean intakes of 1.04 µg/kg bw/day and upper 99.9th percentile exposures of 3.8 µg/kg bw/day. Children 7-18 yrs of age were estimated to have mean intakes of 0.71 µg/kg bw/day and upper 99.99 percentile exposures of 3.2 µg/kb bw/day.

The US FDA estimated exposures for the US population using method 1 and several different surveys for food consumption including data from the USDA's Continuing survey of food intake conducted in 1994-96, 98 (DiNovi and Howard 2003). The assessment has estimated baseline exposures as well as the potential impact on intake if levels of acrylamide change in foods. Exposures have been calculated using different models including Monte Carlo modelling to incorporate more realistic estimates of the probability of occurrence of residues. Depending upon the scenario and the subpopulation, the range of mean exposures reported was 0.32 – 1.26 µg/kg body weight/day; the corresponding range of 90th percentile intakes was 0.66-2.33 µg/kg body weight/day. Figure 1 presents the results of one of the analyses conducted by FDA (DiNovi and Howard, 2003) that is typical of the calculated exposures when all of the data are considered.

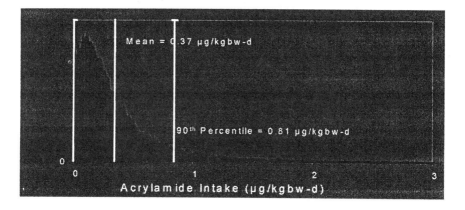

Figure 1. Acrylamide Intake Distribution for the US Population 2 Years and Older[1]

Note: 1) based on FDA's analysis of foodstuffs for acrylamide and the USDA Continuing Survey of Food Intake by Individuals from 1994-96, 1998 (USDA CSFII 1994-96,98) (from DiNovi and Howard 2003)

3.2 Exposure to Acrylamide Based on Method 2

To date, there has been only one Total Diet Study conducted to estimate exposure to acrylamide. That study was conducted by the Swiss government using a duplicate diet protocol (Swiss, BAG-CH, 2002). The results of that study are presented in Table 1. Duplicate diets were collected for 27 participants for 2 days and included all solid food. Beverage consumption was determined from recorded consumption and the acrylamide content separately analysed in the different beverages. The mean daily intake was

0.28 µg/kg bw/day. Table 1 summarizes the contribution of each meal to overall intake.

Table 1. Exposure estimates for Swiss consumers who provided a duplicate diet for analysis (Swiss 2002)

Sources of exposure:
Breakfast 8%
Lunch 21%
Dinner 22%
Snacks 13%
Coffee 36%
Mean exposure: 0.28 µg/kg bw/day

3.3 Available Acrylamide Sampling Data and Associated Uncertainties

The available data on acrylamide in foodstuffs have not been collected by following a statistically designed survey of foods in the marketplace; nor have sampling programs been employed that would allow an understanding of the true variations in levels between food categories, within food categories, among different brands or even within brands on different dates. However, in the past two years many different foodstuffs have been analyzed for acrylamide (JIFSAN, 2004). The available data, as of Jan 2003 are summarized in Tables 2 and 3. Additional information is available by country on individual country websites which can be reached through links on the JIFSAN website (JIFSAN, 2004). In addition to analyzing foodstuffs, research has been conducted that has attempted to identify the sources of the variability that is seen as well as to identify possible approaches that would reduce variability. FDA has presented data showing the extent of variation in different brands of the same food and different lots of the same brand (FDA, 2004).

Based on initial evaluations of the preliminary data, there is a wide variation in of acrylamide levels in foods. The highest detected level in each food category is much higher than the average levels (see Table 3). Thus, attempt to conduct a "worst case" exposure assessment using the highest acrylamide level will dramatically over-estimate exposure, and subsequently, risk.

Table 2. Acrylamide Levels in Foods

Category	European Data	FDA Data
Breads	12-3200	<10-364
Crispbread	<30-1670	
Crackers and Biscuits	<30-2000	26-620
Cereal	<30-2300	11-1057
Other Grains	<30	
Potato Chips	150-1280	117-2762
Other Salty Snacks	122-416	12-1243
French Fries	85-1104	20-1325
Other Potato Products	<20-12400	
Other Vegetable and Fruit Products	10-<50	<10-70
Prepared Foods	<30-30	
Meats	<30-64	<10-116
Candy and Dessert items	<20-110	<10-909
Cookies		36-432
Coffee and Tea	170-700	37-374
Other Nonalcoholic Beverages	<30	
Alcoholic Beverages	30	
Dairy Products	10-100	<10-43
Baby Food and Formula	40-120	<10-130
Dry Soup Mixes		<10-1184
Gravy and Seasonings		38-54
Miscellaneous	70-200	<10-125

Table 3. Acrylamide Residues Found in Foods Tested by FDA (Dec 2002, Feb 2003)[1]

Broad Categories Evaluated by FDA	AVG (PPB)	MIN (PPB)	MAX (PPB)	N
Breads and bakery products	55	<10	364	42
Cereals	133	11	1057	21
Coffee (ground, not brewed)	207	37	374	50
Coffee (brewed)	8	5	11	6
Cookies	222	36	432	7
Crackers	194	26	620	15
Potato Chips	534	117	2762	43
Snack Foods	372	12	1243	16
French Fries	340	20	1325	53

[1] Not including: protein foods, infant formula, baby foods, dried foods, canned fruits and vegetables, sauces and gravies, nuts and nut butters, dairy, frozen vegetables; Acrylamide findings were insignificant

3.4 Dietary sources of exposure

The amount of dietary acrylamide intake is influenced by not only the level of acrylamide in a food, but also, by how much of that food is consumed. While foods that are staples of our diet may be low in acrylamide than less frequently consumed foods, they could be a significant source of acrylamide, because of the large number of the population consuming the food and the large quantities that are consumed on the daily basis. Table 4 provides a summary of average amount consumed per day for major food categories that have been tested positive for acrylamide and the percent of the US population consuming these foods.

Breads and bakery and cereal products, while having much lower acrylamide levels than French fries (see Table 3), contribute appreciably to total acrylamide exposure due to significantly higher daily food intakes. Breads and bakery products and cereals contribute approximately 32% to total estimated acrylamide exposure, while French fries contributed 24% to total exposure (See Figure 2).

Table 4. Reported 2-day Avg Food Intakes (g/day) for US Foods Tested positive for Acrylamide (CSFII 1994-1998)

Food Categories	Mean Per Capita (g/day)	Mean Per User (g/day)	Percent Consuming
Breads and bakery products	89	97	92%
Cereals	16	39	42%
Coffee	249	562	44%
Cookies	9	29	30%
Crackers	4	18	24%
Potato Chips	5	26	18%
Snack Foods	8	29	28%
French Fries	16	57	28%

3.5 Risk reduction and potential impact on nutrition

The ability to change levels has been explored by various researchers. FDA considered the potential impact of modifications to the food supply on average exposure. Based on those analyses, reducing levels of acrylamide in any single food exposure would have relatively little impact on overall exposure (DiNovi and Howard, 2003). Similarly, our own "what if" analysis of reducing level of acrylamide in any single food revealed little impact on the overall population average acrylamide intake and lifetime excess cancer risk (see Table 5).

While the reduction of acrylamide level in each food has minimal effect on total exposure and risk, the foods that have been identified as significant sources of acrylamide are the major sources of macronutrients and micronutrients. Tables 6 and 7 summarize the macronutrient and micronutrient contribution by foods reported by FDA to contain acrylamide, respectively.

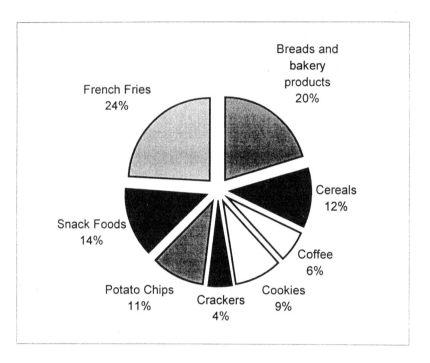

Figure 2. Contribution to Dietary Acrylamide Exposure by Food Categories (FDA2002 and 2003 Tested Foods, USDA 1994-98 CSFII

Table 5. Exposure and risk mitigation by reducing acrylamide level in each food to zero

	Population Mean Acrylamide Intake (μg/kg/day)	Excess lifetime cancer risk[1]
Original Estimates	0.43	1.92E-03
Removal of Acrylamide from		
Breads and bakery products	0.34	1.53E-03
Cereals	0.38	1.70E-03
Coffee	0.40	1.80E-03
Cookies	0.39	1.74E-03
Crackers	0.41	1.84E-03
Potato Chips	0.38	1.72E-03
Snack Foods	0.37	1.66E-03
French Fries	0.33	1.47E-03

Note: 1) Excess lifetime cancer risk is calculated assuming q* = 4.5 per mg/kg/day

Table 6. Contribution to macronutrient intakes for foods reported by FDA to contain acrylamide

Macronutrients	Contribution (%) to total daily intake
Energy (kcals)	38%
Carbohydrates (g/day)	33%
Fiber (g/day)	36%
Fat (g/day)	28%
Protein (g/day)	25%

Table 7. Contribution to micronutrient intakes for foods reported by FDA to contain acrylamide

Micronutrients	Contribution (%) to total daily intake
Iron (mg/day)	47%
Niacin	44%
Thiamin	43%
Folate (mcg/day)	42%
Selenium (mcg/day)	35%
Vitamin E (ATE)	34%
Riboflavin	33%
Vitamin B-6	32%
Magnesium	31%
Copper	30%
Zinc	29%
Phosphorus	26%
Potassium	25%
Vitamin B-12 (mcg/day)	22%
Calcium (mg/day)	20%
Vitamin A (RE)	16%
Vitamin C	10%
Vitamin A (IU)	9%
Carotene (RE)	2%

4. DISCUSSION AND CONCLUSIONS

4.1 Putting Cancer Risks in Context

The hypothetical cancer risk estimate (1.92×10^{-3}) based on estimate of total dietary acrylamide exposure and EPA's cancer slope factor is well above regulatory "brightline" for cancer risks (i.e. 10^{-4} to 10^{-6} excess lifetime cancer risks). Nevertheless, the question of whether dietary exposure to acrylamide poses a real public health risk remain open. While this issue is being examined, several researchers have suggested that dietary exposure to acrylamide has no measurable impact on risks of cancer (Mucci et al, 2003; Erdreich and Friedman, 2004)

Although some food production processes are relatively newer (e.g. dried cereals), humans have eaten foods that are cooked at high temperatures for millennia. As such, acrylamide is not the only cancer-risk-increasing compound formed during cooking and the notion of "carcinogens" in food is not new. Potentially toxic components of most foods have already been identified, including many known or potential carcinogens (NAS, 1996). Several of these identified food carcinogens are much more potent cancer causing agents than acrylamide. Table 8 provides examples of carcinogens found in food with oral cancer slope factor that are or higher potency than acrylamide.

Table 8. Examples of Known Food Carcinogens

Chemical Name	Oral Slope Factor (per mg/kg-day)	Relative Potency
Acrylamide	$4.50E+00^1$	1.00
Di-benzo(a,e)pyrene	$1.2E+02^2$	26.67
Trp-P-1(Tryptophan-P-1)	$2.6E+01^2$	5.78
Gyrometrin	$1.00E+01^2$	2.22
Benzo[a]pyrene(BaP)	$1.20E+01^2$	2.67
Benzo[a]pyrene(BaP)	$7.30E+00^1$	1.62
Glu-P-1(2-Amino-6-methyldipyrido [1,2-a:3',2'-d]imidazole	$4.80E+00^2$	1.07
Dibenz(a,h)anthracene	$4.10E+00^2$	0.91

Note: 1) EPA-IRIS; 2) Cal EPA, 2002

4.2 Balance Acrylamide Risk and Potential Impact on Nutritional Status

The data summarized in Tables 2 and 3 demonstrate that there are multiple sources of acrylamide. As shown in figure 2, some foods with lower levels contribute appreciably to the overall mean population intake because they are commonly consumed. Lowering acrylamide in one or a few foods has little effect on long-term exposure. No one food accounts for the majority of the mean population acrylamide exposure. Overall exposures will not drop by even an order of magnitude without major disruption of the food supply unless the levels in the foods are significantly reduced. It seems highly unlikely that acrylamide can be eliminated from all dietary sources. However, the levels may be reduced.

Since these foods also contribute significantly to the nutrient content of the diet (see Tables 6 and 7), the potential modifications to the diet must be carefully considered. There is a need to maintain a balance between retaining a nutritious and healthy diet with reducing risk from acrylamide exposures. Traditionally risks have been evaluated by assuming "worst case" scenarios, e.g. the most potent toxicity is assumed and the highest potential exposures. Such scenarios often make it difficult to understand the differences between hypothetical and actual risks. As can be seen with acrylamide in foods, there are potential impacts on benefits (e.g. nutritional quality of the diet) as well as potential risks. Therefore, it will be more important to conduct realistic assessments including refining both the toxicological profiles and the exposure estimates. Since there are many bioactive substances present in foods, approaches for evaluating the total diet and all its components are needed in order to achieve a balance between risks and benefits.

Clearly, before any interventions are proposed, we need to fully understand the nature of the low dose hazard to humans, the impact of any proposed interventions and whether there are any unintended consequences to public health. The relative contribution of foods that are home-cooked versus industry prepared is also of interest. Some food categories are primarily industry-prepared while others are more likely to be prepared and/or processed. This information would likely inform if mitigation strategies that focus on industrial processes would likely be effective in reducing exposure and risks.

REFERENCES

California EPA, Office of Environmental Health Hazard Assessment. (2002) Hot Spots Unit Risk and Cancer Potency Values. http://www.oehha.ca.gov/air/hot_spots/pdf/TSDlookup2002.pdf

DiNovi, M and Howard D. (2003) The updated Exposure Assessment for Acrylamide. FDA. Exposure Assessments for Acrylamide presented to the Food Advisory Committee, December Feb 2003. http:www.cfsan.fda.gov

EPA, Integrated Risk Information System. Acrylamide, CASRN 79-06-1 (http://www.epa.gov/iris/subst/0286.htm)

Erdreich. L.S., M.A. Friedman. (2004) Epidemiologic evidence for assessing the carcinogenicity of acrylamide. Regulatory Toxicology and Pharmacology, 39:150–157

FDA. (2004) Exploratory Data on Acrylamide in Food FY 2003 Total Diet Study Results. http:www.cfsan.fda.gov/~dms/acrydat2.html and Exploratory Data on Acrylamide in Food (data through November 15, 2002, data through February 7, 2003 and data through October 1, 203) http://www.cfsan.fda.gov/~dms/acrydata.html

FAO/WHO. (2002) Health Implications of Acrylamide in Food. report of a Joint FAO/WHO Consultation. WHO Headquarters, Geneva, Switzerland June 25-27, 2002

Slimani N, M. Fahey, A. Welch, E. Wirfalt, C. Stripp, E. Bergstrom; EPIC Working Group on Dietary Patterns. (2002) Do dietary patterns actually vary within the EPIC study? Unit of Nutrition and Cancer, IARC, Lyon, France. IARC Sci Publ. 2002;156:49-52.

JIFSAN (2004). Joint Institute of Food Safety and Nutrition. www.jifsan.umd.edu/acrylamide2004.htm

Hagmar L. Tornqvist, M Nordander, C. Rosen I. Bruze M. Kautiainen A, Magnusson A-L, Malmberg B, Aprea P, Axmom A., (2001) Health effects of occupational exposure to acrylamide using Hb adducts as biomarkers of internal dose. Scan J. Work Environ Health 27:219-226.

Konings,E.J.M, A.J. Baars, J.D. van Klaveren, M.C. Spanjer, P.M. Rensen, M. Hiemstra, J.A. van Kooij, P.W.J. Peters. (2003). Acrylamide exposure from foods of the Dutch population and an assessment of the consequent risks. Food and Chemical Toxicology 41:1569–1579

Mucci, L.A., PW Dickman, G Steineck, HO Adami, and K Augustsson. (2003) Dietary acrylamide and cancer of the large bowel, kidney, and bladder: Absence of an association in a population-based study in Sweden. British Journal of Cancer, 88:84-89

National Academy of Sciences, National Research Council. (1996) Carcinogens and Anticarcinogens in the Human Diet, 'A comparison of naturally occurring and synthetic substance. National Academy Press.

Petersen, B.J. (2000). Probabilistic modeling: theory and practice. Food Additives and Contaminants 17(7): 591-599.

Petersen, B.J. (2003) Methodological aspects related to aggregate and cumulative exposures to contaminants with common mechanisms of toxicity. Toxicology Letters 140-141:427-435.

Petersen, BJ and LM Barraj. (1996) Assessing the Intake of Contaminants and Nutrients: An Overview of Methods. J. of Food Composition and Analysis 9:243-254.

Swiss (2002). Swiss Federal Office of Public Health Assessment of acrylamide intake by duplicate diet study. http://www.bag.admin.ch/verbrau/aktuell

Svensson,K, L. Abramsson, W. Becker, A. Glynn, K.E. Hellena, Y. Lind, J. Rose. (2003) Dietary intake of acrylamide in Sweden Food and Chemical. Toxicology 41:1581–1586

SNFA (2002). Acrylamide is formed during the preparation of food and occurs in many foodstu.s. Press Release from Livsmedelsverket, Swedish National Food Administration, Uppsala, April 24, 2002. Internet: www.slv.se.

Swiss, BAG-CH. (2002) Teneur d'acrylamide mesurée dans divers aliments (liste non exhaustive, état au 16.07.02). Office fédéral de la santé publique, OFSP, CH-3003 Berne1/2 Tel +41 (0)31 322 21 11, Fax +41 (0)31 322 95 07 courriel: info@bag.admin.ch

Tareke, E, P. Ryberg, P. Rydberg, P. Karlsson, S. Eriksson, and M. Tornqvist. (2002) Analysis of Acrylamide, carcinogen formed in heated foodstuffs. J. Agric. Food Chem 50:4998-5006.

U.S. Department of Agriculture (USDA). CSFII Data Set and Documentation: The 1994-96, 1998 Continuing Surveys of Food Intakes by Individuals. Food Surveys Research Group. Beltsville Human Nutrition Research Center. Agricultural Research Service. April 2000.

ACRYLAMIDE AND GLYCIDAMIDE: APPROACH TOWARDS RISK ASSESSMENT BASED ON BIOMARKER GUIDED DOSIMETRY OF GENOTOXIC/MUTAGENIC EFFECTS IN HUMAN BLOOD

Matthias Baum[1], Evelyne Fauth[2], Silke Fritzen[1], Armin Herrmann[2], Peter Mertes[1], Melanie Rudolphi[2], Thomas Spormann[1], Heinrich Zankl[2], Gerhard Eisenbrand[1] and Daniel Bertow[1]
[1]*University of Kaiserslautern, Department of Chemistry, Divsion of Food Chemistry and Environmental Toxicology, D-67663 Kaiserslautern, Germany,* [2]*University of Kaiserslautern Department of Biology, Division of Human Genetics, D-67663 Kaiserslautern, Germany; e-mail:mbaum@rhrk.uni-kl.de*

Abstract: Acrylamide (AA) is a carcinogen as demonstrated in animal experiments, but the relevance for the human situation is still unclear. AA and its metabolite glycidamide (GA) react with nucleophilic regions in biomolecules. However, whereas AA and GA react with proteins, DNA adducts are exclusively formed by GA under conditions simulating in vivo situations. For risk assessment it is of particular interest to elucidate whether AA or GA within the plasma concentration range resulting from food intake are "quenched" by preferential reaction with non-critical blood constituents or whether DNA in lymphocytes is damaged concomitantly under such conditions. To address this question dose- and time-dependent induction of hemoglobin (Hb) adducts as well as genotoxic and mutagenic effects by AA or GA were studied in human blood as a model system.

Key words: acrylamide; glycidamide; genotoxicity; mutagenicity; protein adducts

1. INTRODUCTION

Acrylamide (AA) can be formed in heated foodstuffs in substantial amounts depending on heating conditions and type and concentration of precursors (Tareke, 2002; Stadler, 2002; Granvogel, 2004).

AA is a carcinogen as demonstrated in animal experiments with rats after oral consumption (Johnson et al., 1986; Friedman et al., 1995). These studies showed that the carcinogenic potential is relatively low, especially when compared to other genotoxic carcinogens for which there is also evidence for carcinogenicity in humans like mycotoxins, polycyclic aromatic hydrocarbons or N-nitrosamines. For AA, the mechanism of carcinogenicity and its relevance for the human situation is still under debate. Several studies demonstrated the ability of AA to induce chromosomal mutations in vivo, but doses applied to animals in most cases were rather high, covering a range of 30 and 150 mg/kg body weight (Dearfield et al., 1995; Madle, 2003; Paulsson et al., 2002). There is consistent evidence that AA induces sister chromatid exchanges and chromosomal aberrations in mammalian cells in vitro (Adler et al., 2002; Tyl and Friedman, 2003). However, AA was found to be inactive in mutagenicity tests with bacteria in the presence or absence of activating systems (Zeiger et al., 1987; Knaap et al., 1988; Tsuda et al., 1993). In total, there is limited evidence for a direct mutagenic and genotoxic potential of AA. However, AA itself does not show direct reactivity towards DNA, although under forced chemical conditions and after extended reaction time, adducts with DNA bases (guanine) could be synthesized and characterized (Solomon et al., 1985). Taken together, the biological activity of AA under physiological conditions appears to strongly depend on the balance between toxifying and detoxifying mechanisms in the organism.

Glycidamide (2, 3-epoxypropionamide, GA) is generated in vivo from AA primarily by CYP450-2E1-mediated metabolism (Fig. 1). A direct correlation between AA exposure and GA-Hb adduct levels has been observed in exposed Chinese workers. Hb-adduct levels with GA were about 30% of the amounts formed with AA (Calleman et al., 1990). In contrast to AA, GA induces mutations in bacteria (Hashimoto and Tanii, 1985). After exclusive application of 50 mg AA/kg body weight intraperitonally (i.p.), DNA-GA adducts with N-7 of guanine and N-3 of adenosine were formed as major products (da Costa et al., 2003). GA applied i.p. in doses between 10 and 120 mg/kg body weight induced micronuclei in bone marrow from rodents (Paulsson et al., 2003). It can be assumed that GA is the ultimate carcinogen derived from AA.

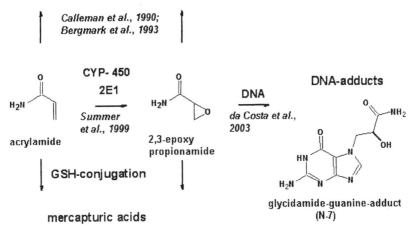

Figure 1. Toxifiying and detoxifying pathways of AA.

2. INDUCTION OF MUTATIONS AT THE HPRT-GENE LOCUS IN V79 MAMALIAN CELLS

The HPRT mutagenicity test in V79 Chinese hamster lung fibroblastes was used to investigate the capacity of AA and GA to induce HPRT forward gene mutations (Bradley et al., 1981; Glatt, 1993).

Cells were treated with AA (100-10 000 µM) or GA (400-2000 µM) for 24 hours. For selection of mutations, 1 million cells were incubated in 6-thioguanine containing medium (Bradley et al., 1981; Glatt, 1993). Mutation frequencies (MF) were determined as mutants per 1 million cells.

AA did not induce HPRT-mutations at all concentrations tested. In contrast, with GA significantly elevated mutation frequencies became detectable from 800 µM upwards (MF: 14±8; solvent-control MF: 4±2) (Fig.2).

3. THE MODEL SYSTEM: HUMAN BLOOD

The extent of DNA damage is expected to be mainly governed by kinetics of activating and deactivating metabolic pathways (Fig. 1). We therefore used human blood as a model system to study concentration-dependent interactions with DNA in lymphocytes as an easily accessible substrate to measure mutagenicity as well as binding of AA or GA to blood proteins (Fig.3). Concentration-response relationships of these reactions were investigated to find out whether a concentration can be determined below which DNA in lymphocytes is not damaged, as a consequence of AA and GA being preferentially used up by "quenching" reactions with blood proteins.

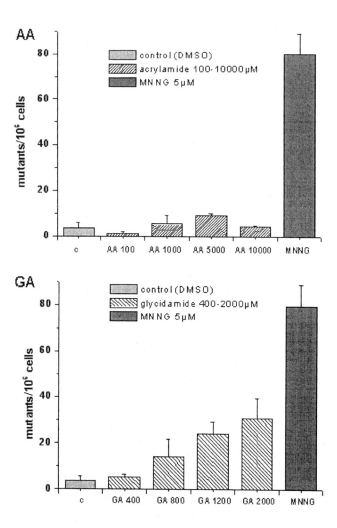

Figure 2. Induction of HPRT mutations in V79 cells by AA and GA (incubation time: 24 h; c: control (DMSO); MNNG: methyl N-nitroso guanidine, positive control) mutations are selected by treatment with 6-thioguanine.

DNA-damage was monitored by single cell gel electrophoresis. To investigate the potential to induce chromosomal mutations, induction of micronuclei, sister chromatid exchange events and chromosomal aberrations in lymphocytes were measured. Blood protein binding was assessed by measuring binding of ^{14}C-AA to blood components. As specific biomarker for AA exposure, formation of adducts with the N-terminal valine in Hb was followed.

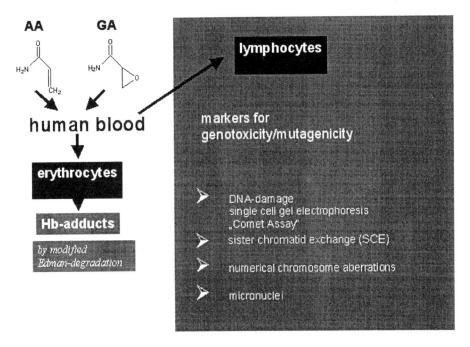

Figure 3. Human blood as a model system: concentration/time dependent induction of DNA damage and mutations by AA and GA in lymphocytes, in relation to measurements of Hb adducts as a biomarker of exposure.

4. DNA DAMAGE (MONITORED BY SINGLE CELL GEL ELECTROPHORESIS) IN LYMPHOCYTES

DNA-damage in lymphocytes was monitored by single cell gel electrophoresis and expressed as tail intensity (TI). Blood was freshly collected from donors, heparinized and incubated for 1, 2, or 4 hours with AA (1000-6000 µM) or GA (100-3000 µM). Blood aliquots in low melting agarose were put onto slides. After alkaline DNA unwinding and electrophoresis, DNA was stained and the TI of the cells was quantified by computer-assisted microscopy (Singh et al., 1988; Tice et al., 2000;

Hartmann et al., 2003). AA did not induce significant DNA-damage at all concentrations tested, whereas GA induced significant DNA-damage, beginning at 300 μM after 4 h (Fig.4).

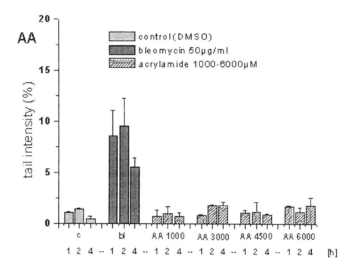

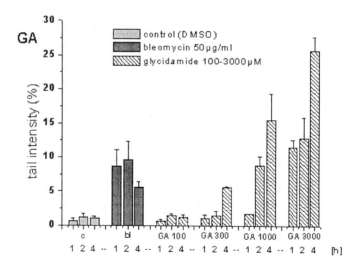

Figure 4. DNA damage induction in lymphocytes after incubation of human blood (1-4 h) with AA (1000-6000μM) or GA (100-3000μM) (c: DMSO as negative control; bl: bleomycin as positive control); DNA damage is expressed as % tail intensity (computer assisted calculation by *Comet II*[R] soft ware).

5. INDUCTION OF MICRONUCLEI

To investigate induction of micronuclei (MN) in lymphocytes we used the cytokinesis block micronucleus assay. Blood was treated with phytohemagglutinin to stimulate mitosis and then incubated with AA (500-5000 µM) or GA (50-1000 µM) for 23 hours. Lymphocytes were treated with cytochalasin B to block cytokinesis. After Giemsa staining, 1000 binucleated lymphocytes were investigated for MN-formation (Fenech et al., 2000).

With AA and GA, significantly enhanced MN frequencies in lymphocytes were only observed at the highest concentrations tested (Fig.5: AA: 5000 µM, GA: 1000 µM).

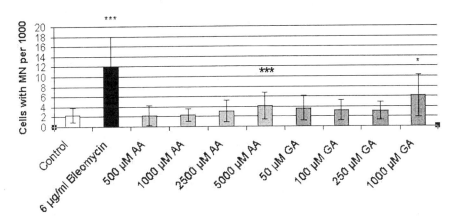

Figure 5. Concentration-dependent induction of MN in lymphocytes from human blood incubated for 23 h with AA or GA (MN-induction expressed as frequency of MN in 1000 binucleates). Statistical analysis by Mann-Whitney-Wilcoxon-test *** $P< 0.001$; * $P< 0.05$; negative control: DMSO; positive control: bleomycin).

6. BINDING OF AA AND GA TO BLOOD PROTEINS

AA and GA are reactive towards nucleophilic centers in peptides and proteins. In range finding studies, we found substantial binding of ^{14}C-AA (30µM) to erythrocytes after 1h.

In studies measuring occupational exposure, adducts formed with N-terminal valines in Hb from erythrocytes are used as sensitive biomarker for the quantification of the internal AA or GA doses in exposed persons (Törnqvist et al., 1986; Calleman et al., 1994; Bergmark et al., 1993;

Bergmark, 1997; Hagmar et al., 2001; Schettgen et al., 2002). Hb-adducts might also be useful to assess nutritional exposure to AA. The finding of a consistent AA-Val background level in the range of 30 pmol/g Hb in apparently otherwise unexposed persons gave rise to the assumption that food might be a source for acrylamide (Bergmark, 1997).

Erythrocytes from blood incubated with AA or GA within a concentration range of 30µM-3mM were spun down and Hb was extracted. Pentafluorophenyl thiohydantoine (PFPTH) derivatives of valine adducts were obtained by treatment of Hb with pentafluorophenyl isothiocyanate (Schettgen et al., 2002). PFPTH-derivatives were measured by HPLC triple MS using atmospheric pressure chemical ionisation (APCI).

Time- and dose-dependent formation of valine adducts with AA and GA is shown in Fig. 6. Incubation with 30 µM AA for 6 h resulted in an adduct level (480 pmol/g Hb) corresponding to about ten times the Hb protein "adduct background" that had been reported by Bergmark (1997). Under the same conditions, Hb protein adduct formation with GA yielded lower adduct levels (230 pmol/g Hb).

In summary, concentrations of GA as low as 30 µM clearly yielded Hb adducts. However, indications for induction of DNA damage in lymphocytes became only detectable at about ten times higher GA concentrations. Under these conditions, AA did not induce genotoxicity up to high mM concentrations.

7. SUMMARY

AA does not exert genotoxic activity in lymphocytes up to a blood concentration of 6mM, as evidenced by negative results obtained with the Comet assay. AA does not exert mutagenic activity even up to 10mM in V79 hamster cells, as evidenced by results obtained with the HPRT-mutagenicity test. Significant MN-induction in lymphocytes was only observed at an extremely high concentration (5mM).

Genotoxic activity induced by GA in lymphocytes became detectable at 300 µM (4 h exposure). Significant MN-induction in lymphocytes was observed at 1 mM GA.

In V79-cells, GA induced HPRT-mutations in a dose-dependent manner, beginning at 800 µM.

Incubation of blood with 30 µM AA or GA resulted in clearly measurable and time-dependent formation of AA/GA-Val adducts. GA at this concentration did, however, not induce detectable DNA damage in lymphocytes. DNA damage became only detectable at a 10-fold higher GA-concentration (300µM; Comet assay).

As a first conclusion, our data suggest that AA and GA in whole blood at concentrations up to 30 μM preferentially reacted with non-critical blood constituents, resulting in protection against lymphocytic DNA damage. It remains to be shown whether such "quenching" reactions by blood proteins are fast enough to also protect other organs from being damaged, especially the liver, the first organ reached by the blood stream through the portal vein after AA uptake from the gastrointestinal tract.

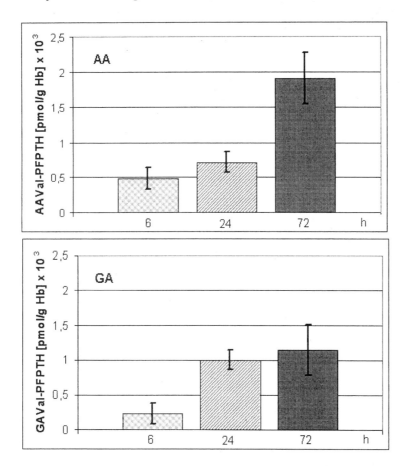

Figure 6. Formation of AA/GA-Val Hb adducts in human blood incubated (6-72 h) with 30 μM AA or GA (Val-AA/GA adducts were derivatized by treatment of Hb with PFPITC into PFPTHs and analyzed by HPLC-triple MS (APCI)).

ACKNOWLEDGEMENTS

We thank Mario Demary for his help with the Comet assay measurements, Nathalie Avice for help in the preparation of Hb-Val adduct derivatives, Karl Heinz Merz for help with the synthesis of GA and AA/Ga-Val-PFPTHs and Ingrid Hemm for help in preparing the manuscript.
This work was supported by AiF 108 ZBG/5

REFERENCES

Adler, I.D., Schmid, T. E., and Baumgartner, A., 2002, Induction of aneuploidy in male mouse germ cells detected by the sperm-FISH assay: a review of the present data base, *Mutat. Res.* **504**:173-182.

Bergmark, E., Calleman, C. J., He, F., and Costa, L. G., 1993, Determination of hemoglobin adducts in humans occupationally exposed to acrylamide, *Toxicol. Appl. Pharmacol.* **120**:45-54.

Bergmark, E., 1997, Hemoglobin adducts of acrylamide and acetonitrile in laboratory workers, smokers and nonsmokers, *Chem. Res. Toxicol.* **10**(1):78-84.

Bradley, M. O., Bhuyan, B., Francis, M. C., Langenbach, R., Peterson, A., and Huberman E., 1981, Mutagenisis by chemical agents in V79 Chinese hamster cells: A review and analysis of the literature; A report of the Gene-Tox Program, *Mutat. Res.* **87**:81-142.

Calleman, C. J., Bergmark, E., and Costa, L. G., 1990, Acrylamide is metabolized to glycidamide in the rat: evidence from hemoglobin adduct formation, *Chem. Res. Toxicol.* **3**:406-412.

Calleman, C. J., Wu, J., He, F., Tian, G., Bergmark, E., Zhang, S., Deng, H., Wang, Y., Crofton, K. M., Fennell, T., and Costa, L. G., 1994, Relationship between biomarkers of exposure and neurological effects in a group of workers exposed to acrylamide, *Toxicol. Appl. Pharmacol.* **126**:197-201.

Da Costa, G., Churchwell, M. I., Hamilton, P., von Tungeln, L. S., Beland, F. A., Marques, M. M., and Doerge, D. R., 2003, DNA adduct formation from acrylamide via conversion to glycidamide in adult and neonatal mice, *Chem. Res. Toxicol.* **16**:1328-1337.

Dearfield, K. L., Douglas, G. R., Ehling, U. H., Moore, M. M., Sega, G. A., and Brusick, D. J., 1995, Acrylamide: a review of its genotoxicity and assessment of heritable genetic risk, *Mutat. Res.* **330**:71-99.

Fenech, M., 2000, The in vitro micronucleus technique; *Mutat. Res.* **455**: 81-95

Friedman, M. A., Dulak, L. H., and Stedham, M. A., 1995, A lifetime oncogenicity study in rats with acrylamide, *Fundam. Appl.. Toxicol..* **27**:95-105.

Glatt, H.-R., 1993, HPRT-Genmutationstest in V79-Zellen des Chinesischen Hamsters, in: *Mutationsforschung und genetische Toxikologie*, R. Fahrig, ed., Wissenschaftliche Buchgesellschaft, Darmstadt, pp. 243-262.

Granvogl, M., Jezussek, M., Koehler, P., and Schieberle, P., 2004, Quantitation of 3-aminopropionamide in potatoes - A minor but potent precursor in acrylamide formation; *J. Agric. Food Chem*, **52**:4751-4757.

Hagmar, L., Törnqvist, M., Nordander, C., Rosén, I., Bruze, M., Kautiainen, A., Magnusson, A.-L., Malmberg, B., Aprea, P., Granath, F., and Axmon, A., 2001, Health effects of occupational exposure to acrylamide using hemoglobin adducts as biomarkers of internal dose, *Scand. J. Work Environ. Health* **27**(4):219-226.

Hartmann, A., Plappert, U., Poetter, F., and Suter, W., 2003, Comparative study with the alkaline Comet assay and the chromosome aberration test, *Mutat. Res.* **536**(1-2):27-38.

Hashimoto, K., and Tanii, H., 1985, Mutagenicity of acrylamide and its analogues in Salmonella typhimurium, *Mutat. Res.* **158**:129-133.

Johnson, K. A., Gorzinski, S. J., Bodner, K. M., Campbell, R. A., Wolf, C. H., Friedman, M. A., and Mast, R. W., 1986, Chronic toxicity and oncogenicity study on acrylamide incorporated in the drinking water of Fischer 344 Rats, *Toxicol. Appl. Pharmacol.* **85**:154-168.

Knaap, A., Kramers P., Voogd C., Bergkamp, W., Langebroek, P., Mout, H., van der Stel, J., and Verharen, H., 1988, Mutagenic Activity of Acrylamide in Eucaryotic Systems but not in Bacteria, *Mutagenesis* **3**:263-268.

Madle, S., Broschinski, L., Mosbach-Schulz, O., Schönig, G., and Schulte, A., 2003, Zur aktuellen Risikobewertung von Acrylamid in Lebensmitteln, *Bundesgesundheitsblatt-Gesundheitsforschung-Gesundheitsschutz* **46**:405-415.

Paulsson, B., Grawe, J., and Törnqvist, M., 2002, Hemoglobin adducts and micronucleus frequencies in mouse and rat after acrylamide or N-methyloacrylamide treatment, *Mutat. Res.* **516**(1-2):101-111.

Paulsson, B., Athanassiadis, I., Rydberg, P., and Tornqvist, M., 2003, Hemoglobin adducts from glycidamide: acetonization of hydrophilic groups for reproducible gas chromatography/ tandem mass spectrometric analysis, *Rapid Commun. Mass Spectrom.* **17**(16):1859-1865.

Schettgen, T., Broding, H. C., Angerer, J., and Drexler, H., 2002, Hemoglobin adducts of ethylene oxide, propylene oxide, acrylonitrile and acrylamide-biomarkers in occupational and environmental medicine, *Toxicol. Lett.* **134**(1-3):65-70.

Singh, N. P., McCoy, M. T., Tice, R. R., and Schneider, E. L., 1988, A simple technique for quantitation of low levels of DNA damage in individual cells., *Exp. Cell Res.* **175**:184-191.

Solomon, J. J., Fedyk, J., Mukai, F., and Segal A., 1985, Direct alkylation of 2'-desoxynucleosides and DNA following in vitro reactions with acrylamide, *Cancer Res.* **45**:3465-3470.

Stadler, R. H., Blank, I., Varga, N., Robert, F., Hau, J., Guy, P. A., Robert, M.-C., and Riediker, S., 2002, Acrylamide from Maillard reaction products, *Nature* **419**:449.

Sumner, S. C. J., Fennell, T. R., Moore, T. A., Chanas, B., Gonzalez, F., and Ghanayem, B. I., 1999, Role of cytochrom P450 2E1 in the metabolism of acrylamide and acetonitrile in mice, *Chem. Res. Toxicol.* **12**:1110-1116.

Tareke, E., Rydberg, P., Karlsson, P., Eriksson, S., and Törnqvist, M., 2002, Analysis of acrylamide, a carcinogen formed in heated foodstuffs, *J. Agric. Food Chem.* **50**:4998-5006.

Tice, R. R., Agurell, E., Anderson, D., Burlinson, B., Hartmann, A., Kobayashi, H., Miyamae, Y., Rojas, E., Ryu, J.-C., and Sasaki, Y. F., 2000, Single cell gel/Comet assay: guidelines for in-vitro and in-vivo genetic toxicology testing, *Environ. Mol.. Mutagen.* **35**:206-221.

Törnqvist, M., Mowrer, J., Jensen, S., and Ehrenberg, L., 1986, Monitoring of enviromental cancer initiators through hemoglobin adducts by modified Edman degradation method, *Anal. Biochem.* **154**:255-266.

Tsuda, H., Shimizu, C. S., Taketomi, M. K., Hasegawa, M. M., Hamada, A., Kawata, K. M., and Inui, N., 1993, Acrylamide: induction of DNA damage, chromosomal aberrations and cell transformation without gene mutations, *Mutagenesis* **8**(1):23-29.

Tyl, R. W., and Freidman, M. A., 2003, Effects of acrylamide on rodent reproductive performance, *Reprod. Toxicol.*, **17**:1-13.

Zeiger, E., Anderson, B., Haworth, S., Lawlor, T., Mortelmans, K., and Speck, W., 1987, Salmonella mutagenicity tests. III. Results from the testing of 255 chemicals, *Envrion. Mutagen.*, **9**(Suppl. 9):1-110.

PILOT STUDY ON THE IMPACT OF POTATO CHIPS CONSUMPTION ON BIOMARKERS OF ACRYLAMIDE EXPOSURE

Hubert W. Vesper, Hermes Licea-Perez, Tunde Meyers, Maria Ospina, and Gary L. Myers
National Center for Environmental Health, Centers for Disease Control and Prevention, 477 Buford Hwy NE MS F25, Atlanta, GA 30341; e-mail: HVesper@cdc.gov

Abstract: Food is assumed to be one major source of acrylamide exposure in the general population. Acrylamide exposure is usually assessed by measuring hemoglobin adducts of acrylamide and its primary metabolite glycidamide as biomarkers. Little is known about the impact of acrylamide in food on biomarkers of acrylamide exposure. Therefore, CDC is conducting a feeding study to investigate the effect of consumption of endogenous acrylamide in food on biomarkers of acrylamide exposure. As part of this study, we performed a pilot study to obtain further information on the magnitude of the changes in biomarker levels after consumption of high amounts of potato chips (21 ounces) over a short period of time (1 week) in non-smokers. After 1 week, biomarkers levels increased up to 46% for acrylamide adducts and 79% for glycidamide adducts. The results indicate that changes in biomarker levels due to consumption of potato chips can be detected. However, because of the design of this pilot study, the observed magnitude of change cannot be generalized and needs to be confirmed in the main study.

Key words: acrylamide, glycidamide, hemoglobin adducts, LC-MS/MS, potato chips, pilot study

1. INTRODUCTION

The assessment of human exposure to acrylamide is commonly performed by measuring the reaction product of acrylamide and its primary metabolite glycidamide with hemoglobin (Farmer, 1999; Friedman, 2003; Törnqvist et al., 2002). These so called hemoglobin adducts reflect the amount of acrylamide present in the circulation over the lifetime of the erythrocyte. Methods have been developed to specifically measure acrylamide that reacted with the N-terminal valine of the α- and β-chain of hemoglobin (Bergmark et al., 1993; Fennell et al., 2003; Paulsson et al., 2003; Perez et al., 1999). These methods were developed for the assessment of occupational and smoking exposure or the toxicological effects of acrylamide. Most of these methods use gas chromatography coupled with mass spectrometry (GC/MS or GC/MS/MS) to analyze acrylamide and glycidamide adducts after cleavage of the modified N-terminal valine through Edman reaction and liquid-liquid extraction to isolate the Edman products. Analysis of these Edman products by GC/MS requires derivatization of the glycidamide adduct. To overcome this additional derivatization step, a method based on liquid chromatography coupled with electrospray mass spectrometry was described recently. In addition, this method employed solid-phase extraction to purify the Edman products.

While these hemoglobin adducts are well characterized markers of exposure, they do not provide any information about the source of exposure, such as occupational exposure, smoking and, as recently discovered, food sources (Bergmark, 1997; Schettgen et al., 2002; Tareke et al., 2002). Highest biomarkers levels (up to 34,000 pmol/g globin for acrylamide adducts) were found in workers involved in an occupational accident that exposed them to very high concentrations of acrylamide (Bergmark et al., 1993). Other investigations have identified cigarette smoking as another source of acrylamide exposure and strong correlations between the number of cigarettes smoked and biomarkers of acrylamide exposure have been described (Bergmark, 1997; Schettgen et al., 2002). The biomarker levels observed in smokers are much lower than those in occupational accidents (13 - 294 pmol/g globin for acrylamide adducts (Schettgen et al., 2002)). Finally, very low levels of these biomarkers have been found in people who were not occupationally exposed and were not smoking (Bergmark et al., 1993; Bergmark, 1997; Perez et al., 1999; Schettgen et al., 2002). This so called 'background exposure' is assumed to be caused from food and ranges from 12 - 70 pmol/g globin for acrylamide adducts.

Until recently, the presence of acrylamide in food had not been anticipated because it is not generally used in food processing (Friedman, 2003). However, as confirmed by an increasing number of studies, this compound is found in a wide variety of processed food (FDA, 2004). The levels are generally highest in potato products that have been subjected to heat such as frying, grilling or baking. Food items containing high amounts of acrylamide (up to 2,500 µg/kg) include French fries, potato chips, and certain snack food. The amino acid asparagine in its free (non-protein-bound) form and reducing sugars are the major precursors for acrylamide formation in food (Mottram et al., 2002; Stadler et al., 2002). The high amounts of acrylamide in potato products can be explained with high concentrations of naturally occurring free asparagine in untreated potatoes. Therefore, acrylamide in food is not a food contaminant like pesticides but rather an endogenously formed compound through food processing. Thus, people have been exposed to this compound through food for a long time.

The amount of acrylamide exposure through food was estimated using food consumption data and acrylamide content data measured in certain food items. Based on the European Prospective Investigation into Cancer and Nutrition (EPIC) food intake database and data on acrylamide in food provided by Swedish researchers, the World Health Organization (WHO) estimated that the exposure from food would range from 0.08 to 0.3 µg/kg/day (WHO, 2002). The exposure from food in the U.S. populations was estimated using the Continuing Survey of Food Intakes by Individuals (CSFII) food consumption data and FDA data on acrylamide in food. The estimated average acrylamide intake was 0.4 µg/kg/day (90th percentile: 0.8 µg/kg/day) for the general population and 0.1 µg/kg body weight and day (90th percentile: 2.15 µg/kg/day) for children 2 to 5 years of age (Robie, 2003).

Currently, it is not known how the exposure from food affects the levels of biomarkers in blood, because the bioavailability of acrylamide in food is not known and the metabolism of acrylamide at such low exposure concentrations is not fully understood. Furthermore, studies on smokers and non-smokers showed a considerable overlap in these biomarker concentrations making the differentiation between background exposure and exposure from smoking difficult.

To assess the impact of acrylamide exposure from food on biomarkers of acrylamide exposure, a focused study is needed that takes other sources of acrylamide exposure into consideration. As a first step, we performed a pilot study assessing the biomarker levels in individuals using a liquid chromatography-mass spectrometry method after Edman reaction and liquid-liquid chromatography before and after consumption of potato chips to obtain preliminary information on whether current methodologies can detect background levels of acrylamide exposure and to obtain information about intra- and inter-individual variability. The findings of this pilot study will be used in the design of a larger food exposure study.

2. MATERIALS AND METHODS

Six volunteers (age 18 years and older) were enrolled in this study. They were asked to eat 3 ounces of potato chips per day for 7 days while maintaining their regular dietary behavior. EDTA-whole blood was collected at the beginning of the study and on the end (eighth day) of the study. The blood was used to determine hemoglobin adducts of acrylamide and glycidamide as markers of acrylamide exposure. The blood was processed according to methods described previously (Perez et al., 1999; Törnqvist, et al., 1986). In brief, erythrocytes were isolated from 1 mL EDTA-whole blood and lyzed. Globin was isolated by precipitation out of a solution of 1-propanol/hydrochloric acid using ethylacetate. The N-terminal valine adducts (N-(2-carbamoylethyl)valine for acrylamide adducts and N-(2-carbamoyl-2-hydroxyethyl)-valine for glycidamide adducts) were removed from globin (50 mg) by modified Edman reaction in 1.5 mL formamide using pentafluorophenyl-isothiocyanate (PFPITC) for 16 hours at room temperature and then for 1.5 hours at 45°C. The Edman products (pentafluorophenylhydantion derivatives) were isolated using liquid-liquid extraction with diethylether after adding 1 mL of water to the reaction mixture. After evaporation of the diethylether, the samples were redissolved in 200 µL of a methanol/water mixture (1:1 v/v) and analyzed by HPLC/MS/MS (ThermoFinnigan TSQ Quantum MS with ThermoFinnigan Surveyor HPLC). Calibrators, blanks and quality control materials were processed together with the samples. The calibrators were octapeptides (VHLTPEEK) with acrylamide or glycidamide added at the N-terminal valine. The same peptides with a $^{13}C_5^{15}N$-labeled valine were used as internal standards (Bachem). The chromatographic separation was obtained using a C18 column (2.1 x 100 mm, 3 µ, endcapped, Luna, Phenomenex, Torrance, CA) and isocratic conditions of 60% methanol in water with 0.025% trifluoroacetic acid at a flow rate of 300 µL/min. The following transitions were used (ESI positive ion mode, 4 kV, CE: 10 eV): AA-Val-PFPTH: m/z 396 → m/z 379, AA-Val($^{13}C_5^{15}N$)-PFPTH: m/z 402 → m/z 385, GA-Val-PFPTH: m/z 412 → m/z 395, GA-Val($^{13}C_5^{15}N$)-PFPTH: m/z 418 → m/z 401.

The detection limit for each peptide was 0.5 pmol on column. The intra-assay coefficient of variation for AA-Edman reaction products was 4% at 154 pmol/g globin and 9% at 41 pmol/g globin and for GA-Edman reaction products 15% at 151 pmol/g globin and 11% at 21 pmol/g globin. It was assumed that changes in biomarker levels in the magnitude of ±2 x SE were non-detectable. Given the determined variability, changes for acrylamide adducts of up to ±3 pmol/g globin and for glycidamide adducts of up to ±2 pmol/g globin are considered non-detectable. Three bags of potato chips were analyzed for their acrylamide content by Covance Laboratories (Madison, WI).

3. RESULTS

The average acrylamide content of the potato chips used in this study was 1,373 µg/kg (SD 45). Thus, a daily consumption of 3 ounces (84 g) of these potato chips results in an exposure of 115 µg acrylamide/person/day or about 1.9 µg acrylamide/kg/day.

The concentrations of the acrylamide and glycidamide adducts at the beginning of the study (Table 1) were on average 43 pmol/g globin (SD 8.4) and 26 pmol/g globin (SD 5.3), respectively. The ratio between glycidamide adducts and acrylamide adducts averaged 0.64 (SD 0.22). After 1 week, acrylamide adduct levels were on average 41 pmol/g globin (SD 6.8) and glycidamide adduct levels averaged 32 pmol/g globin (SD 9.3) and the average ratio between glycidamide and acrylamide levels was 0.81 (SD 0.32). The inter-subject variability was 20% for both the acrylamide and glycidamide adducts at baseline and 16% and 29% for acrylamide adducts and glycidamide adducts after one week, respectively.

Table 1. Concentrations of biomarkers of acrylamide exposure at the beginning and the end of the study

Subject	AA pmol/g globin		AA Change from baseline		GA pmol/g globin		GA Change from baseline		GA/AA ratio	
	Start	1 week	pmol/g globin	%	Start	1 week	pmol/g globin	%	Start	1 Week
#1	46	50	-8	-17	26	37	11	41	0.6	1.0
#2	38	51	0	1	23	41	18	79	0.6	1.0
#3	39	33	-1	-2	30	43	13	44	0.9	1.3
#4	40	48	1	2	30	25	-4	-15	0.6	0.5
#5	35	54	15	44	31	25	-5	-17	0.9	0.5
#6	34	36	-18	-33	17	21	4	21	0.3	0.6

None of the participants showed an increase in both acrylamide and glycidamide adducts. Increase in acrylamide adducts was observed in only one person, while an increase in glycidamide adducts was seen in 4 people. The median increase in biomarker values was 14 pmol/g globin and the median percent increase from baseline was 43%. One person did not show any changes in acrylamide adducts and a slight decrease in glycidamide adducts. A decrease in acrylamide adducts was observed in 2 subjects and in 2 different people for glycidamide adducts. The median decrease was 7 pmol/g globin and the median change from baseline was 17%.

4. DISCUSSION

The exposure of the study participants to acrylamide from potato chips was 1.9 µg/kg/day, which was higher than the FDA estimated average exposure of 0.4 µg/kg/day in adults but still within in the range estimated for children (2.15 µg/kg/day). The participants were asked to eat 3 ounces of potato chips per day, an amount about 3 times higher than the estimated average daily potato chips consumption (USDA, 2004). Furthermore, the determined average acrylamide content in these potato chips was about 3 times higher than the average acrylamide content of this type of food. Therefore, the exposure in this study does not reflect the average exposure of people to acrylamide from food, but is within the range for certain subgroups of the population.

With the method applied, acrylamide and glycidamide adducts could be detected in all subjects. The analytical variability allows the detection of differences in adducts levels of ± 3 pmol/g globin, which is smaller than the observed intra- and inter-subject variability. The sensitivity of the methods allows the reliable detection of background levels of acrylamide exposure.

The observed adduct levels were within the range reported as background levels in other studies (Bergmark et al., 1993; Bergmark, 1997; Perez et al., 1999;, Schettgen et al., 2002). The same applies for the ratio of glycidamide to acrylamide adduct levels. None of the adduct levels were within the range that would suggest high exposure to tobacco smoke (adduct levels of >100 pmol/g globin). At the end of the study, the observed biomarker concentrations remained below those reported for people highly exposed to tobacco smoke.

At low exposure doses as used in this study, acrylamide and glycidamide adducts do not change consistently. The observed decreases in biomarker levels of some subjects in this study may be attributed to non-steady state conditions or intra-subject variability. Though the daily acrylamide exposure from potato chips was higher than estimated average intakes, the duration of one week was short and thus the overall acrylamide exposure through potato chips was low. Longer durations and thus higher cumulative exposure may result in more pronounced and consistent results across subjects. Because acrylamide exposure through food can be considered chronic or long-term, low-dose exposure, the assessment of exposure changes reaching from one steady-state to a new steady-state might provide more comprehensive information on the effect of acrylamide in food on biomarkers of acrylamide exposure, than short term high dose exposure scenarios. Acrylamide is metabolized by the P450 2E1 to glycidamide. The conversion seems to depend on the acrylamide concentration and the induction of this P450 (Calleman et al., 1992). Because all the parameters affecting the metabolism of acrylamide to glycidamide are not fully understood, it is necessary to determine acrylamide as well as glycidamide biomarkers to obtain a comprehensive information about the total acrylamide exposure. Steady-state conditions, smoking exposure and other factors that may affect the metabolism of acrylamide to glycidamide need to be taken into consideration in a larger, main food exposure study.

ACKNOWLEDGEMENT

We would like to thank Dr. John Osterloh for his valuable suggestions on the document.

REFERENCES

Bergmark, E., 1997, Hemoglobin adducts of acrylamide and acrylonitrile in laboratory workers, smokers and nonsmokers, *Chem. Res. Toxicol.* **10**:78-84.

Bergmark, E., Calleman, C. J., He, F., and Costa, L. G., 1993, Determination of hemoglobin adducts in humans occupationally exposed to acrylamide, *Toxicol. Appl. Pharmacol.* **120**:45-54.

Calleman, C. J., Stern, L. G., Bergmark, E., and Costa, L. G., 1992, Linear versus nonlinear models for hemoglobin adduct formation by acrylamide and its metabolite glycidamide: implications for risk estimation, *Cancer Epidemiol. Biomark. Prev.* **1**:361-368.

Farmer, P. B., 1999, Studies using specific biomarkers for human exposure assessment to exogenous and endogenous chemical agents, *Mutat. Res.* **428**:69-81.

FDA, 2004, Exploratory Data on Acrylamide in Food (April 27, 2004), http://www.cfsan.fda.gov/~dms/acrydata.html.

Fennell, T. R., Snyder, R. W., Krol, W. L., and Sumner, S. C., 2003, Comparison of the hemoglobin adducts formed by administration of N-methylolacrylamide and acrylamide to rats, *Toxicol. Sci.* **71**:164-175.

Friedman, M., 2003., Chemistry, biochemistry, and safety of acrylamide: A review, *J. Agric. Food Chem.* **51**:4504-4526.

Mottram, D. S., Wedzicha, B. L., and Dodson, A. T., 2002, Acrylamide is formed in the Maillard reaction, *Nature* **419**:448-449.

Paulsson, B., Athanassiadis, I., Rydberg, P., and Törnqvist, M., 2003, Hemoglobin adducts from glycidamide: acetonization of hydrophilic groups for reproducible gas chromatography/tandem mass spectrometric analysis, *Rapid Comm. Mass Spec.* **17**:1859-1865.

Perez, H. L., Cheong, H. K., Yang, J. S., and Osterman-Golkar, S., 1999, Simultaneous analysis of hemoglobin adducts of acrylamide and glycidamide by gas chromatography-mass spectrometry, *Anal. Biochem.* **274**:59-68.

Robie, D., 2003, The Exposure Assessment for Acrylamide, (July 27, 2004), http://www.cfsan.fda.gov/~dms/acryrob2.html.

Schettgen, T., Broding, H. C., Angerer, J., and Drexler, H., 2002, Hemoglobin adducts of ethylene oxide, propylene oxide, acrylonitrile and acrylamide-biomarkers in occupational and environmental medicine, *Toxicol. Lett.* **134**:65-70.

Stadler, R. H., Blank, I., Varga, N., Robert, F., Hau, J., Guy, P. A., Robert, M. C., and Riediker, S., 2002, Acrylamide from Maillard reaction products, *Nature* **419**:449-450.

Tareke, E., Rydberg, P., Karlsson, P., Eriksson, S., and Törnqvist, M., 2002, Analysis of acrylamide, a carcinogen formed in heated foodstuffs, *J. Agric. Food Chem.* **50**:4998-5006.

Törnqvist, M., Fred, C., Haglund, J., Helleberg, H., Paulsson, B., and Rydberg, P., 2002, Protein adducts: quantitative and qualitative aspects of their formation, analysis and applications, *J. Chromatogr* **778**:279-303.

Törnqvist, M., Mowrer, J., Jensen, S., and Ehrenberg, L., 1986, Monitoring environmental cancer initiatiors through hemoglobin adducts by a modified Edman degradation method, *Anal. Biochem.* **154**:255-266.

USDA, 2004, Food Consumption data system. (May 17, 2004), http://ers.usda.gov/Data/FoodConsumption/DataSystem.asp?ERSTab=2.

WHO, 2002, Health implications of acrylamide in food, *WHO Report of a Joint FAO/WHO Consultation*, Geneva, Switzerland.

LC/MS/MS METHOD FOR THE ANALYSIS OF ACRYLAMIDE AND GLYCIDAMIDE HEMOGLOBIN ADDUCTS

Maria Ospina, Hubert W. Vesper, Hermes Licea-Perez, Tunde Meyers, Luchuan Mi, and Gary Myers
National Center for Environmental Health, Centers for Disease Control and Prevention, 477 Buford Highway, MS-F25, Atlanta, GA 30341; e-mail: MOspina@cdc.gov

Abstract: Hemoglobin adducts of acrylamide and its primary metabolite, glycidamide are used as biomarkers of acrylamide exposure. Several methods for analyzing these biomarkers in blood have been described previously. These methods were developed to analyze small numbers of samples, not the high sample throughput that is needed in population screening. Obtaining data on exposure of the US population to acrylamide through food and other sources is important to initiate appropriate public health activities. As part of the Centers for Disease Control and Prevention biomonitoring activities, we developed a high throughput liquid chromatography tandem mass spectrometry (LC/MS/MS) method for hemoglobin adducts of acrylamide. The LC/MS/MS method consists of using the Edman reaction and isolating the reaction products by protein precipitation and solid-phase extraction (SPE). Quantitation is achieved by using stable-isotope labeled peptides as internal standards. The method is performed on an automated liquid handling and SPE system. It provides good sensitivity in the low-exposure range as assessed in pooled samples and enables differentiation between smokers and non smokers.

Key words: acrylamide, glycidamide, hemoglobin adducts, LC/MS/MS

1. INTRODUCTION

A great number of diseases might result from exposure to chemicals. Blood protein adduct analysis has been used to assess and estimate the dose associated with exposure to chemicals, as well as the potential effects of these exposures (Meyer and Bechtold, 1996). The measurements of protein adducts are more useful than DNA-adducts since they are not subjected to repair mechanisms in the body and therefore can provide a more precise measure of dose integrated over the lifespan of the protein (Skipper and Tannenbaum, 1990). Although a wide variety of proteins form adducts, only a few proteins are practical for chemical, biological or epidemiological studies. Blood is a fairly accessible material and the abundance of hemoglobin in blood makes this a great protein for very sensitive assays (Törnqvist et al., 2002). N-terminal valine, cysteines, histidine and carboxylic groups from hemoglobin can react with electrophiles and their metabolites to form chemically stable compounds under biological conditions. Also, the lifespan of hemoglobin is the same as the red blood cells, approximately 120 days. Adducts accumulate with repetitive exposure and reach a steady state when exposure exceeds the cells lifespan. Since the kinetics and turnover rates of hemoglobin are known, an integrated in-vivo concentration of an electrophile during the 4-month prior to blood sampling can be calculated (Calleman et al., 1978; Ehrenberg and Osterman-Golkar, 1980; Skipper and Tannenbaum, 1990; Törnqvist et al., 2002).

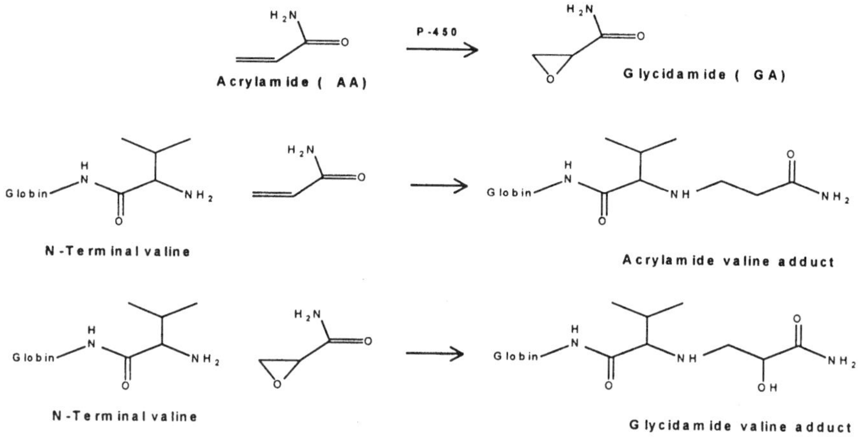

Figure 1. Metabolism of acrylamide and adduct formation with the N-terminal valine of hemoglobin. Adapted from Licea-Perez et al., 1999.

Acrylamide (AA) is a known neurotoxin shown to be carcinogenic in animals and therefore classified as a potential human carcinogen (IARC, 1994). AA is metabolized to glycidamide (GA), an epoxide reactive towards DNA. GA is assumed to be the mutagenic and cancer-initiating species, whereas AA is probably the main cause of neurotoxic effects of AA exposure. Both acrylamide and glycidamide are alkylating agents that react with hemoglobin, to form adducts (Calleman et al., 1990; Bergmark et al.,

1993). The reactions of both compounds with the n-terminal valine of hemoglobin are shown in Figure 1.

Several methods have been developed to measure hemoglobin adducts of AA and GA. The main methods are summarized in Table 1. One of the methods is based on acid hydrolysis of globin, isolation of a cystein adduct by ion-exchange chromatography and derivatization with HCL/methanol and HFBA (Calleman et al., 1990). The other methods are based on a modified Edman degradation (Törnqvist et al., 1988) in which adducts to the N-terminal valine of hemoglobin are detached from the protein as phenylthiohydantoin (PTH) derivatives and analyzed by either gas chromatography or liquid chromatography coupled to mass spectrometry (Bergmark et al., 1993; Licea-Perez et al., 1999; Tareke et al., 2000; Fennell et al., 2003).

Table 1. Methods for Analysis of Hemoglobin-Acrylamide Adducts

Analytical Method	Quantitation Species	Internal Standard	Calibration	Reference
GC/MS Pos CI	AA-Cys-HFBA, GA-Cys-HFBA	(^{2}H$_2$)AA-Cys, (^{2}H$_2$)GA-Cys	Mix of alkylated globins	Calleman et al. 1990
GC/MS/MS Neg CI	AA-Val-PFPTH, GA-Val-PFPTH (acetylated)	(^{2}H$_3$)AA-Val-PFPTH	AA[^{14}C]ValGly Gly ethyl ester	Licea-Perez et al. 1999
GC/MS Neg CI	AA-Val-PFPTH, GA-Val-PFPTH	Globin alkylated with (^{2}H$_4$)ethylene oxide	Ratio of in-vitro alkylated globin to IS	Bergmark et al. 1993
GC/MS/MS Neg C	AA-Val-PFPTH	(^{2}H$_7$)AA-Val-PFPTH	Globin adduct	Tareke et al. 2000
LC/MS/MS Neg ESI	AA-Val-PTH, GA-Val-PTH	AA-Val(^{13}C$_5$)PTH, GA-Val(^{13}C$_5$)PTH	AA-Val-Leu-anilide	Fennell et al. 2003

The method using the modified Edman degradation has been the favorite since it is less time consuming than the method that involves total acid hydrolysis and ion exchange. Some modifications such as acetonization and acetylation of the GA-valine adduct have been introduced to improve the chromatography of the GA adduct affording more reproducible results in the quantitation (Licea-Perez et al., 1999; Paulsson et al., 2003).

The impact of acrylamide on public health is uncertain and there is a clear need for human data. This paper describes a liquid chromatography-tandem mass spectrometry method that we are developing to measure acrylamide and glycidamide hemoglobin adducts and that will be used for population studies. The method is based on Edman degradation chemistry and uses solid phase extraction (SPE) as sample clean-up. The method is automated for high throughput analysis.

2. MATERIALS AND METHODS

2.1 Chemicals

AA-VHLTPEEK, GA-VHLTPEEK and the corresponding AA-Val ($^{13}C_5$ ^{15}N)-HLTPEEK and GA-Val ($^{13}C_5$ ^{15}N)-HLTPEEK were synthesized by Bachem and used as standards for quantitation. Methanol, acetic acid, pentafluorophenylisothiocyanate (PFPITC), phenylisothiocyanate (PITC), ethyl acetate, pentane and formamide were purchased from Sigma. Sodium hydroxide was obtained from Aldrich.

2.2 Isolation of red blood cells

A conventional method was used for isolation of red blood cells. Briefly, 400 µL of EDTA whole blood was centrifuged and the plasma removed. The red blood cells were isolated and washed three times with saline solution. Water (200 µL deionized) was then added to washed cells and stored in 10 mL tubes.

2.3 Isolation of globin

Globin isolation was also done with a well-known procedure (Mowrer, et al., 1986) in which 1200 µL of a 2-propanol/HCl mixture was added to the isolated red blood cells and mixed vigorously. The sample was centrifuged at 3,000 *g* and 4°C for 45 min. The supernatant (containing the globin) was transferred into 10 mL Pyrex glass vial and 7 mL ethyl acetate were added. The globin sample was centrifuged at 900 *g* and 4°C for 5 min and the supernatant was removed. The washing with ethyl acetate was repeated two more times. Globin was washed also with 7 mL pentane and centrifuged at 900 g and 4°C for 5 min. The supernatant was removed and the precipitate (purified globin) was dried in a SpeedVac at room temperature for 10 min. The sample was stored at −20°C for later use.

2.4 Edman Reaction

A Tecan Genesis Fredom 200 was used for handling all the solvents and solutions. The modified Edman method was used for derivatization of the globin samples (Törnqvist et al., 1988), and the calibration solutions. 50 mg of globin were mixed with 1.5 mL of formamide and 40 µl of 1 M NaOH. Internal standard solution (20 µL) and PFPITC (20 µL) were also added to the mixture. The sample was incubated for 16 hours overnight at room temperature and then for 1.5 h at 45°C. The samples were cooled at room temperature and the excess globin was precipitated with acetonitrile. The sample was then vortexed and centrifuged and the supernatant was transferred to a 10 mL tube. Acetonitrile was evaporated at 45°C in a SpeedVac and the remaining formamide solution was transferred to a 96 well plate for sample clean-up.

2.5 Solid phase extraction clean-up

Solid phase extraction (SPE) was done in a Gilson 215 SPE system to isolate the pentafluorophenylthiohydantoin (PFPTH) derivatives from the Edman reaction. The sample was applied to a Strata X SPE 96 well plate (Phenomenex, Inc., Torrance, CA), washed with 1 mL water and eluted with 1 mL of acetonitrile-ethyl acetate 1:1 mixture. The eluate was concentrated in a SpeedVac for 2 h at 45°C and redissolved in 300 µL of methanol: water 1:1. Samples were centrifuged and 200 µl were transferred to another well plate and analyzed by LC/MS/MS

2.6 LC/MS/MS analysis

Analysis was carried out using a Surveyor HPLC system (ThermoFinnigan, San Jose, CA) equipped with a Surveyor autosampler and a Surveyor pump. Samples (100 µL) were injected and separated on a Luna C18 (2) column, (10 cm x 2 mm, 3 um, Phenomenex, Torrance, CA) at a temperature of 45°C. The chromatographic separation was obtained with isocratic conditions of methanol (solvent A) and 0.1 % acetic acid (solvent B) 60:40 at a flow rate of 300 µL/min. The samples were maintained at 10°C in the autosampler tray.

Calibrators (0-100 nM), blanks and quality control materials were processed together with the samples. The detection of the PFPTH derivatives was carried out in a TSQ Quantum mass spectrometer (ThermoFinnigan, San Jose, CA) fitted with an electrospray ionization (ESI) source. The mass spectrometer was operated in positive ion mode with multiple reaction monitoring (MRM). The following MRM transitions were monitored for the target analytes and their corresponding internal standards: AA-Val-PFPTH, m/z 396 → m/z 379; AA-Val ($^{13}C_5$ ^{15}N)-PFPTH, m/z 402 → m/z 385; GA-Val-PFPTH, m/z 412 → m/z 395; and for GA-Val ($^{13}C_5$ ^{15}N)-PFPTH: m/z 418 → m/z 401.

3. RESULTS AND DISCUSSION

3.1 Fluorinated *vs.* non-fluorinated Edman reagent

PFPITC and PITC were tested to assess the sensitivity of the method of both reaction products. Calibration curves (Figure 2) were linear with both reagents and showed acceptable correlation coefficients for the concentration range fron 5-100 nM for the peptide solutions.

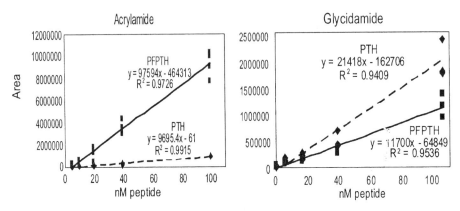

Figure 2. Calibration curves for fluorinated and non-fluorinated PTH derivatives of AA and GA hemoglobin adducts.

The sensitivity for fluorinated AA derivatives was approximately seven times higher compared to the non-fluorinated ones, as determined from the slopes of the calibration curves. For glycidamide, the sensitivity of the assay with the non-fluorinated Edman reagent was approximately twice that the fluorinated derivatives. Therefore, the fluorinated Edman reagent was used for subsequent experiments.

3.2 Mobile phase additives

The analysis of hemoglobin adducts requires careful optimization of all the parameters to obtain the sensitivity necessary to measure the low concentration of the adducts in the body. We were successful in using 0.025% TFA as a modifier for the aqueous phase., This was, therefore, our starting point for this determination. Although we obtained good signals for the AA-valine-PFPTH adduct, the signal for the GA-valine-PFPTH was very weak. Since the ion-suppression characteristics of TFA are well known, formic acid and acetic acid were tested as alternate modifiers. An aqueous phase with no modifiers was also included in this comparison (Figure 3).

The absolute areas were lower for both, AA- and GA-Val-PFPTH derivatives when 0.1% formic acid was used as the aqueous phase. In fact, a decrease in about 50% of the signal was observed in both cases. When 0.1% acetic acid was used, the signal for AA-Val-PFPTH was almost the same as when 0.025% TFA was used, but the signal for GA-Val-PFPTH, was almost two times bigger than when 0.025% TFA was used and almost four times than when 0.1% formic acid was added as modifier. An increase in the ionic strength of the acetic acid solution decreased the area counts in both cases. No peaks for GA-Val-PFPTH and a dramatic decrease in the signal for AA-Val-PFPTH were observed when the modifier was removed from the mobile phase. Therefore, acetic acid (0.1%) was chosen as the modifier for the separation and analysis of the Edman derivatives.

3.3 Characterization of quality control materials

The method presented here was used to determine the level of AA and GA hemoglobin adducts in four blood pools, two pools of non-smokers, non-coffee drinkers and two pools from smokers, coffee drinkers. Four replicate samples of each pool were prepared according to the procedure described and each replicate was injected only once (Figure 4).

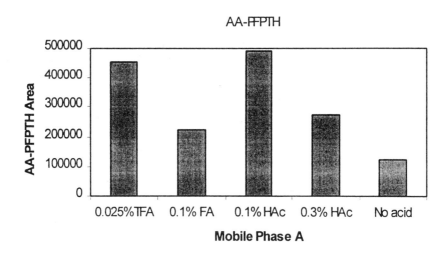

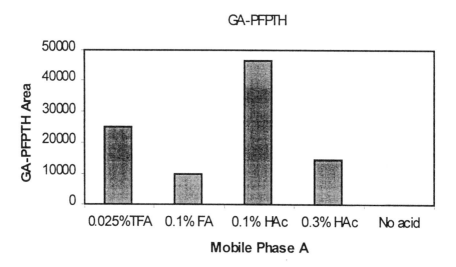

Figure 3. Effect of the mobile phase additive on the peak areas of AA- and GA-Val PFPTH adducts.

Analysis of Acrylamide and Glycidamide Adducts

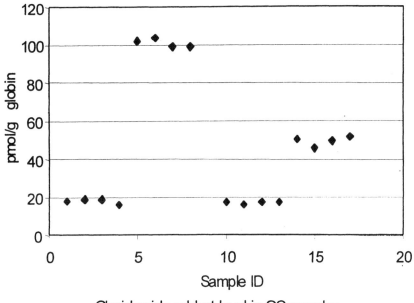

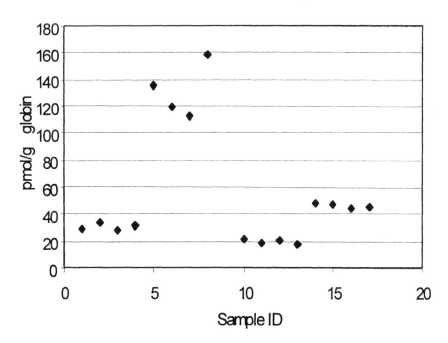

Figure 4. QC pool characterization

A clear distinction between the smoker, coffee drinker and non-smoker, non-coffee drinker pools can be observed. The precision ranged from 2 to 7% for the acrylamide adducts and from 4 to 16% for glycidamide depending on the concentration of the samples. These coefficients of variation are expected since the signal for GA adducts is weaker than the signal for the AA adducts. The values obtained were comparable to results reported previously (Bergmark, 1997; Törnqvist, this Symposium).

4. CONCLUSIONS AND FUTURE WORK

A high throughput method to analyze acrylamide and glycidamide hemoglobin adducts has been developed that can be used for population studies. The method is based on currently used methodology to assure comparability of results. The major modifications introduced here are: precipitation of protein after the Edman reaction, SPE for sample clean up and analysis by LC/MS/MS using electrospray ionization in the positive ion mode.

The results demonstrated acceptable linearity and precision for the method. Validation using NCCLS requirements is currently in progress.

Studies are also being planned that focus on acrylamide exposure not only derived from food sources but also from smoking and occupational exposures. This should make it possible to obtain information on the overall exposure to acrylamide.

REFERENCES

Calleman, C.J., Bergmark, E., and Costa, L.G., 1990, Acrylamide is metabolized to glycidamide in the rat: Evidence from hemoglobin adduct formation, *Chem. Res. Toxicol.* **3**: 406-412.

Bergmark, E., 1997, Hemoglobin adducts of acrylamide and acrylonitrile in laboratory workers, smokers and nonsmokers, *Chem. Res. Toxicol.* **10**: 78-84.

Bergmark, E., Calleman, C. J., He, F., and Costa, L. G., 1993, Determination of hemoglobin adducts in humans occupationally exposed to acrylamide, *Toxicol. Appl. Pharmacol.* **120**: 45-54

Fennell, T. R., Snyder, R. W., Krol, W. L., and Sumner, S. C., 1993, Comparison of the hemoglobin adducts formed by administration of N-methylolacrylamide and acrylamide to rats, *Toxicol. Sci.* **71**: 164-175.

IARC, 1994, *Monographs on the evaluation of of carcinogen risks to humans: some industrial chemicals*, International Agency for Research on Cancer, Lyon. **60**: 389-433.

Licea Perez, H., Cheong, HK., Yang, J. S., and Osterman-Golkar, S., 1999, Simultaneous analysis of hemoglobin adducts of acrylamide and glycidamide by gas chromatography-mass spectrometry, *Anal. Biochem.* **274**: 59-68.

Meyer, M. J., and Bechtold, W. E., 1996, Protein adduct biomarkers:State of the art, *Environ. Health Perspect.* **104** Suppl 5: 879-882.

Mowrer, J., Törnqvist, M., Jensen, S., and Ehrenberg, L., 1986, Modified Edman degradation applied to hemoglobin for monitoring occupational exposure to alkylating agents, *Toxicol. Environ. Chem.* **11**: 215-231.

Paulsson, B., Athanassiadis, I., Rydberg, P., and Törnqvist, M., 2003, Hemoglobin adducts for glycidamide:acetonization of hydrophylic groups for reproducible gas chromatography /tadem mass spectrometric analysis, *Rapid Comm. Mass Spec.* **17**: 1859-1865.

Tareke, E., Rydberg, P., Karlsson, P., Eriksson, S., and Törnqvist, M., 2000, Acrylamide: a cooking carcinogen?, *Chem Res. Toxicol.* **13**: 517-522.

Törnqvist, M., Fred, C., Haglund, J., Helleberg, H., Paulsson, B., and Rydberg, P., 2002, Protein adducts: quantitative and qualitative aspects oftheir formation, analysis and applications, *J. Chromatogr. B* **778**: 279-308.

COMPARISON OF ACRYLAMIDE METABOLISM IN HUMANS AND RODENTS

Timothy R. Fennell[1] and Marvin A. Friedman[2]
[1]RTI International, Research Triangle Park, North Carolina; [2]University of Medicine and Dentistry (UMDNJ), Newark, New Jersey; e-mail: Fennell@rti.org

Abstract: Acrylamide is metabolized by direct conjugation with glutathione or oxidation to glycidamide, which undergo further metabolism and are excreted in urine. In rats administered 3 mg/kg 1,2,3-$^{13}C_3$ acrylamide, 59 % of the metabolites excreted in urine was from acrylamide-glutathione conjugation, whereas 25% and 16% were from two glycidamide-derived mercapturic acids. Glycidamide and dihydroxypropionamide were not detected at this dose level. The metabolism of acrylamide in humans was investigated in a controlled study with IRB approval, in which sterile male volunteers were administered 3 mg/kg 1,2,3-$^{13}C_3$ acrylamide orally. Urine was collected for 24 h after administration, and metabolites were analyzed by ^{13}C NMR spectroscopy. At 24 h, urine contained 34 % of the administered dose, and 75 % of the metabolites were derived from direct conjugation of acrylamide with glautathione. Gycidamide, dihydroxypropionamide and one unidentified metabolite were also detected in urine. This study indicated differences in the metabolism of acrylamide between humans and rodents.

Key words: Acrylamide, glycidamide, metabolism

1. INTRODUCTION

The discovery of acrylamide in food has caused considerable concern in the potential toxicity of acrylamide in humans at low levels of exposure over a long period of time. Prior to this finding, the main concern about human exposure has been from workplace exposure to acrylamide and exposure to residual acrylamide monomer in polyacrylamide used in consumer products (European Union, 2002). The routes of exposure of concern for these exposures are dermal and inhalation, whereas the route of concern for ingestion in food is oral.

Acrylamide is known to exert a number of toxic effects in rodents, including neurotoxicity, heritable genetic effects, cancer, heritable translocations, and dominant lethal mutations (Dearfield et al., 1988; Dearfield et al., 1995; IARC, 1994). In humans, exposure to acrylamide can cause peripheral neuropathy (Spencer and Schaumburg, 1974a, 1974b, 1975). Of particular interest is whether low levels of exposure to acrylamide over a long period in the diet can cause any adverse effects in exposed people. An important element in evaluating the potential risk for people is in understanding the role of metabolism in the disposition and removal of acrylamide, the potential role of metabolism in the toxicity of acrylamide, and the differences between species in the metabolism of acrylamide.

The intention of this paper is to review work on the metabolism of acrylamide in rodents, and to summarize some recent studies comparing the metabolism of acrylamide in rodents and humans.

2. METABOLISM IN RODENTS

Acrylamide is reactive, and can react with nucleophilic sites of protein, and with the sulphydryl group of glutathione. A number of early studies on the metabolism of acrylamide in vitro and in rats identified glutathione conjugation as a pathway of metabolism, with excretion of N-acetyl-S-(3-amino-3-oxopropyl)cysteine as a metabolite in urine (Dixit et al., 1982; Edwards, 1975; Miller et al., 1982).

The oxidation of acrylamide to glycidamide was initially identified by the investigation of hemoglobin adducts in rats administered acrylamide (Calleman et al., 1990). After administration of acrylamide, globin was isolated and hydrolyzed with 6 N HCl. S-Carboxyethylcysteine, formed by the reaction of acrylamide with cysteine and subsequent hydrolysis, was the main adduct detected from acrylamide. A second adduct, S-2-carboxy-2-hydroxyethylcysteine, was identified. This was postulated to arise from reaction of glycidamide with globin, and identified the pathway of acrylamide oxidation.

The metabolism of acrylamide was extensively investigated in rodents (Sumner et al., 1992), using a new method for characterization of metabolites by ^{13}C NMR spectroscopy. A single dose of 1,2,3-^{13}C labeled acrylamide was administered by gavage. Urine samples were collected in metabolism cages, and after centrifugation and addition of D_2O, the samples were analyzed directly by ^{13}C NMR spectroscopy. The presence of adjacent labeled carbon atoms in the acrylamide molecule provided a means of identifying the carbon atoms derived from acrylamide in the various urinary metabolites. The low natural abundance of ^{13}C (1.1%) results in signals for each carbon atom that are essentially singlets, with little contribution from carbon-carbon coupling to adjacent labeled carbon atoms (1.1% of 1.1%). In the labeled acrylamide molecules, each labeled carbon has one or two labeled neighbors, resulting in carbon-carbon coupling, and the presence of a doublet, a doublet of doublets and a doublet for the three carbon atoms. These multiplet carbon signals can readily be distinguished from the unlabeled singlets. The dispersion of the carbon signals over a wide frequency range enabled the separation of the metabolite signals. Application of several two-dimensional techniques, HET2DJ (heteronuclear 2-D J-resolved spectroscopy) and INADEQUATE (incredible natural abundance double quantum transfer experiment) enabled the determination of the proton multiplicity of each protonated carbon signal, and the connectivity between two labeled carbon signals, respectively. The amount of each metabolite was measured based on comparison of the peak area with that of an added standard (dioxane) in spectra acquired under conditions in which nuclear Overhauser enhancement was reduced.

The metabolites identified in rat and mouse urine are shown in Figure 1. The major metabolite was N-acetyl-S-carboxamidoethylcysteine (metabolite 1), derived from direct glutathione conjugation of acrylamide. Two products derived from the reaction of glutathione with glycidamide were also detected: N-acetyl-S-(3-amino-2-hydroxy-3-oxopropyl)cysteine (metabolite 2) and N-acetyl-S-(1-carbamoyl-2-hydroxyethyl)cysteine (metabolite 3). The hydrolysis product of glycidamide, glyceramide (metabolite 5), was also found in urine. Interestingly, glycidamide itself was also detected in urine, indicating that the epoxide was sufficiently stable to undergo excretion in the urine. Acrylamide was also detected in urine, but was not quantitated.

Summation of the metabolites derived from direct conjugation vs. the metabolites derived from oxidation and subsequent metabolism of glycidamide (Table 1) provided a measure for comparing the metabolism of acrylamide in rats and mice. The extent of oxidation of acrylamide was considerably higher in the mouse than the rat.

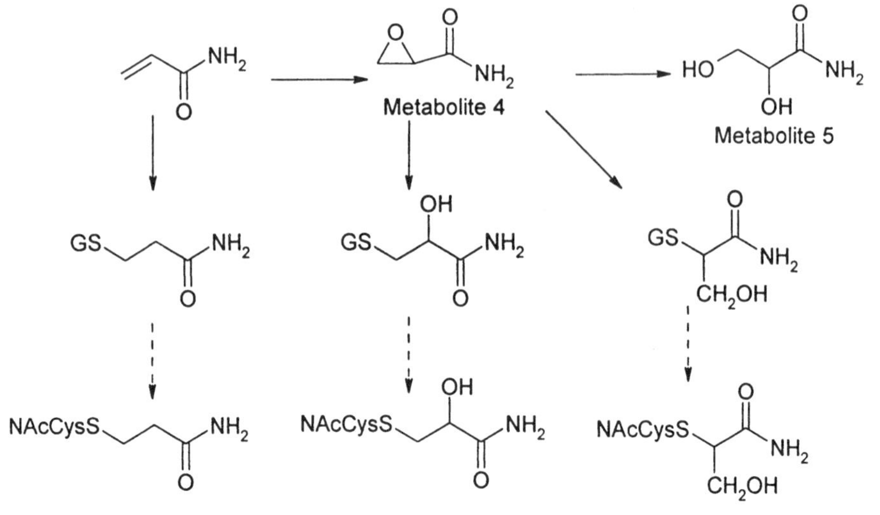

Figure 1. Metabolism of acrylamide in rodents. The numbering system is that used by Sumner et al. (1992). GS represents a glutathionyl residue, and NAcCysS represents an *N*-acetylcysteine residue

Table 1. Urinary metabolites of acrylamide in the rat and mouse following gavage administration of 50 mg/kg acrylamide (from Sumner et al. (1992)).

		Rat	Mouse
GSH-AA	1,1'	71 ± 3.8[a]	41 ± 2.2
GSH-GA	2,2'	13 ± 2.1	21 ± 2.6
	3,3'	6.8 ± 1.1	12 ± 0.6
GA	4	7.3 ± 1.6	17 ± 2.1
GA hydrolysis	5	1.2 ± 0.5	5.3 ± 1.2
Σ GA		28 ± 3.8	59
% of dose in urine		53 ± 7.6	51

[a] Average ± S.D., % of urinary metabolites.
AA, acrylamide; GSH, glutathione; GA, glycidamide

Calleman et al. suggested from analysis of acrylamide and glycidamide hemoglobin adducts that the extent of acrylamide oxidation was greater at lower doses than at higher doses (Calleman et al., 1992). On administration of a dose of 3 mg/kg 1,2,3-$^{13}C_3$ acrylamide to rats, 59 ± 1.5 % of the urinary metabolites were derived from glycidamide, and 41 ± 1.5 % from oxidation via glycidamide (Fennell et al., 2004). This is considerably different from the extent of oxidation at 50 mg/kg with a ratio of direct conjugation:oxidation of approximately 1.43 at 3 mg/kg, compared with 2.54

at the 50 mg/kg. Almost all of the material metabolized via glycidamide at the low dose was present as mercapturic acid conjugates, with little material present as glycidamide or glyceramide.

3. METABOLISM IN HUMANS

Many areas of uncertainty have existed about the metabolism of acrylamide in humans that are now considered of great importance. What is the extent of uptake from ingestion orally of acrylamide? What is the extent of metabolism by direct conjugation of acrylamide with GSH, *versus* metabolism via oxidation? Are there significant differences between individuals in the rate of metabolism of acrylamide via the two main pathways? Are the metabolites of acrylamide that are observed in rodents also observed in humans? Are there additional pathways of metabolism in humans?

A number of studies have been conducted in humans exposed to acrylamide in the workplace to examine acrylamide and its metabolites in plasma and urine, and adducts in hemoglobin. Acrylamide was detected in plasma from people exposed to acrylamide in the workplace. Calleman et al. (1994) measured the amount of acrylamide present in plasma at the end of shift. Detectable acrylamide was reported in 17 out of 41 of the workers exposed to acrylamide, the concentrations detected ranged from 0.6 to 3.5 µmol/L in plasma. Interestingly acrylamide was reported in plasma of three (of a total of 10) of the control subjects without occupational exposure to acrylamide (0.9 µmol/liter).

Acrylamide was reported in the urine of people who ingested potato chips containing acrylamide (Sorgel et al., 2002).

In a recent study, we investigated the metabolites of acrylamide excreted in urine from people administered a single oral dose of $1,2,3-^{13}C_3$ acrylamide by ^{13}C NMR spectroscopy (Fennell et al., 2004). We also measured the hemoglobin adducts derived from exposure to acrylamide. A complete description of the study and the results obtained is beyond the scope of this report, and we will confine the description to the results of the urinary metabolite analysis in a small part of the overall study.

Briefly, $1,2,3-^{13}C_3$ acrylamide was administered orally as a solution in water to sterile male non-smoking volunteers. The study was conducted in a clinical laboratory setting, in which the volunteers were closely monitored, with a physical exam prior to the study, and with careful monitoring by medical staff during the course of the study. The protocol was reviewed and approved by Institutional Review Boards both at the Clinical Research Facility conducting the administration of acrylamide, and at the laboratory conducting the analysis of samples. The oral administration of acrylamide was conducted in a dose-escalating manner, with 0.5 mg/kg administered initially to 5 people in a group of 6 volunteers. The additional volunteer in each group received water as a control. After a two-week period, a second group of volunteers received a dose of 1.0 mg/kg. Finally, after a further two week period, a third group received a dose of 3.0 mg/kg. Blood samples were collected prior to administration and at 24 hr following administration of acrylamide. A pre-exposure urine sample was collected, and urine samples were collected at 0-2, 2-4, 4-8, 8-16, and 16-24 hr following exposure.

Urine sample analysis by ^{13}C NMR was conducted in only the highest dose group, at 3 mg/kg. From each individual, a pooled sample was generated, combining aliquots from each time point in proportion to their volume. Each sample was centrifuged, and D_2O containing dioxane was added. Each sample was analyzed by ^{13}C NMR spectroscopy on a Varian 500 MHz NMR spectrometer operating at 125 MHz for ^{13}C. Samples were prepared by adding D_2O, or D_2O containing dioxane at a known concentration (200 µl) to an aliquot of a urine sample, a composite urine sample, or a concentrated composite urine sample (800 µl). Carbon-Carbon connectivity was established using two-dimensional incredible natural abundance double quantum transfer spectra (INADEQUATE) using the Varian-supplied pulse sequence.

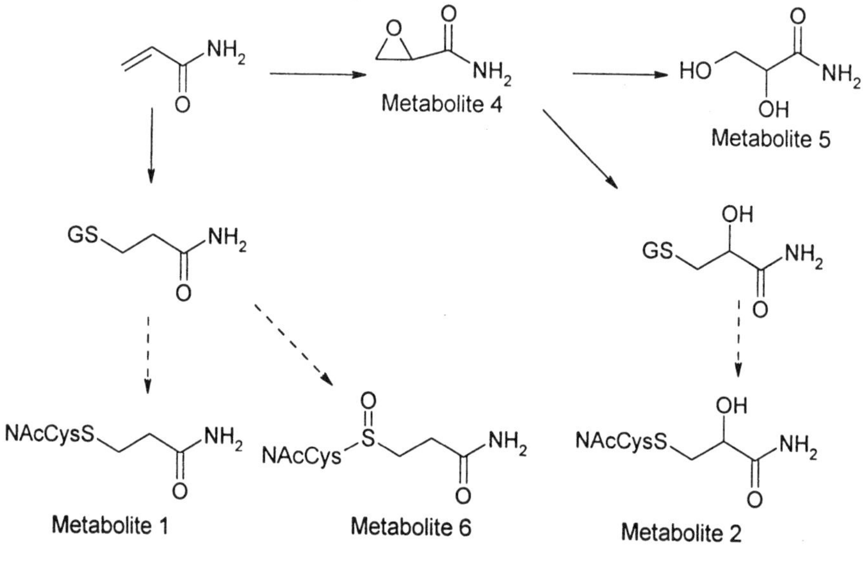

Figure 2. Metabolism of acrylamide in humans

The analysis of acrylamide metabolites in human urine indicated the presence of metabolites indicated in Figure 2. The major metabolite was the same as that in rodent urine, namely *N*-acetyl-*S*-(3-amino-3-oxopropyl)cysteine (metabolite 1). This metabolite accounted for approximately 72 % of the urinary metabolites detected. Other metabolites were present at low levels. Of the metabolites identified in rodents, glyceramide (metabolite 5, 11 %) and glycidamide (metabolite 4, 2.6 %) were readily detected. Only one of the mercapturic acid metabolites of glycidamide was detected (metabolite 2), at levels that were too low to quantitate. The other mercapturic acid, metabolite 3, was not detected. An additional metabolite not previously detected in rodent urine was detected in human urine. It has been assigned to *N*-acetyl-*S*-carbamoylethylcysteine-*S*-oxide, and accounted for 14 % of the metabolites detected in urine. Acrylamide was detected in urine, but was not quantitated by NMR spectroscopy.

4. CONCLUSIONS

Species differences were observed in the fraction of acrylamide that is metabolized via direct conjugation with glutathione, compared with oxidation via glycidamide. The extent of oxidation of acrylamide was mouse > rat > human. In the mouse and rat, the majority of the metabolism of glycidamide resulted from conjugation with glutathione. In contrast, in humans, the majority of the metabolism of glycidamide was via hydrolysis, with little via glutathione conjugation.

ACKNOWLEDGEMENTS

We would like to acknowledge the contributions of Dr. Susan Sumner in conducting many of the rodent studies, and in the genesis of the human metabolism study. We would like to acknowledge Dr. William E. Bridson, and Ms Rebecca Spicer of the Covance Clinical Research Unit, Inc, Madison WI. We would also like to acknowledge the support of Mr. Rodney Snyder and Dr. Jason Burgess in the NMR analysis of human urine.

REFERENCES

Calleman, C. J., Bergmark, E., and Costa, L. G., 1990, Acrylamide is metabolized to glycidamide in the rat: evidence from hemoglobin adduct formation, *Chem Res Toxicol* **3**:406-412.

Calleman, C. J., Stern, L. G., Bergmark, E., and Costa, L. G., 1992, Linear versus nonlinear models for hemoglobin adduct formation by acrylamide and its metabolite glycidamide: implications for risk estimation, *Cancer Epidemiol Biomarkers Prev* **1**:361-368.

Dearfield, K. L., Abernathy, C. O., Ottley, M. S., Brantner, J. H., and Hayes, P. F., 1988, Acrylamide: its metabolism, developmental and reproductive effects, genotoxicity, and carcinogenicity, *Mutat Res* **195**:45-77.

Dearfield, K. L., Douglas, G. R., Ehling, U. H., Moore, M. M., Sega, G. A., and Brusick, D. J., 1995, Acrylamide: a review of its genotoxicity and an assessment of heritable genetic risk, *Mutat Res* **330**:71-99.

Dixit, R., Seth, P. K., and Mukjtar, H., 1982, Metabolism of acrylamide into urinary mercapturic acid and cysteine conjugates in rats, *Drug Metab Dispos* **10**:196-197.

Edwards, P. M., 1975, The distribution and metabolism of acrylamide and its neurotoxic analogues in rats, *Biochem Pharmacol* **24**:1277-1282.

European Union, 2002, European Union Risk Assessment Report
Acrylamide, Luxembourg, pp. 210 pp.

Fennell, T. R., Sumner, S. C., Snyder, R. W., Burgess, J., Spicer, R., Bridson, W. E., and Friedman, M. A., 2004, Metabolism and Hemoglobin Adduct Formation of Acrylamide in Humans, *Tox. Sci.* submitted.

IARC, 1994, Acrylamide, *IARC Monogr Eval Carcinog Risks Hum* **60**:389-433.

Miller, M. J., Carter, D. E., and Sipes, I. G., 1982, Pharmacokinetics of acrylamide in Fisher-344 rats, *Toxicol Appl Pharmacol* **63**:36-44.

Sorgel, F., Weissenbacher, R., Kinzig-Schippers, M., Hofmann, A., Illauer, M., Skott, A., and Landersdorfer, C., 2002, Acrylamide: increased concentrations in homemade food and first evidence of its variable absorption from food, variable metabolism and placental and breast milk transfer in humans, *Chemotherapy* **48**:267-274.

Spencer, P. S., and Schaumburg, H. H., 1974a, A review of acrylamide neurotoxicity. Part I. Properties, uses and human exposure, *Can J Neurol Sci* **1**:143-150.

Spencer, P. S., and Schaumburg, H. H., 1974b, A review of acrylamide neurotoxicity. Part II. Experimental animal neurotoxicity and pathologic mechanisms, *Can J Neurol Sci* **1**:152-169.

Spencer, P. S., and Schaumburg, H. H., 1975, Nervous system degeneration produced by acrylamide monomer, *Environ Health Perspect* **11**:129-133.

Sumner, S. C., MacNeela, J. P., and Fennell, T. R., 1992, Characterization and quantitation of urinary metabolites of [1,2,3-^{13}C]acrylamide in rats and mice using ^{13}C nuclear magnetic resonance spectroscopy, *Chem Res Toxicol* **5**:81-89.

KINETIC AND MECHANISTIC DATA NEEDS FOR A HUMAN PHSIOLOGICALLY BASED PHARMACOKINETIC (PBPK) MODEL FOR ACRYLAMIDE

Pharmacokinetic Model for Acrylamide

Melvin E. Andersen[1], Joseph Scimeca[2], and Stephen S. Olin[3]
[1]*CIIT Centers for Health Research, Six Davis Drive PO Box 12137, Research Triangle Park, NC 27709-2137, Tel: +1-919-558-1205, Fax: +1-919-558-1404, e-mail: MAndersen@ciit.org;* [2]*Cargill, Inc., PO Box 9300, MS 56, Minneapolis, MN 55440-9300, Tel: +1-952-742-7276, Fax: +1-952-742-6678, e-mail: joseph_scimeca@cargill.com;* [3]*International Life Sciences Institute, Risk Science Institute, One Thomas Circle, Ninth Floor, Washington, DC 20005-5802, Tel: +1-202-659-3306, Fax:+1-202-659-3617, e-mail: solin@ilsi.org*

Abstract: A pharmacokinetic (PBPK) model has been developed for acrylamide (AMD) and its oxidative metabolite, glycidamide (GLY), in the rat based on available information. Despite gaps and limitations to the database, model parameters have been estimated to provide a relatively consistent description of the kinetics of acrylamide and glycidamide using a single set of values (with minor adjustments in some cases). Future kinetic and mechanistic studies will need to focus on the collection of key data for refining certain model parameters and for model validation, as well as for conducting studies that elucidate the mechanism of action. Development of a validated human AMD/GLY PBPK model capable of predicting target tissue doses at relevant dietary AMD exposures, in combination with expanding data on modes of action, should allow for a substantive improvement in the risk assessment of acrylamide in food.

Key words: Acrylamide, Glycidamide, metabolism, pharmacokinetics, risk assessment

1. **INTRODUCTION**

Risk assessments generally require a number of extrapolations from laboratory data in order to assess human health implications from chemical exposure, including:
- high doses to low doses
- across route of exposure
- continuous versus intermittent exposure
- animals to humans

These extrapolations are best accomplished when based on tissue dose rather than administered dose (Clewell and Andersen, 1985). Physiologically based pharmacokinetic (PBPK) modeling is an ideal tool for determining tissue doses. The underpinnings of PBPK models are a series of mass based differential equations that are numerically integrated to simulate kinetic behavior. For example, the dose of a chemical in a particular tissue compartment, like the liver, can be calculated as such:

$$dA_l/dt = Q_l (C_a - C_{vl}) - V_m * C_{vl}/(K_m + C_{vl}) \qquad (1)$$

where the rate of change of the amount in the liver (dA_l/dt) is equal to the rate of uptake in arterial blood (Q_l*C_a) minus loss in venous blood (Q_l*C_{vl}) minus the rate of loss by metabolism ($V_m*C_{vl}/K_m +C_{vl}$). Q_l is liver blood flow; C_a, arterial blood concentration; C_{vl}, the free concentration of compound in the venous blood exiting the liver; V_m, the maximum rate of metabolism in the liver; and K_m, the dissociation constant for the enzyme(s) involved in metabolism.

In cells, acrylamide (AMD) and glycidamide (GLY) react with various macromolecules to irreversibly alter the structure of the macromolecule. Some adducts formed by these reactions may have little functional consequence, as is believed to be true with hemoglobin or albumin adducts. Other adducts may have significant biological consequences. Adverse response to adduction may occur when a large proportion of a group of critical macromolecules are adducted or when a small proportion of critical macromolecules (such as DNA-bases) are adducted leading to impairment of a critical biological process or increasing probabilities of mutation during cell division.

Ideally, risk assessments with AMD and GLY would be based on a measure of the proportion of specific molecules that are altered by adduction. However, at present we do not know which macromolecules are directly involved in the various responses, i.e., in the neurotoxicity, carcinogenicity, or reproductive/developmental toxicity of these compounds. Absent the identification of specific macromolecular targets, risk assessments will have to rely on a measure of internal dose that correlates with the potential for formation of these adducts. The dose metric for reactive compounds that form stable adducts will be net exposure to tissues, expressed as an area under the concentration curve (AUC) rather than administered dose or peak concentration.

The sections below outline the expected relationships of adducts and toxicity.

1.1 Protein Adducts

We can consider a case where toxicity is related to the proportion of a specific macromolecule, MM, lost through adduction. For this example, MM reacts with AMD or GLY with a second-order rate constant, k_{mm}. The rate equation for the rate of change of concentration of macromolecule with time (dMMt/dt) has terms for the synthesis rate (k_o), the basal degradation rate (k_{de}) and the loss by adduction. In this example the loss is related to AMD.

$$d\,MMt/dt = k_o - k_{de}*MMt - k_{mm}*AMD*MMt \qquad (2)$$

The steady-state concentration of MMt as a function of the average concentration of AMD becomes:

$$MMs\text{-}s = K_o/(k_{de} + k_{mm}*AMD) \qquad (3)$$

Although the concentration of AMD varies with time after dosing, the equation can be recast as the average concentration within a dosing interval. This value would be the AUC in the dose interval/dose interval. This average inter-dose interval concentration is depicted by the italicized *AMD* in the equation. This curve for MMs-s is a rectangular hyperbola with a value of K_o/k_{de} in the absence of AMD exposure and falling asymptotically to zero as average AMD concentration increases (**Figure 1**). When $k_{mm}*AMD$ equals k_{de}, the steady-state concentration of MM is half its value in the absence of any reaction with AMD.

The risk model used with protein adduction would likely be based on a non-linear threshold risk model with safety/uncertainty factors (US EPA, 2002). The goal in PBPK modeling would be to insure accurate estimation of the daily area under the blood concentration curve for AMD (or GLY) for all relevant dose routes in laboratory animals and humans.

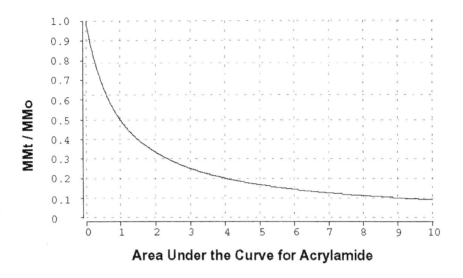

Figure 1. The relationship expected for the proportion of macromolecule remaining (MMt/MMo) in relation to average daily AUC for acrylamide where the w/axis AUC is normalized to multiples of k_{de}.

1.2 DNA-Adducts

In some cases, linear risk models may be appropriate for direct acting mutagens/carcinogens where no other modes of action influence the carcinogenic responses (US EPA, 2003). While we can outline a risk strategy for adducts in a linear risk model, it bears emphasis that the database with AMD and GLY does not necessarily argue for a linear model for cancer risk at this time. For DNA-adducts, where there is no background concentration of adduct, the rate of change of adduct concentration with time would be:

$$d(DNA\text{-}GLY)/dt = k_{mm}*DNA*GLY - k_{repair}*DNA\text{-}GLY \qquad (4)$$

The steady-state concentration of adduct would be linearly related to the average interdose interval concentration of *GLY*:

$$[DNA\text{-}GLY] = (k*DNA*GLY) / k_{repair} \qquad (5)$$

If risk were related directly to the adduct concentration, a linear risk model would be applied based on an average concentration of *GLY* (or *AMD*) depending on specific modes of action.

1.3 PBPK Modeling for AMD and GLY

There are several clear design criteria to keep in mind in developing a PBPK model for AMD and GLY. The PBPK model should accurately predict circulating concentrations of AMD and GLY for multiple exposure routes and over a wide range of doses in experimental animals and people. The model should have three main purposes.

First, the model developed for the laboratory animal will be used to determine the relationship between tissue exposures to AMD and GLY. In this application, the PBPK model output can be used to estimate a benchmark dose (BMD) of AMD that is associated with a 10% increase in incidence of specific adverse responses. In this case the BMD is determined in terms of *AMD* or *GLY*, as noted in the equations above. Thus, the BMDs might have units of mg AMD (or mg GLY)/liter blood, while average AUCs would have units of (mg substance/liter) x hours.

Second, the PBPK models for the rat and/or mouse are used to analyze the full kinetic data sets in these animals to determine whether there are significant non-linearities in metabolism, first-pass losses that affects tissue doses at low doses, or any large route-to-route differences in kinetics. These latter issues affect the low dose extrapolations that will be eventually required for the human risk assessment.

Third, a human model that predicts tissue concentrations of AMD and GLY is used in the risk assessment based on the risk model chosen for the toxic endpoint under evaluation. In this application, the tissue dose-BMD from the animal studies is adjusted depending on the risk model. These adjustments consist of either an extrapolation to zero for a linear response with an acceptable risk at 10^{-5} or 10^{-6} or use of uncertainty factors to adjust the rodent tissue dose-BMD to an acceptable human tissue dose. Once the 'acceptable' human internal tissue dose is obtained from the risk model, the human PBPK model becomes the tool to determine the acceptable exposures. An acceptable exposure situation is one where the tissue dose achieved from the particular simulated human exposure is lower than the 'acceptable' human tissue dose from the risk model. PBPK models permit evaluation of multi-dose and multi-route exposures and support sensitivity and variability analysis to assist in identifying potentially susceptible groups in the human population based on such factors as metabolic polymorphisms or specific dietary habits. Some good examples of use of PBPK models to support risk assessments have appeared in recent years (Andersen *et al.*, 2000; Gentry *et al.*, 2002).

Several other model applications are also likely, and include: (1) use of the PBPK model to insure consistency of results across multiple studies, (2) use of the model to provide more quantitative tools to assist human biomonitoring studies, and (3) use of the model to develop improved dose-response models as the molecular targets associated with specific toxic endpoints become more clearly established. Interspecies extrapolations of the risk models related to alterations in concentrations of molecular targets require knowledge or assumptions about synthesis rates of the macromolecules and the repair or turnover rates for the adducted molecules. More empirical risk models that are based on the average daily concentration would not require this information.

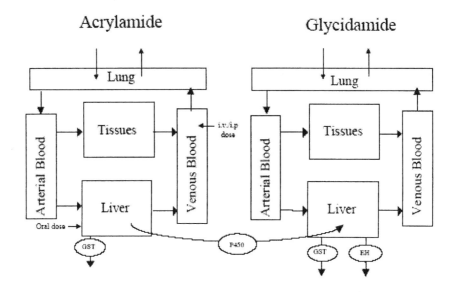

Figure 2. PBPK model structure for acrylamide.

Kirman *et al.* (2003) developed the sole AMD PBPK model, which contains a component for GLY. The model was developed based on three kinetic and disposition studies (Raymer *et al.*, 1993; Miller, *et al.*, 1982; and Sumner, *et al.*, 1992). In the Kirman *et al.* model, AMD is distributed within five compartments (arterial blood, venous blood, liver, lung and all other tissues lumped together) and is linked to the GLY portion of the model via metabolism in the liver (**Figure 2**). GLY is distributed within the same compartments as AMD. Although additional compartments corresponding to sites of toxicity and carcinogenicity could be added (peripheral nerves, central nervous system, testes, adrenal gland, thyroid, mammary gland, uterus, and oral cavity), limitations in available data did not justify separating them from the lumped compartments in the initial model. The arterial and venous blood compartments were further divided into serum and blood cell sub-compartments to allow for modeling certain data sets (e.g., chemical adducted with hemoglobin in red blood cells). This initial model also included a description of synthesis, utilization and depletion of glutathione, following modeling strategies from the literature (D'Souza *et al.*, 1988).

1.4 AMD and GLY Kinetic and Mechanistic Research Needs

Development of the Kirman *et al.* rat PBPK model allows for the: (1) estimation of tissue dose to AMD and GLY under various exposure conditions, (2) identification of gaps in the available knowledge base, and (3) evaluation of measures of tissue dose for various proposed modes of action for AMD carcinogenicity. Proposed carcinogenicity modes of action include mutagenicity, interactions with cell surface receptors, and reactivity with glutathione and tissue sulfhydryls. This model provided a base upon which to build an interspecies (rodent and human) PBPK model as data become available from AMD/GLY research. In particular, future studies linking kinetic studies with the various modes of action should now be more seriously considered. Such studies would define target protein and binding sites and allow PBPK modeling to predict the concentrations of AMD and GLY adducts over time. For example:

1. AMD and GLY may act as carcinogens after AMD exposures by direct reaction with DNA bases leading to mutations during cell division and ultimately to cancer. In this case, the ability to simulate DNA adduct concentrations with the PBPK model over time in animals with specific AMD-associated tumor burdens would form the basis for risk assessment calculations for these compounds. Interspecies extrapolation of DNA and protein reactivity from *in vitro* studies would support cancer risk assessment calculations.
2. A second mode of action related to GSH depletion, or GSH depletion associated with reactivity toward other proteins, could also be explored with the PBPK model already developed. This model could be improved by new studies of GSH status in various tissues following dosing with AMD.
3. A more concerted effort is also likely to be necessary to determine the nature of specific binding sites associated with AMD/GLY in target tissues. These protein reactivity studies can now be pursued due to continuing developments in protein mass-spectrometry that should permit more comprehensive evaluation of the interaction of AMD/GLY with multiple tissue constituents.

4. Another possibility for mode of action is that the reactivity of AMD/GLY with specific cellular targets, either intracellular molecules or surface molecules, is ultimately responsible for the effects in these tissues. In this new generation of studies on AMD/GLY disposition, it would be valuable to more fully catalog cellular targets of AMD/GLY reactivity.
5. Further elaboration of the interactions of these compounds in target tissues from *in vivo* toxicity studies may provide new directions for looking at specific AMD/GLY interactions *in vitro*. Cell lines from target tissues – thyroid, mammary tissues, uterus, scrotal mesothelial cells, and specific brain cell types – could also be used to assess specific binding, or alternatively, *in vivo* studies could be performed with radiolabeled AMD. Specific dopamine-responsive cells/tissues might also be evaluated to test AMD agonist activity and sites of reactivity with tissue protein since these receptors have been implicated in some responses to AMD/GLY.

2. DISCUSSION

PBPK modeling has become an important tool in risk assessment to assist in extrapolations and data integration. General advantages of simulation modeling in physiology, toxicology, and risk assessment have been well established (Yates, 1978; Andersen *et al.*, 1995). Some generic advantages available from the development of PBPK models as a core part of a research strategy include:
- codification of facts and beliefs (organize available information)
- expose contradictions in existing data/beliefs
- explore implications of hypotheses about the chemical
- expose data gaps limiting use of the model
- predict response under new conditions
- identify essentials of system structure
- provide representation of present state of knowledge
- suggest and prioritize new experiments

It is best to conduct model development in concert with data acquisition for any compound. First, the PBPK model can be used to assess the correlation between various measures of tissue dose and toxic response to evaluate possible causal relationships among various measures of tissue exposure and outcome. Second, these models can serve to test the consistency of various studies describing diverse aspects of acrylamide metabolism and disposition. And, thirdly, critical data gaps for model development can be used as part of the design criteria for new studies. Some of these attributes are evident in the work done to date with PBPK modeling for AMD/GLY.

REFERENCES

Andersen, M. E., Clewell, H.J., III, and Frederick, C. B., 1995, Applying simulation modeling to problems in toxicology and risk assessment: a short perspective, *Toxicol. Appl. Pharmacol.* **133**: 181-187.

Andersen, M. E., Sarangapani, R., Gentry, P.R., Clewell, H. J.,III, Covington, T. R. and Frederick, C.B., 2000, Application of a hybrid CFD-PBPK nasal dosimetry model in an inhalation risk assessment: an example with acrylic acid, *Toxicol. Sci.* **57**: 312-325.

Clewell, H. J., III and Andersen, M. E., (1985, Risk assessment extrapolations and physiological modeling, *Toxicol. Ind. Health* **1**: 111-131.

D'Souza, R. W., Francis, W. R., and Andersen, M. E., 1988 A physiologic model for tissue glutathione depletion and increased resynthesis following ethylene dichloride exposure, *J. Pharmacol. Exp. Ther.* **245**: 563-568.

Gentry, P. R., Covington, T. R., Banton, M I., Clewell, H. J., III, and Andersen, M. E., 2002, Application of a physiologically based pharmacokinetic model for isopropanol in the derivation of a reference dose and reference concentration, *Reg. Tox, Pharm.* **36**: 51-68.

Kirman, C. R., Gargas, M. L., Deskin, R., Tonner-Navarro, L., and Andersen, M. E., 2003, A physiologically based pharmacokinetic model for acrylamide and its metabolite, glycidamide, in the rat, *J. Toxicol. Environ. Health Part A* **66**: 253-274.

Miller, M., Carter, D., and Sipes, I., 1982, Pharmacokinetics of acrylamide in Fischer 344 rats, *Toxicol. Appl. Pharmacol.* **63**: 36-44.

Raymer, J. H., Sparacino, C.M., Velez, G. R., Padilla, S., MacPhail, R. C., and Crofton, K. M., 1993, Determination of acrylamide in rat serum and sciatic nerve by gas chromatography-electron-capture detection, *J. Chromatogr.* **619**(2): 223-34.

Sumner, S., MacNeela, J., and Fennell, T., 1992, Characterization and quantitation of urinary metabolites of [1,2,3-^{13}C] acrylamide in rats and mice using ^{13}C nuclear magnetic resonance spectroscopy, *Chem. Res. Toxicol.* **5**: 81-89.

US EPA, 2002, A review of the reference dose and reference concentration processes. Risk assessment forum. EPA/630/P-02/002F. December 2002.

US EPA, 2003, Draft final guidelines for carcinogen risk assessment. Risk assessment forum. EPA/630/P-003/001A. NCEA-F-0644A. February 2003.

Yates, F.E., 1978, Good manners in good modeling: mathematical models and computer simulations of physiological systems, *Am. J. Physiol.* **234**: R159-R160.

IN VITRO STUDIES OF THE INFLUENCE OF CERTAIN ENZYMES ON THE DETOXIFICATION OF ACRYLAMIDE AND GLYCIDAMIDE IN BLOOD

Birgit Paulsson[1], Margareta Warholm[2], Agneta Rannug[2], and Margareta Törnqvist[1]
[1]Dept. of Environmental Chemistry, Stockholm University, SE-106 91 Stockholm, Sweden;
[2]Institute of Environmental Medicine, Karolinska Institutet, SE-171 77 Stockholm, Sweden; e-mail: birgit.paulsson@mk.su.se;

Abstract: Several enzymes involved in the metabolism of xenobiotic substances are polymorphic in humans. Inter-individual differences in response to certain chemicals, such as acrylamide, as a result of such genetic polymorphisms might affect health-risk assessments. Detoxification by, for example, conjugation with glutathione (GSH) will decrease the concentration. The dose of the compound and enzymes that enhance the conjugation with GSH will increase the detoxification rate. The dose of acrylamide or glycidamide has been measured in blood samples from individuals with defined genotypes for the glutathione transferases GSTT1 and GSTM1 after *in vitro* incubation with these compounds. The results indicate that these enzymes have no significant effect on the blood dose, measured as Hb adducts over time, after exposure to acrylamide or glycidamide.

Key words: Acrylamide; glycidamide; polymorphism; detoxifying enzymes.

1. INTRODUCTION

Chemicals entering the body are subjected to biotransformation, often enzymatically catalyzed, in order to facilitate their excretion. Several of the enzymes involved in this metabolism are known to be polymorphic in humans, due to common genetic variants including deletions of specific genes or differences in the coding sequences. Such genetic polymorphisms may result in an inter-individual difference in response to certain chemicals. Information on biotransformation genotypes might clarify dose-effect and dose-response relationships in epidemiological and biomonitoring studies and thus improve health-risk assessments.

Polymorphic enzymes that might play a role in the metabolism of acrylamide and/or glycidamide are: cytochrome P450 2E1 (CYP2E1), which has been shown to catalyze the transformation of acrylamide to glycidamide (Sumner et al., 1999), glutathione transferases (GSTs) such as the isozymes GSTT1, GSTM1 and GSTP1, which catalyze conjugation with glutathione (GSH) (Hayes and Strange, 2000), and epoxide hydrolase (EH), which catalyzes hydrolysis of epoxides (Fretland and Omiecinski, 2000). GSH metabolites from acrylamide and glycidamide as well as the hydrolysis product from glycidamide have been detected in urine of rodents exposed to acrylamide (Sumner et al., 1997).

In the *CYP2E1* gene, several polymorphisms in the promoter region have been described. These may affect the expression of the CYP2E1 enzyme and contribute to inter-individual variations in the enzyme activity resulting in differences in doses from acrylamide and the metabolite glycidamide. CYP2E1 is mainly a hepatic enzyme but is also expressed in for example lung, kidney, bone marrow and white blood cells (Lucas et al., 2001).

The GSTT1 enzyme is involved in the metabolism of several important industrial halogenated chemicals or epoxides such as methylene chloride, diepoxybutane, and ethylene oxide (Hayes and Strange, 2000; Landi, 2000). It is expressed mainly in liver and kidney, but also in red blood cells. The most important polymorphism is a deletion of the entire *GSTT1* gene. Homozygotic individuals (*GSTT1*O/*O*) totally lack GSTT1 enzyme activity (Warholm et al., 1995). The frequency of this GSTT1-null phenotype is in general lower in Caucasians (10–25%) than in Asians (20–65%) (refs. in Alexandrie, 2003).

The GSTM1 enzyme can catalyze GSH conjugation with several carcinogenic epoxides, e.g. aflatoxin B1-8,9-epoxide (Hayes and Strange, 2000) and is expressed in many organs including liver, testis, adrenals and white blood cells. In GSTM1-null individuals (*GSTM1*O/*O*), the enzyme is lacking as a result of a gene deletion. The functional *GSTM1*A* and *GSTM1*B* alleles, differing by a single base in exon 7 (Lys173Asn), appear to give similar enzymatic activities. The frequency of GSTM1-null individuals is in general higher in Caucasians and Asians (50%) than in Africans (25%) (refs. in Alexandrie, 2003).

The GSTP1 enzyme is active in the detoxification of acrolein, benzyl isocyanate and diol epoxides of PAH (Hayes and Strange, 2000). It is widely expressed in tissues and is the major GST enzyme in the blood (white and red blood cells). Two single nucleotide polymorphisms in the coding region of the *GSTP1* gene have been described (Hu et al., 1998) which give alterations in enzyme activity.

Microsomal epoxide hydrolase (EH) is an inducible enzyme expressed in all tissues studied, including white blood cells. The enzyme catalyzes the hydrolysis of reactive epoxides to their corresponding dihydrodiols, and thus plays an important role in detoxification of epoxides. Two polymorphisms due to point mutations resulting in differences in enzymatic activity have been described (Fretland and Omiecinski, 2000).

The concentration over time (the dose) in blood of a reactive compound and/or metabolite can be measured by analysis of the corresponding hemoglobin (Hb) adducts. Biotransformation will diminish the concentration and thus the dose of the compound. Enzymes that enhance conjugation with GSH will increase the detoxification rate and decrease the measured adduct levels. This was shown by Föst et al. (1995) in a study where blood from individuals with and without functional GSTT1 enzyme was exposed to ethylene oxide.

In an ongoing study, we are investigating whether genetically based differences in the expression of certain detoxifying enzymes, such as GSTs, may affect the relationship between exposure and the dose in blood of acrylamide or glycidamide. This paper describes *in vitro* experiments with blood from individuals with different genotypes for the *GSTT1* and *GSTM1* genes. Blood samples were incubated with known initial concentrations of acrylamide and glycidamide and the formed Hb-adduct levels were analyzed at several time periods as a measure of the resulting dose in the blood.

2. MATERIAL AND METHODS

2.1 Chemicals

Acrylamide (acrylamide for electrophoresis, 99%) was obtained from Merck, ethylene oxide from Fluka, ethacrynic acid and 1-chloro-2,4-dinitrobenzene (CDNB) from Sigma. Glycidamide was synthesized according to Paulsson et al. (2003a). Solvents used were of analytical grade.

2.2 Experiments and analysis

In the first experiment blood samples (2 mL) from four individuals with defined genotypes for *GSTT1* and *GSTM1* were incubated with acrylamide (40 µM), glycidamide (10 µM), and with ethylene oxide (10 µM) as a positive control. Two individuals were carrying *GSTM1* but lacking *GSTT1*

(+/−), one was carrying both *GSTM1* and *GSTT1* (+/+) and the last one was lacking *GSTM1* but carrying *GSTT1* (−/+). In the second experiment, blood from one individual carrying the *GSTM1* but lacking *GSTT1* (+/−) was incubated with acrylamide (40 µM) or glycidamide (10 µM). In parallel samples, ethacrynic acid was also added (80 µM) as an inhibitor to GSTs (Ploemen et al., 1993).

The incubations were performed in a warming cupboard (37°C) for 4 or 5 hours, while tilting. Cooling on ice followed by centrifugation (~ 3000 g for 5 min) and rinsing the red blood cells with ice-cold saline terminated the incubations. Globin precipitation as well as derivatization and isolation of adducts to the N-terminal valine in Hb were performed according to the N-alkyl Edman procedure described elsewhere (Paulsson et al., 2003b). Gas chromatographic/mass spectrometric analysis of acrylamide- and glycidamide-adducts was performed according to Bergmark (1997) and Paulsson et al. (2003b), respectively. Ethylene oxide adducts were analyzed according to Törnqvist (1994). The effect of ethacrynic acid as an inhibitor of GST activity was tested with CDNB as a substrate according to Habig et al. (1974). Individuals with homozygous deletions of *GSTT1* or *GSTM1* were identified using a PCR method previously described (Warholm et al., 1995; Carstensen et al., 1999).

3. RESULTS

The result from the first experiment with blood from four individuals with defined GST genotypes showed no significant differences in Hb-adduct levels after incubation with acrylamide or glycidamide. Incubation with ethylene oxide (positive control) showed higher adduct levels in the GSTT1-null individuals after equal incubation time. Adduct levels after 4 hours of incubation are shown in Table 1.

Table 1. Hemoglobin-adduct levels in blood samples from four individuals with defined GSTM1 and GSTT1 genotypes after incubation with acrylamide (40 µM), glycidamide (10 µM) or ethylene oxide (10 µM)

Individual genotype	Hemoglobin-adduct level (nmol/g globin)		
GSTM1/GSTT1	Acrylamide	Glycidamide	Ethylene oxide
+/−	0.86	1.35	2.41
+/−	0.79	1.44	2.21
+/+	0.83	1.48	1.06
−/+	0.79	1.64	0.95

In the second experiment, blood from an individual carrying GSTM1 but lacking GSTT1 (+/−) was incubated with acrylamide or glycidamide with or without a GST inhibitor (ethacrynic acid). No differences between the samples were observed with and without inhibitors. Table 2 shows adduct levels after incubation for 4 hours with acrylamide and for 5 hours with glycidamide.

Table 2. Hemoglobin-adduct levels in blood samples with and without GST inhibitor (ethacrynic acid), after incubation with acrylamide (40 µM) or glycidamide (10 µM)

Substance	Hemoglobin-adduct level (nmol/g globin)	
	With inhibitor	Without inhibitor
Acrylamide	0.76	0.76
Glycidamide	1.62	1.54

4. SUMMARY/CONCLUSIONS

The first experiment using blood from individuals with different *GSTT1* and *GSTM1* genotypes showed no differences in Hb-adduct levels after treatment with acrylamide or glycidamide. In contrast, the positive control treatment with ethylene oxide showed lower Hb-adduct levels in blood from individuals carrying *GSTT1* compared to blood from individuals lacking *GSTT1*, in agreement with previous studies (Föst et al., 1995). Recombinant human GSTT1 has recently been shown to catalyze the glutathione conjugation with acrylamide (Kjuus et al., 2002). The results from the present experiments suggest that acrylamide and glycidamide are less effective substrates for the GSTT1 enzyme in erythrocytes compared to ethylene oxide.

In the second experiment, the blood incubated with acrylamide or glycidamide was from a GSTT1-null individual. Ethacrynic acid inhibits the GST activity catalyzed by the other GSTs present in blood, GSTP1 and GSTM1. Adding this inhibitor to the blood sample resulted in loss of GST activity. No differences in Hb-adduct levels could be observed between samples with and without inhibitor.

The results from the described *in vitro* experiments suggest that the presence of the enzymes GSTT1, GSTM1, and GSTP1 in blood does not have any effect on the Hb-adduct levels after incubation with acrylamide or glycidamide. The results indicate that conjugation of acrylamide and glycidamide with GSH in blood proceeds mainly via uncatalyzed reactions and that polymorphisms in the *GSTT1* or *GSTM1* genes do not affect the blood dose of Hb-adducts after exposure to acrylamide or glycidamide.

ACKNOWLEDGEMENTS

The authors acknowledge the Swedish Council for Work Life Research and the Swedish National Institute for Working Life for financial support.

REFERENCES

Alexandrie, A.-K., 2003, Significance of polymorphisms in human xenobiotic metabolising enzymes, Doctoral Thesis, Institute of Environmental Medicine, Karolinska Institutet, Stockholm, Sweden.

Bergmark, E., 1997, Hemoglobin adducts of acrylamide and acetonitrile in laboratory workers, smokers, and nonsmokers, *Chem. Res. Toxicol.* **10**(1):78–84.

Carstensen, U., Hou, S.-M., Alexandrie, A.-K., Högstedt, B., Tagesson, C., Warholm, M., Rannug, A., Lambert, B., Axmon, A., and Hagmar, L., 1999, Influence of genetic polymorphisms of biotransformation enzymes on gene mutations, strand breaks of deoxyribonucleic acid, and micronuclei in mononuclear blood cells and urinary 8-hydroxydeoxyguanosine in potroom workers exposed to polyaromatic hydrocarbons, *Scand. J. Work Environ. Health* **25**(4):351–360.

Fretland A.J., and Omiecinski, C.J., 2000, Epoxide hydrolases: biochemistry and molecular biology, *Chem. Biol. Interact.* **129**(1-2):41–59.

Föst, U., Törnqvist, M., Leutbecher, M., Granath, F., Hallier, E., and Ehrenberg, L., 1995, Effects of variation in detoxification rate on dose monitoring through adducts, *Hum. Env. Toxicol.* **14**(12):201–203.

Habig, W.H., Pabst, M.J., and Jakoby, W.B., 1974, Glutathione transferases. The first enzymatic step in mercapturic acid formation, *J. Biol. Chem.* **249**(22):7130–7139.

Hayes, J.D., and Strange, R.C., 2000, Glutathione S-transferase polymorphisms and their biological consequences, *Pharmacology* **61**(3):154–166.

Hu, X., Xia, H., Srivastava, S.K., Pal, A., Awasthi, Y.C., Zimniak, P., and Singh, S.V., 1998, Catalytic efficiencies of allelic variants of human glutathione S-transferase P1-1 toward carcinogenic anti-diol epoxides of benzo[c]phenanthrene, *Cancer Res.* **58**(23):5340–5343.

Kjuus, H., Goffeng, L.-O., Skard Heier, M., Hansteen, I.-L., Øvrebø, S., Skaug, V., Ryberg, D., Sjöholm, H., Törnqvist, M., Paulsson, B., Langeland, B.T., and Brudal, S., *Examination of nervous system effects and other health effects in tunnel workers exposed to acylamide and N-methylolacrylamide in Romeriksporten, Norway*, National Institute of Occupational Health, Oslo, Norway, STAMI-rapport 5, 2002.

Landi, S., 2000, Mammalian class theta GST and differential susceptibility to carcinogens: a review, *Mutat. Res.* **463**(3):247–283.

Lucas, D., Ferrara, R., Gonzales, E., Albores, A., Manno, M., and Berthou, F., 2001, Cytochrome CYP2E1 phenotyping and genotyping in the evaluation of health risks from exposure to polluted environments, *Toxicol. Letters* **124**(1-3):71–81.

Paulsson, B., Kotova, N., Grawé, J., Granath, F., Henderson, A., Golding, B., and Törnqvist, M., 2003a, Induction of micronuclei in mouse and rat by glycidamide, the genotoxic metabolite of acrylamide, *Mutat. Res.* **535**(1):15–24.

Paulsson, B., Athanassiadis, I., Rydberg, P., and Törnqvist, M., 2003b, Hemoglobin adducts from glycidamide: acetonization of hydrophilic groups for reproducible gas chromatography tandem mass spectrometric analysis, *Rapid Commun. Mass Spectrom.* **17**(16):1859–1865.

Ploemen, J.H.T.M., van Ommen, B., Bogaards, J.J.P., and van Bladeren, P.J., 1993, Ethacrynic acid and its glutathione conjugate as inhibitors of glutathione S-transferases, *Xenobiotica* **23**(8):913–923.

Sumner, S.C.J., Fennell, T.R., Moore, T.A., Chanas, B., Gonzalez, F., and Ghanayem, B.I., 1999, Role of cytochrome P450 2E1 in the metabolism of acrylamide and acrylonitrile in mice, *Chem. Res. Toxicol.* **12**(11):1110–1116.

Sumner, S.C.J., Selvaraj, L., Nauhaus, S.K., and Fennell, T.R., 1997, Urinary metabolites from F344 rats and B6C3F1 mice coadministered acrylamide and acrylonitrile for 1 or 5 days, *Chem. Res. Toxicol.* **10**(10):1152–1160.

Törnqvist, M., 1994, Epoxide adducts to N-terminal valine of hemoglobin, in: *Methods in Enzymology, Vol. 231*, J.Everse, K.D.Vandegriff, and R.W.Winslow, eds., Academic Press, San Diego, pp. 650–657.

Warholm, M., Rane, A., Alexandrie, A.-K., Monaghan, G., and Rannug, A., 1995, Genotypic and phenotypic determination of polymorphic glutathione transferase T1 in a Swedish population, *Pharmacogenetics* **5**(4):252–254.

BIOLOGICAL EFFECTS OF MAILLARD BROWNING PRODUCTS THAT MAY AFFECT ACRYLAMIDE SAFETY IN FOOD
Biological Effects of Maillard Products

Mendel Friedman
Western Regional Research Center, Agricultural Reseach Service, USDA, 800 Buchanan Street, Albany, CA 94710; e-mail:mfried@pw.usda.gov

Abstract: The heat-induced reaction of amino groups of amino acids, peptides, and proteins with carbonyl groups of reducing sugars such as glucose results in the concurrent formation of so-called Maillard browning products and acrylamide. For this reason, reported studies of adverse biological effects of pure acrylamide may not always be directly relevant to acrylamide in processed food, which may contain Maillard and other biologically active products. These may either antagonize or potentiate the toxicity of acrylamide. To stimulate progress, this paper presents an overview of selected reported studies on the antiallergenic/allergenic, antibiotic, anticarcinogenic/carcinogenic antimutagenic/mutagenic, antioxidative/oxidative, clastogenic (chromosome-damaging), and cytotoxic activities of Maillard products, which may adversely or beneficially impact the toxicity of acrylamide. The evaluation of biological activities of Maillard products and of other biologically active food ingredients suggests that they could both enhance and/or ameliorate acrylamide toxicity, especially carcinogenicity, but less so neurological or reproductive manifestations. Future studies should be directed to differentiate the individual and combined toxicological relationships among acrylamide and the Maillard products, define individual and combined potencies, and develop means to prevent the formation of both acrylamide and the most toxic Maillard products. Such studies should lead to safer foods.

Key words: Acrylamide; Maillard products; beneficial effects; adverse effects; food safety.

1. INTRODUCTION

The relationship between dietary content and human diseases such as cancer and atherosclerosis has become increasingly a major concern for human health. A need exists to more precisely define the relationship between specific diet components and disease and to devise strategies to minimize the formation of the harmful compounds. The potential for formation of mutagens and carcinogens in foods during processing is a major area of concern for human health and safety. Adverse and beneficial effects of Maillard products formed during the processing of food may occur concurrently with the formation of acrylamide.

Reactions of amino with carbonyl groups of food constituents involve those changes commonly termed browning reactions. These include reactions of amines, amino acids, peptides, proteins with reducing sugars and vitamin C (non-enzymatic browning, often called Maillard reactions). These reactions cause deterioration of food during storage and processing. The loss of nutritional quality is attributed to the destruction of essential amino acids, a decrease in digestibility, and inhibition of proteolytic and glycolytic enzymes. The production of both antitoxic and toxic compounds may further impact the safety of heated foods. Studies in this area include influence of damage to essential amino acids on nutrition and food safety, nutritional damage as a function of processing conditions, and simultaneous formation of deleterious and beneficial compounds. These compounds include carcinogens, mutagens, antimutagens, antioxidants, antibiotics, and allergens, and antiallergens (Friedman, 1973, 1974, 1975a, 1975b, 1977a, 1977b, 1977c, 1978, 1982, 1984, 1986, 1989a, 1989b, 1991, 1992, 1994, 1996, 1997, 1999a, 1999b, 2003, 2004a, 2004b). Mutidisciplinary studies are needed to reveal the complex interplay between the chemistry, biochemistry, nutrition, pharmacology, and toxicology of food ingredients.

2. BIOLOGICAL EFFECTS OF ACRYLAMIDE

Heat induces the formation of acrylamide ($CH_2=CH-CO-NH_2$) in food under conditions that also induce the formation of Maillard browning products. This observation stimulated interest in the underlying chemistry that may be responsible for the formation of acrylamide as well as the chemical and biochemical basis of the toxicological effects of this animal carcinogen, neurotoxin, and reproductive toxin (Friedman, 2003). A recent study showed that even extremely low doses of acrylamide adversely affected the lifespan of nematodes (Hasegawa et al., 2004). Because most of the reported studies on the biological effects of acrylamide in animals were carried out with pure acrylamide, it is not known how the concurrent formation of Maillard products and the presence of other biologically active compounds in food impacts the dietary significance of acrylamide. The main of objective of this paper is to present an overview of these effects in order to stimulate needed research on the safety of acrylamide in processed foods. Research needs in this area include the following:

1. What impact does prevention of acrylamide formation have on the concurrent formation of Maillard products?
2. Do Maillard browning products and other food ingredients affect the safety of concurrently formed, heat-induced acrylamide after consumption? How do antitoxic and toxic compounds formed under these conditions modulate the safety of acrylamide in the human diet?
3. Does prevention of food browning also prevent acrylamide formation?
4. Can the manual ninhydrin reaction be used to study the extent of alkylation of amino groups in amino acids, peptides, and protein by acrylamide *in vitro* and *in vivo* (Friedman, 2004a; Pearce et al., 1988)?

2.1 Beneficial and adverse effects of Maillard products

2.1.1 Antioxidative Maillard products

Antioxidants exert their effect by donating electrons or hydrogen atoms to free-radical-containing lipids and by forming antioxidant-lipid complexes (Friedman, 1997). Any substance that inhibits the propagation step in the chain-reaction, decomposes lipid hydroperoxides, chelates heavy metal ions, or prevents light- and/or radiation-induced initiation of the chain reaction can, in principle, serve as an antioxidant.

Controlled browning is often used to develop desirable flavor, odor, and color properties in foods including coffee, bread, and soybean sauce (Schwimmer, 1981; Schwimmer and Friedman, 1972). Such browning reactions often lead to the formation of naturally occurring antioxidants. For example, antioxidants are formed during the Maillard reaction between tryptophan and fructose or glucose (Chiu et al., 1991). The most potent antioxidative effect was achieved by reaction of 8.2 mM fructose and 6.0 mM tryptophan at 65°C for 7 days. Advanced stage Maillard products were stronger antioxidants than those formed at the early stages of the reaction.

2.1.2 Additional observations relevant to antioxidative effects

1. Honey-lysine Maillard products exhibited antioxidative activities in a linoleic acid emulsion (Antony et al., 2000). The authors suggest that

combining honey with meat proteins and adding them to foods prior to thermal processing could prepare antioxidative compounds.

2. Maillard products resulting from the reaction of β-lactoglobulin glycated with arabinose, galactose, glucose, lactose, rhamnose, and ribose exhibited free radical scavenging activity in *in vitro* cell assays, with the arabinose derivative being the most active (Chevalier et al., 2001). None of the products were cytotoxic to Caco-2 cells.

3. Aged garlic extracts contain the strong antioxidant Nα-(1-deooxy-D-fructos-1-yl)-L-arginine, presumably formed from arginine and fructose (Ryu et al., 2001) (Figure 1). A minimum of four months of aging was required to generate the Maillard product in garlic.

4. Compared to bread crumb, dark brown bread crust fractions showed much higher antioxidative activity, elevated glutathione S-transferase (GST) activity, and decreased phase I NADPH-cytochrome C reductase (CCR) activity in intestinal Caco-2 cells (Lindenmeier et al., 2002). Protein-bound pronyl-lysine derivatives in the bread crust may be responsible for the observed beneficial effects.

5. Maillard products derived from cysteine and histidine exhibited greater antiradical activity in an *in vitro* assay than analogous products derived from lysine or glycine (Morales and Babbel, 2002).

6. Several pyrrole compounds derived from the Maillard reaction in coffee inhibited hexanal oxidation by >80% at concentrations of ~10 μg/mL (Yanagimoto et al., 2002). Among furan derivatives tested, unsubstituted furan had the greatest antioxidative activity.

7. A study of the antioxidative activities of eight pyrrole compounds, some of which are formed during food browning, both individually and as mixtures, revealed that the observed activities are the sum of the activities of the individual compounds present in the mixture (Hidalgo et al., 2003).

8. Maillard products formed from the reaction of D-glucose with L-arginine, glycine, and L-lysine inhibited copper-induced *in vitro* oxidation of human low-density lipoprotein (LDL). The Maillard reductones, 3-hydroxy-4- (morpholino)-3-buten-2-one and an amino hexose reductone, showed similar dose-dependent inhibition of human LDL oxidation (Dittrich et al., 2003).

9. Lysine-glucose and lysine-fructose Maillard products inhibited formation of free radicals produced in the Fenton reaction. Higher molecular weight products exhibited greater antioxidative activities as well as toxicities in Caco-2 cells than lower molecular ones (Jing and Kitts, 2004).

2.1.3 Antimutagenic Maillard products

The antimutagenic effect in *Salmonella typhimurium* (Ames test) of Maillard products prepared by heating lysine and xylose resulted from the reaction of the product with metabolites of the mutagenic/carcinogenic heterocyclic amine (IQ) to form inactive adducts (desmutagenic effect), not by direct inhibition of hepatic microsomal activation, which transforms inactive IQ to a DNA biological alkylating agent (bioantimutagenic effect) (Yen et al., 1992). The inhibitory effect of glucose-tryptophan Maillard product can also be described as a desmutagenic effect. The antimutagenicity of pyrazines present in Maillard products correlated with the inhibition in microsomes of the mutagen-activating enzyme cytochrome P-450 IA2-linked ethoxycoumarin deethylase (Jenq et al., 1994).

2.1.4 Activation of detoxifying enzymes

Water-soluble nonenzymatic browning products (melanoidins) isolated from roasted malt induced the liver detoxifying enzymes NADPH-cytochrome c-reductase and glutathione-S-transferase in intestinal Caco-2 cells (Faist et al., 2002). This observation suggests that melanoidins may facilitate the conjugation, metabolism, and elimination of toxic xenobiotics catalyzed by these enzymes.

2.1.5 Antibiotic Maillard products

Extensive studies have been carried on the inhibition of bacterial growth by Maillard reaction products (Einarsson et al., 1988; Stecchini et al., 1993). The products were prepared by refluxing solutions containing either arginine and xylose or histidine and glucose. The effects were measured by determining the minimum concentration that inhibits the growth of the microorganisms. The bacteria include pathogenic *Salmonella* and *Streptococcus faecalis* strains. The results demonstrate a wide range of susceptibility among the 20 strains tested for growth inhibition by the Maillard product. The higher molecular weight fractions had a greater inhibitory effect than the lower ones.

Evaluation of antibiotic potencies of Maillard products revealed that (a) products formed from arginine were more potent than those derived from histidine; (b) arginine mixtures with either glucose or xylose had the same inhibitory effects; (c) formation of antibacterial compounds was favored when the reaction between amino acid and carbohydrate was carried out at slightly alkaline pH of 9; and (d) the antibacterial effect increased with increase in reaction time in the mixture containing arginine.

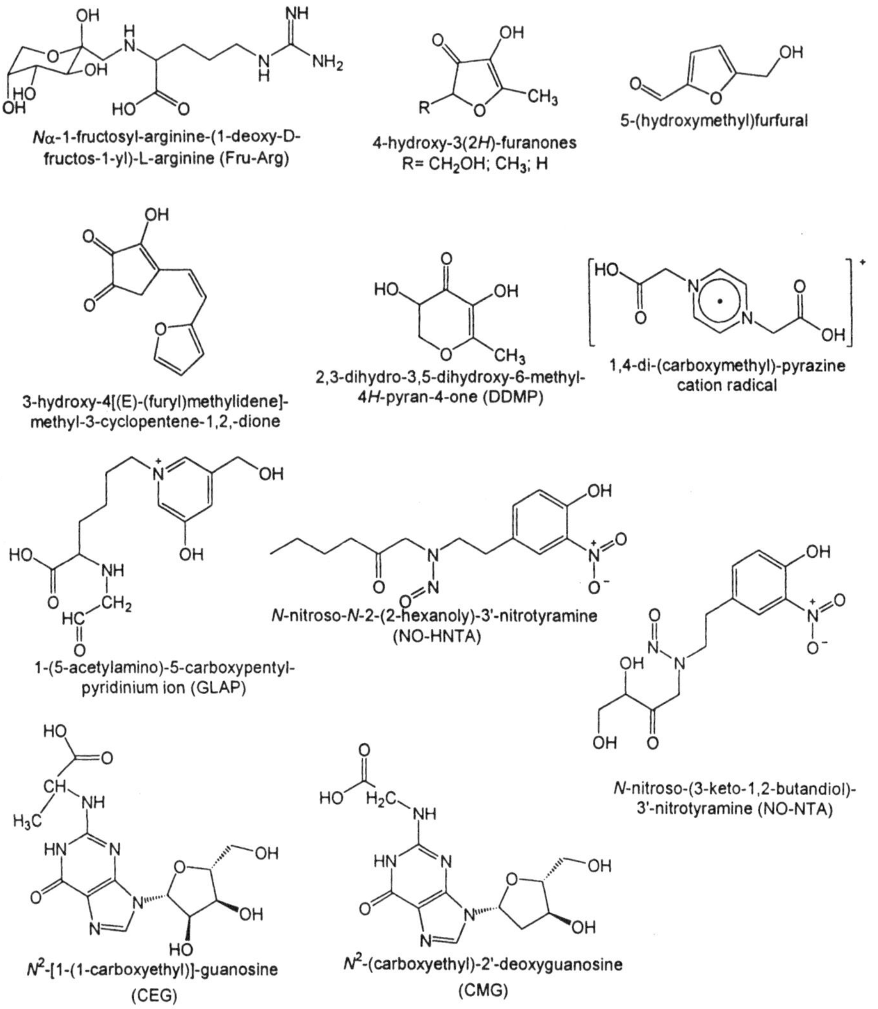

Figure 1. Structures of some characterized Maillard reaction products. See text.

Related studies showed that (a) antibacterial activities of lysozyme-dextran Maillard products against the Gram-negative bacteria *Bacillus cereus* and *Staphylococcus aureus* appear to be associated with the surfactant and emulsifying properties of the conjugates (Nakamura et al., 1996); (b) Lysozyme-galactomannan conjugates prepared through a controlled Maillard reaction protected carp fish against the virulent Gram-negative pathogen *Edwardsiella tarda* (Nakamura et al., 1996); (c) Maillard melanoidins stimulated the growth of anaerobic fecal bacteria (Ames et al., 1999); and (d) Maillard products inhibited the growth of the hyperthermophilic organism *Aeropyrum pernix* (Kim and Lee, 2003) and lessened the virulence of gene expression in the foodborne pathogen *Listeria monocytogenes* (Sheikh-Zeinoddin et al., 2000); and (e) Maillard products may be formed during pasteurization (heating) of fruit juices designed to kill microorganisms (Alwazeer et al., 2003; Ekasari et al., 1989; Friedman et al., 1990a).

It is not clear whether antibiotic Maillard products potentiate the action of naturally occurring antimicrobial plant compounds and whether they can kill antibiotic-resistant pathogens (Friedman et al., 2004a; 2004b; 2002; 2003b).

2.1.6 Allergenicity/antiallergenicity of Maillard products

Relatively mild conditions of heating food proteins with carbohydrates can reduce the antigenicity of and possibly modify sites known to elicit allergenic responses. For example, we (Öste et al., 1990; Öste and Friedman, 1990) heated a solid mixture of soybean trypsin inhibitor (KTI) and carbohydrates in an oven at 120°C and analyzed the dialyzed product by enzyme-linked immunosorbent assays (ELISA). Glucose, lactose, and maltose decreased the antigenicity of KTI by 60-80%, compared to a control sample heated without carbohydrate. Starch was less effective than the three reducing sugars. The decrease was rapid, occurring within 10 min when glucose was heated with KTI, with retention of 60% of the chemically available lysine. Longer heating times increased browning and reduced the level of available lysine in KTI, without further reducing antigenicity. These reactions can also introduce new antigens into a food protein (Friedman and Brandon, 2001).

Studies on lactose-protein Maillard adducts (Matsuda et al., 1992) showed that (a) different components of the lactose-bovine serum albumin adduct produced as the reaction proceeds could serve as antigenic determinants; (b) the carbohydrate residue of the Amadori compound ε-deoxylactulosyl-lysine formed in the early stage of the Maillard reaction was specific for the haptenic antigen of the lactose-proteins Maillard complex; and (c) the monoclonal antibodies did not react with milk glycoproteins of fresh raw milk, but did react with proteins in pasteurized milk. The cited observations suggest that monoclonal antibodies could detect lactose-protein Maillard adducts as possible milk allergens in pasteurized milk and milk products.

The glycated end products N^ε-(carboxymethyl)lysine, malondialdehyde, and 4-hydroxynonanal appear to be associated with the greater allergenicity, measured by immunoglobulin E (IgE) binding of roasted peanuts compared to raw nuts. The inhibition of the IgE binding by immunoglobulin G (IgG) antibodies to the glycated products may ameliorate the allergic response to roasted peanuts (Chung and Champagne, 2001).

2.1.7 Mutagenicity – genotoxic effects of Maillard products

Mutagenic and carcinogenic products in cooked protein-rich foods are formed by several mechanisms, including carbohydrate caramelization, protein pyrolysis, amino acid/creatinine reactions, and amino-carbonyl (Maillard) reactions, in which free amino groups condense with reducing sugars to produce brown melanoidins, furans, carbolines, and a variety of other products (Friedman, 1996; Friedman and Cuq, 1988; Friedman and Henika, 1991).

Of the numerous compounds formed in the Maillard reaction, some are mutagenic and some are anti-mutagenic (Taylor et al., 2004). Unless these are characterized and tested individually, the assays provide the net genotoxic effect. They also suggest that a battery of tests should be used, including the Ames bacterial test, to detect gene mutations and the micronucleus test to detect chromosomal aberrations.

We heated gluten, carbohydrates, and gluten-carbohydrates blends in a simulation of low-moisture crust baking (Friedman and Finot, 1990; Friedman et al., 1990b). The baked materials were then assayed by the Ames *Salmonella* his-reversion test in order to evaluate the formation of mutagenic browning products (Figure 2). Baked D-glucose, maltose, lactose, sucrose, wheat starch, potato amylose, cellulose, microcrystalline hydrocellulose, sodium ascorbate, L-ascorbic acid, (carboxymethyl)cellulose, and (hydroxy-propyl)methylcellulose were moderately mutagenic in strain TA98 with microsomal (S9) activation and were weakly mutagenic without microsomal activation in strains TA100, TA102, and TA1537. Heated blends of gluten with 20% of these carbohydrates were also mutagenic, but the total activity recovered did not exceed levels of the individual ingredients baked separately.

Biological Effects of Maillard Browning Products

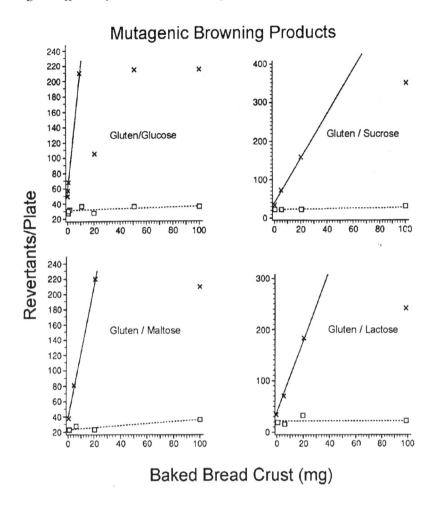

Figure 2. Heat-induced formation of mutagenic Maillard products. □ = unheated controls; x = heated. Adapted from (Friedman et al., 1990b).

2.1.8 Additional observations relevant to mutagenicity

1. Several aromatic amine mutagens formed by heating amino acids with arginine may be responsible for the reported mutagenicity of baked grain-based foods (Knize et al., 1994a; Knize et al., 1994b).
2. The Maillard mutagen 5-(hydroxymethyl)furfural (Figure 1) present in many foods exerts genotoxicity only after metabolic activation *via* sulfonation and chlorination (Surh et al., 1994).
3. Maillard products of 1:1 solid glucose-amino acid mixtures heated at 200°C for 5 min induced single-strand breaks of DNA incubated

overnight at 37°C and pH 7.4 (Hiramoto et al., 1997). The Maillard products did not induce double-strand breaks in the DNA. See Figure 3.

4. The active DNA breaking and mutagenic component of soy sauce was characterized as 4-hydroxy-5-methyl-3(2H)-furanone (Figure 1), a fragrant compound formed from the Maillard reaction of amino acids with pentoses (Hiramoto et al., 1997).

5. Casein-ketose (fructose and tagatose) Maillard products exhibited greater mutagenicity in the Ames test than did casein-aldose (galactose and glucose) derivatives. These differences in the formation of direct-acting mutagens may be due to differences in reaction mechanisms between ketose and aldoses (Brands et al., 2000).

6. Pyrazine cation radicals (Figure 1) are active mutagens formed from the Maillard reaction of glucose, glycine, and creatinine (Kikugawa et al., 2000).

7. Lysine-fructose, lysine-glucose, and lysine-xylose Maillard products damaged DNA of human lymphocytes in the concentration range ~0.1-0.4 mg/mL (comet assay); induced the formation of free radicals and 8-hydroxy-2'-deoxyguanosine, a direct marker of oxidative damage; and decreased glutathione (GSH) content and activities of the antioxidant enzymes GSH reductase and catalase. These observations unequivocally demonstrated that under the test conditions, the Maillard products strongly damaged a human cell system (Yen et al., 2002).

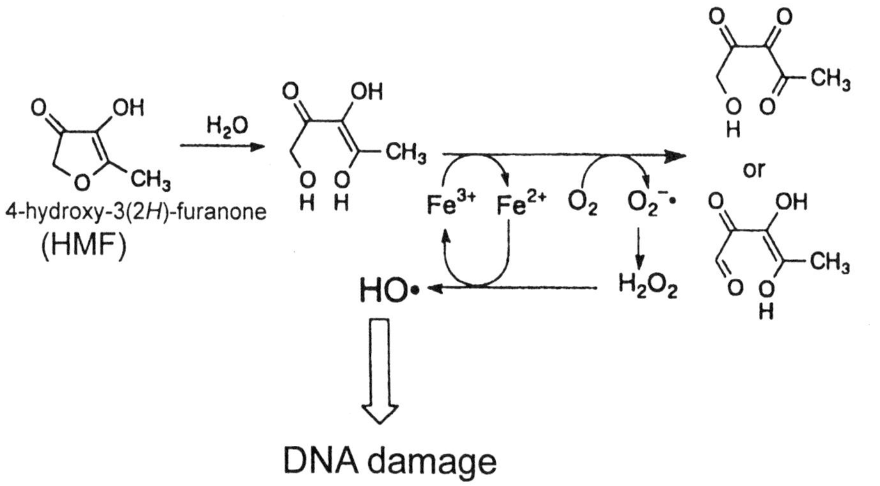

Figure 3. Mechanism of generation of DNA-damaging free hydroxyl radicals (HO·) from the Maillard product HMF via the Fenton reaction. Adapted from (Hiramoto et al., 1997).

2.1.9 Clastogenicity – chromosome damaging effects

Heated sugar/amino acid reaction mixtures known to contain products clastogenic (chromosome-damaging) and/or mutagenic (DNA-damaging) to cells were evaluated for clastogenic activity in mice using the erythrocyte micronucleus assay (MacGregor et al., 1989). Heated fructose/lysine reaction mixtures were also evaluated in the *Salmonella* his-reversion assay and the Chinese hamster ovary cell (CHO) chromosomal aberration assay to confirm and extend previous *in vitro* observations. Significant mutagenicity of fructose/lysine mixtures was observed in *Salmonella* strains TA100, TA2637, TA98, and TA102, with greater activity in mixtures heated at pH 10 than at pH 7. Both pH 7 and pH 10 reaction mixtures of the fructose/lysine browning reaction were highly clastogenic in CHO cells (Table 1). However, heated mixtures of fructose and lysine, and of glucose or ribose with lysine, histidine, tryptophan, or cysteine, did not increase the frequency of micronucleated erythrocytes in mice when administered orally. This indicates the absence of chromosomal aberrations in erythrocyte precursor cells. Evidently, the genotoxic components of the browned mixtures are not absorbed and distributed to bone marrow cells in amounts sufficient to induce micronuclei when given orally, or they are metabolized to an inactive form. It is imperative to further evaluate the *in vivo* genotoxicity of browning products in cell populations other than bone marrow tissue.

Table 1. Effect of heated fructose/lysine mixtures on the incidence of chromosomal aberrations in Chinese hamster ovary cells (adapted from MacGregor et al., 1989).

	Dose (mM)	Mitotic index	Normal cells	Abnormal cells	Chromatid deletions	Chromosome deletions	Chromatid exchanges	Chromosome exchanges
Control	0	0.194	94	6	3	1	0	2
pH 7	3.12	0.190	96	4	1	2	1	0
	6.25	0.324	94	6	3	1	2	0
	12.50	0.080	85	15	4	3	8	0
	25.00	0.014	59	41	30	5	34	4
pH 10	1.60	0.012	64	36	10	10	37	2
	3.12	0.026	97	3	1	4	0	0
	6.25	0.000	no metaphases					
	12.50	0.017	98	2	1	0	1	0

2.1.10 Cytotoxic Maillard products

Reaction products of glyceraldehydes with bovine serum albumin or casein were cytotoxic, induced the formation of reactive oxygen species, and depressed intracellular glutathione levels in HL-60 cells. The active cytotoxic component was characterized as 1-(5-acetylamino-5-carboxypentyl)-3-hydrfoxy-5-hydroxymethyl pyridinium (GLAP) (Figure 1) (Usui et al., 2004).

2.1.11 Carcinogenic Maillard products

Heating nitrite-containing foods can result in the formation of genotoxic compounds, as evidenced by the transformation of tyrosine and glucose to the DNA-damaging *N*-nitroso-*N*-(3-keto-1,2-butanediol)-3'-nitrotyramine (Figure 1). The compound initiated carcinogenesis in mouse embryo C3H10T1/e cells (Tseng et al., 1998; Wang et al., 1998).

Depending on structure, furanones derived from the Maillard reaction can either promote or inhibit the growth *in vitro* of human gastric carcinoma cells GXF251L (Marko et al., 2003). Inhibition of cell proliferation appears to be the result of apoptosis. A related study (Chen et al., 2001) showed that a Maillard product formed from lysine and glucose in the presence of sodium nitrite contained a tumor promoter that induced promotion by two-stage carcinogenesis in mouse embryo fibroblasts (Dedon and Tannenbaum, 2004).

2.1.12 Vitamin browning and degradation

Ascorbic acid (vitamin C) has been reported to play multiple roles in nutrition, human health, and food chemistry. Dietary deficiency of the vitamin causes the human disease scurvy due to formation of abnormal collagen resulting in skin and gum lesions and fragility of blood vessel. Vitamin C is widely reported to protect both plants and animals against oxidative stress induced by potentially toxic as well as cancer- and ateriosclerosis-inducing reactive oxygen species (ROS) including hydroxyl radicals, superoxide anions, singlet oxygen, and hydrogen peroxide. In foods, the vitamin may prevent formation of carcinogenic nitrosamines in cured meats, to protect against toxic metal toxicity, to inhibit oxidized fat-rancidity and food browning, and to improve the baking quality of doughs.

Recently we (Han et al., 2004) analyzed the content of ascorbic acid in tubers of four Korean potato cultivars, in baked, boiled, braised, fried, microwaved, pressure-cooked, and sautéed potatoes slices from the Dejima cultivar, and in 14 commercial Korean and 14 processed potato foods sold in the United States (chips, snacks; mashed potatoes; fries). The distribution of vitamin C in each of the eight potato slices (sticks, plugs) cut horizontally from the stem end of the Dejima potato ranged from 6.8 to 19.3% of the total. The corresponding distribution in seven sticks cut vertically was much narrower, ranging from 11.7 to 17.5% of the total. Losses observed during home-processing of three varieties with low (Dejima, 16 mg/100g), intermediate (Sumi, 32 mg/100g), and high (Chaju, 42 mg/100g) vitamin C content were on average (in %): boiling in water, 82; boiling in water containing 1-3% NaCl, 70; frying in oil, 67; sautéing 64; pressure-cooking in water, 58; braising, 56; baking, 42, and microwaving, 27. The content of the Korean foods ranged from trace amounts to 25-mg/100 g and that of the US foods from 0.4 to 46 mg/100g. These results permit optimizing the vitamin C content of the diet by using high-vitamin C potato varieties, selecting sticks cut horizontally for frying; baking or microwaving rather than boiling or frying; and selecting commercial potato foods with a high vitamin C content.

Concomitant degradation pathways of the vitamin involve oxidation of ascorbic acid to dehydroascrobic acid followed by hydrolysis to 2,3-diketogulonic acid and further oxidation, dehydration, polymerization, and reaction with amino acids and proteins to generate up to 50 nutritionally inactive products. It is not known to what extent, if any, these degradation products adversely affect the safety of the diet and whether vitamin C degradation parallels acrylamide formation.

Related earlier studies revealed that when a nutritionally complete, low-protein basal diet containing 10% casein was supplemented with 20% protein from unheated casein, wheat gluten, or soybean, test mice gained weight (Friedman et al., 1990b; Öste et al., 1990; Öste and Friedman, 1990; Ziderman and Friedman, 1985; Ziderman et al., 1987; Ziderman et al., 1989). In contrast, weight gain was markedly reduced when the supplement was soy protein or gluten heated at 200° or 215°C for 72 min in the dry state (simulated crust baking). After heating with sodium ascorbate (but not L-ascorbic acid), soy protein and gluten completely prevented growth when added to the basal diet. A heated casein-ascorbate mixture, but less than with the other proteins, also aggravated growth inhibition. The toxic effect increased sharply with heating temperature in the range 180 to 215°C, and with sodium ascorbate concentration in the range 1 to 20%. It was also found that sodium ascorbate heated with tryptophan results in the formation of toxic compounds. These considerations suggest the need to use ascorbic acid rather than sodium ascorbate in baking formulations.

Studies by other investigators showed that (a) an ELISA method can detect N^2-[1-(1-carboxy)ethyl]guanosine (Figure 1), formed by glycation of DNA with glucose or ascorbic acid (Seidel and Pischetsrieder, 1998); and (b) the amino group linked to the pyrimidine ring of folic acid reacted with lactose and maltose at 100°C in a phosphate buffer to form N^2-[1-(carboxyethyl)]folic acid in yields of 43 and 49%, respectively. Whether folic acid is degraded by the Maillard reaction in folate-enriched cereals and fruit juices is not known (Schneider et al., 2002).

2.1.13 Metallo complexes

The peptide-bound Maillard products N^α-hippuryl-N-fructosyllysine and N^α-hippuryl-N-carboxymethyllysine form complexes with copper ions (Seifert et al., 2004). These observations suggest that such binding by dietary metal-chelating compounds *in vivo* could adversely impact the biological and nutritional utilization of essential trace elements and inhibit active sites of metalloenzymes. Related observations are described in (Furniss et al., 1989; Wijewickreme and Kitts, 1998).

Other studies revealed that lysinoalanine, a crosslinked amino acid also formed during food processing, is also a strong chelator of essential metal ions (Friedman and Pearce, 1989; Pearce and Friedman, 1988; Sarwar et al., 1999).

2.1.14 Inhibition of browning

Because many individuals are sensitive to the anti-browning compound sodium sulfite, we explored the potential of sulfur amino acids to prevent browning (Friedman, 1996; Friedman and Bautista, 1995; Friedman and Molnar Perl, 1990; Friedman et al., 1992; Molnar Perl and Friedman, 1990a, 1990b). We found that SH-containing amino acids were nearly as effective as sodium sulfite in preventing browning in apples, potatoes, fruit juices, and protein-containing foods such as nonfat dry milk and barley and soy flours.

Related studies showed that: (a) L-cysteine and *N*-acetyl-L-cysteine inhibited the formation of the Maillard product hydroxymethylfurfural (Haleva-Toledo et al., 1999); (b) *N*-acetyl-L-cysteine inhibited the formation of the Maillard product N^ε-(carboxymethyl)lysine (Nakayama et al., 1999); and (c) a rutin derivative inhibited the generation of N^ε-fructoselysine (Nagasawa et al., 2003).

The following compounds have been evaluated for the ability to inhibit Maillard reactions *in vivo*: (a) glutathione (Ortwerth and Olesen, 1988); (b) aminoguanidine (Requena et al., 1993), (c) 3-(2-thienyl)-2-piperizone (tenilsetam) (Shoda et al., 1997); (d) green tea catechins (Song et al., 2002); and (e) 3,5-dimethylpyrazole-1-carboxyamidine (Miller et al., 2003).

Because the mechanisms of formation of Maillard products and of acrylamide are linked, it is worth exploring whether some of the above-mentioned antibrowning agents also inhibit the formation of acrylamide in foods.

3. CONCLUSIONS

In summary, the main objective of this overview was to showcase aspects of the Maillard reaction, which may impact the reported biological/toxicological aspects of dietary acrylamide. Since almost nothing is known to what extent this interaction actually occurs in complex food systems, predictions of the safety of acrylamide in foods based on the extensively reported studies with pure acrylamide may not always be justified. Based on the cited biological/toxicological/antitoxicological effects, it may be predicted that antioxidative and genotoxic Maillard products may modulate (enhance or suppress) possible carcinogenic and other adverse effects of acrylamide. However, the situation may be more complicated because other food ingredients could also modulate the safety of dietary acrylamide. These include embryotoxic (Friedman et al., 2003a; Friedman et al., 1991) and cancer cell-inhibiting (Lee et al., 2004) glycoalkaloids and glycosidase-inhibiting calystegine alkaloids (Friedman et al., 2003c) present in potatoes; biologically active enzyme inhibitors present in soybeans (Brandon et al., 2004; Brandon and Friedman, 2002; Friedman, 2004a; Friedman and Brandon, 2001); crosslinked and D-amino acids formed during food processing (Boschin et al., 2003; Friedman, 1999a, 1999b; Liardon et al., 1991); the processing-induced non-genotoxic carcinogen, 3-monochloroprpanediol, formed from the reaction of chloride with fats (Tritscher, 2004), and the liver carcinogen aflatoxin B_1, which alkylates DNA by a mechanism similar to that proposed for acrylamide (Friedman, 2003; Friedman et al., 1982). Antioxidative, antimutagenic, and anticarcinogenic polyphenolic compounds (anthocyanins, catechins, flavonoids) present in fruits and vegetables are also expected to affect the toxic manifestations of dietary acrylamide. Understanding the complex chemistry, metabolism, pharmacology, and toxicology of dietary acrylamide, Maillard products, and other biological active food ingredients should make it possible to design food processes to minimize the formation of both toxic Maillard products and acrylamide.

ACKNOWLEDGEMENTS

I thank Carol E. Levin for assistance with the preparation of this paper and Sigmund Schwimmer for helpful comments including the suggestion that chimney smoke and engine emissions should be tested for the presence of acrylamide (Zou and Atkinson, 2003).

REFERENCES

Alwazeer, D., Delbeau, C., Divies, C., and Cachon, R., 2003, Use of redox potential modification by gas improves microbial quality, color retention, and ascorbic acid stability of pasteurized orange juice, *Int. J. Food Microbiol.* **89**:21-29.

Ames, J. M., Wynne, A., Hofmann, A., Plos, S., and Gibson, G. R., 1999, The effect of a model melanoidin mixture on faecal bacterial populations in vitro, *Br. J. Nutr.* **82**:489-495.

Antony, S. M., Han, I. Y., Rieck, J. R., and Dawson, P. L., 2000, Antioxidative effect of maillard reaction products formed from honey at different reaction times, *J. Agric. Food Chem.* **48**:3985-3989.

Boschin, G., D'Agostina, A., Rinaldi, A., and Arnoldi, A., 2003, Lysinoalanine content of formulas for enteral nutrition, *J. Dairy Sci.* **86**:2283-2287.

Brandon, D. L., Bates, A. H., and Fridman, M., 2004, Immunoassays of Bowman-Birk and Kunitz soybean inhibitors in infant formula, *J. Food Sci.* **69**:FCT11-15.

Brandon, D. L., and Friedman, M., 2002, Immunoassays of soy proteins, *J. Agric. Food Chem.* **50**:6635-6642.

Brands, C. M., Alink, G. M., van Boekel, M. A., and Jongen, W. M., 2000, Mutagenicity of heated sugar-casein systems: effect of the Maillard reaction, *J. Agric. Food Chem.* **48**:2271-2275.

Chen, C. C., Tseng, T. H., Hsu, J. D., and Wang, C. J., 2001, Tumor-promoting effect of GGN-MRP extract from the Maillard reaction products of glucose and glycine in the presence of sodium nitrite in C3H10T1/2 cells, *J. Agric. Food Chem.* **49**:6063-6067.

Chevalier, F., Chobert, J. M., Genot, C., and Haertle, T., 2001, Scavenging of free radicals, antimicrobial, and cytotoxic activities of the Maillard reaction products of beta-lactoglobulin glycated with several sugars, *J. Agric. Food Chem.* **49**:5031-5038.

Chiu, W. K., Tanaka, M., Nagashima, Y., and Taguch, T., 1991, Prevention of sardine lipid oxidation of by antioxidative Maillard reaction products prepared from fructose-tryptophan., *Nippon Suisan Gakaishi* **57**:1773-1781.

Chung, S. Y., and Champagne, E. T., 2001, Association of end-product adducts with increased IgE binding of roasted peanuts, *J. Agric. Food Chem.* **49**:3911-3916.

Dedon, P. C., and Tannenbaum, S. R., 2004, Reactive nitrogen species in the chemical biology of inflammation, *Arch. Biochem. Biophys.* **423**:12-22.

Dittrich, R., El-Massry, F., Kunz, K., Rinaldi, F., Peich, C. C., Beckmann, M. W., and Pischetsrieder, M., 2003, Maillard reaction products inhibit oxidation of human low-density lipoproteins *in vitro*, *J. Agric. Food Chem.* **51**:3900-3904.

Einarsson, H., Eklund, T., and Nes, I. F., 1988, Inhibitory mechanisms of Maillard reaction products, *Microbios* **53**:27-36.

Ekasari, I., Bonestroo, M. H., Jongen, W. M. F., and Pilnik, W., 1989, Mutagenicity and possible occurrence of flavonol aglycones in heated orange juice, *Food Chem* **31**:289-294.

Faist, V., Lindenmeier, M., Geisler, C., Erbersdobler, H. F., and Hofmann, T., 2002, Influence of molecular weight fractions isolated from roasted malt on the enzyme activities of NADPH-cytochrome c-reductase and glutathione-S-transferase in Caco-2 cells, *J. Agric. Food Chem.* **50**:602-606.

Friedman, M., 1973, *The Chemistry and Biochemistry of the Sulfhydryl Group in Amino Acids, Peptides, and Proteins,* Pergamon Press, Oxford, UK, pp. 485.

Friedman, M., ed., 1974, *Protein-Metal Interactions-Advances in Experimental Medicine and Biology* (AEMB) Vol. 40, Plenum, New York, pp. 692.

Friedman, M., ed., 1975a, *Protein Nutritional Quality of Food and Feeds-Part 1. Assay Methods - Biological, Biochemical, Chemical*, Marcel Dekker, New York, pp. 626.

Friedman, M., ed., 1975b, *Protein Nutritional Quality of Foods and Feeds-Part 2. Quality Factors-Plant Breeding, Composition, Processing and Antinutrients*, Marcel Dekker, New York, pp. 674.

Friedman, M., 1977a, Effects of lysine modification on chemical, physical, nutritive, and functional properties of proteins, in: *Food Proteins*, J. R. Whitaker, and S. R. Tannenbaum, eds., AVI, Westport, CT, pp. 465-483.

Friedman, M., ed., 1977b, *Protein Crosslinking: Biochemical and Molecular Aspects*-AEMB Vol. 86A, Plenum, New York, pp. 760.

Friedman, M., ed., 1977c, *Protein Crosslinking: Nutritional and Medical Consequences*-AEMB Vol. 86B, Plenum, New York, pp. 740.

Friedman, M., ed., 1978, *Nutritional Improvement of Food and Feed Proteins*-AEMB Vol. 105, Plenum, New York, pp. 882.

Friedman, M., 1982, Chemically reactive and unreactive lysine as an index of browning, *Diabetes* **31**:5-14.

Friedman, M., ed., 1984, *Nutritional and Toxicological Aspects of Food Safety*-AEMB Vol. 177, Plenum, New York, pp. 584.

Friedman, M., ed., 1986, *Nutritional and Toxicological Significance of Enzyme Inhibitors in Foods*-AEMB Vol. 199, Plenum, New York, pp. 572.

Friedman, M., ed., 1989a, *Absorption and Utilization of Amino Acids*, 3 Volumes, CRC Press, Boca Raton, FL.

Friedman, M., 1989b, Nutritional and toxicological consequences of browning during simulated crust baking, in: *Protein Quality and Effects of Processing*, R. D. Phillips, and J. W. Finley, eds., Marcel Dekker, New York, pp. 189-217.

Friedman, M., ed., 1991, *Nutritional and Toxicological Consequences of Food Processing*-AEMB Vol. 289, Plenum, New York, pp. 540.

Friedman, M., 1992, Dietary impact of food processing, *Annu. Rev Nutr.* **12**:119-137.

Friedman, M., 1994, Improvement in the safety of foods by SH-containing amino acids and peptides. A review, *J. Agric Food Chem.* **42**:3-20.

Friedman, M., 1996, Food browning and its prevention: an overview, *J. Agric Food Chem.* **44**:631-653.

Friedman, M., 1997, Chemistry, biochemistry, and dietary role of potato polyphenols. A review, *J Agric. Food Chem.* **45**:1523-1540.

Friedman, M., 1999a, Chemistry, biochemistry, nutrition, and microbiology of lysinoalanine, lanthionine, and histidinoalanine in food and other proteins, *J. Agric. Food Chem.* **47**:1295-1319.

Friedman, M., 1999b, Chemistry, nutrition, and microbiology of D-amino acids, *J. Agric. Food Chem.* **47**:3457-3479.

Friedman, M., 2003, Chemistry, biochemistry, and safety of acrylamide. A review, *J. Agric.Food Chem.* **51**:4504-4526.

Friedman, M., 2004a, Applications of the ninhydrin reaction for analysis of amino acids, peptides, and proteins to agricultural and biomedical sciences, *J. Agric. Food Chem.* **52**:385-406.

Friedman, M., 2004b, Effects of food processing, in: *Encyclopedia of Grain Science*, C. Wrigley, H. Corke, and C. E. Walker, eds., Elsevier, Oxford, UK, Vol. 2, pp. 328-340.

Friedman, M., and Bautista, F. F., 1995, Inhibition of polyphenol oxidase by thiols in the absence and presence of potato tissue suspensions, *J. Agric. Food Chem.* **43**:69-76.

Friedman, M., and Brandon, D. L., 2001, Nutritional and health benefits of soy proteins, *J. Agric. Food Chem.* **49**:1069-1086.

Friedman, M., Buick, R., and Elliott, C., 2004a, Antibacterial activities of naturally occurring compounds against antibiotic-resistant *Bacillus cereus* vegative cells and spores, *Escherichia coli*, and *Staphylococcus aureus*, *J. Food Protection* **67**:1774-1778.

Friedman, M., and Cuq, J. L., 1988, Chemistry, analysis, nutritional value, and toxicology of tryptophan in food. A review, *J. Agric. Food Chem.* **36**:1079-1093.

Friedman, M., and Finot, P. A., 1990, Nutritional improvement of bread with lysine and γ-glutamyl-lysine, *J. Agric. Food Chem.* **38**:2011-2020.

Friedman, M., and Henika, P. R., 1991, Mutagenicity of toxic weed seeds in the Ames test: jimson weed (*Datura stramonium*), velvetleaf (*Abutilon theophrasti*), morning glory (*Ipomoea* spp.), and sicklepod (*Cassia obtusifolia*), *J. Agric. Food Chem.* **39**:494-501.

Friedman, M., Henika, P. R., Levin, C. E., and Mandrell, R. E., 2004b, Antibacterial activities of plant essential oils and their components against *Escherichia coli* O157:H7 and *Salmonella enterica* in apple juice, *J. Agric. Food Chem.* **52**:6042-6048.

Friedman, M., Henika, P. R., and Mackey, B. E., 2003a, Effect of feeding solanidine, solasodine, and tomatidine to non-pregnant and pregnant mice, *Food Chem. Toxicol.* **41**:61-71.

Friedman, M., Henika, P. R., and Mandrell, R. E., 2002, Bactericidal activities of plant essential oils and some of their isolated constituents against *Campylobacter jejuni*, *Escherichia coli*, *Listeria monocytogenes*, and *Salmonella enterica*, *J. Food Protection* **65**:1545-1560.

Friedman, M., Henika, P. R., and Mandrell, R. E., 2003b, Antibacterial activities of phenolic benzaldehydes and benzoic acids against *Campylobacter jejuni*, *Escherichia coli*, *Listeria monocytogenes*, and *Salmonella enterica*, *J. Food Protection* **66**:1811-1821.

Friedman, M., and Molnar Perl, I., 1990, Inhibition of browning by sulfur amino acids. Part 1. Heated amino acid-glucose systems, *J. Agric. Food Chem.* **38**:1642-1647.

Friedman, M., Molnar-Perl, I., and Knighton, D. R., 1992, Browning prevention in fresh and dehydrated potatoes by SH-containing amino acids, *Food Addit. Contam.* **9**:499-503.

Friedman, M., and Pearce, K. N., 1989, Copper(II) and cobalt(II) affinities of LL- and LD-lysinoalanine diastereomers: implications for food safety and nutrition, *J. Agric. Food Chem.* **37**:123-127.

Friedman, M., Rayburn, J. R., and Bantle, J. A., 1991, Developmenal toxicology of potato alkaloids in the frog embryo teratogenesis assay-*Xenopus* (FETAX), *Food Chem. Toxicol.* **29**:537-547.

Friedman, M., Roitman, J. N., and Kozukue, N., 2003c, Glycoalkaloid and calystegine contents of eight potato cultivars, *J. Agric. Food Chem.* **51**:2964-2973.

Friedman, M., Wehr, C. M., Schade, J. E., and MacGregor, J. T., 1982, Inactivation of aflatoxin B_1 mutagenicity by thiols, *Food Chem. Toxicol.* **20**:887-892.

Friedman, M., Wilson, R. E., and Ziderman, I. I., 1990a, Effect of heating on mutagenicity of fruit juices in the Ames test, *J. Agric. Food Chem.* **38**:740-743.

Friedman, M., Wilson, R. E., and Ziderman, I. I., 1990b, Mutagen formation in heated wheat gluten, carbohydrates, and gluten/carbohydrate blends, *J. Agric. Food Chem.* **38**:1019-1028.

Furniss, D. E., Vuichoud, J., Finot, P. A., and Hurrell, R. F., 1989, The effect of Maillard reaction products on zinc metabolism in the rat, *Br. J. Nutr.* **62**:739-749.

Haleva-Toledo, E., Naim, M., Zehavi, U., and Rouseff, R. L., 1999, Effects of L-cysteine and N-acetyl-L-cysteine on 4-hydroxy-2, 5-dimethyl-3(2H)-furanone (furaneol), 5-(hydroxymethyl)furfural, and 5-methylfurfural formation and browning in buffer solutions containing either rhamnose or glucose and arginine, *J. Agric. Food Chem.* **47**:4140-4145.

Han, J. S., Kozukue, N., Young, K. S., R., L. K., and Friedman, M., 2004, Distribution of ascorbic acid in potato tubers and in home-processed and commercial potato foods, *J. Agric. Food Chem.* **52**:6516-6521.

Hasegawa, K., Miwa, S., Tsutsumiuchi, K., Taniguchi, H., and Miwa, J., 2004, Extremely low dose of acrylamide decreases lifespan in *Caenorhabditis elegans*, *Toxicol. Lett.* **152**:183-189.

Hidalgo, F. J., Nogales, F., and Zamora, R., 2003, Effect of the pyrrole polymerization mechanism on the antioxidative activity of nonenzymatic browning reactions, *J. Agric. Food Chem.* **51**:5703-5708.

Hiramoto, K., Nasuhara, A., Michikoshi, K., Kato, T., and Kikugawa, K., 1997, DNA strand-breaking activity and mutagenicity of 2,3-dihydro-3,5-dihydroxy-6-methyl-4H-pyran-4-one (DDMP), a Maillard reaction product of glucose and glycine, *Mutat. Res.* **395**:47-56.

Jenq, S. N., Tsai, S. J., and Lee, H., 1994, Antimutagenicity of Maillard reaction products from amino acid/sugar model systems against 2-amino-3-methylimidazo-[4,5-f]quinoline: the role of pyrazines, *Mutagenesis* **9**:483-488.

Jing, H., and Kitts, D. D., 2004, Redox-related cytotoxic responses to different casein glycation products in Caco-2 and int-407 cells, *J. Agric. Food Chem.* **52**:3577-3582.

Kikugawa, K., Hiramoto, K., Kato, T., and Yanagawa, H., 2000, Effect of food reductones on the generation of the pyrazine cation radical and on the formation of the mutagens in the reaction of glucose, glycine and creatinine, *Mutat. Res.* **465**:183-190.

Kim, K. W., and Lee, S. B., 2003, Inhibitory effect of Maillard reaction products on growth of the aerobic marine hyperthermophilic archaeon *Aeropyrum pernix*, *Appl. Environ. Microbiol.* **69**:4325-4328.

Knize, M. G., Cunningham, P. L., Avila, J. R., Jones, A. L., Griffin, E. A., Jr., and Felton, J. S., 1994a, Formation of mutagenic activity from amino acids heated at cooking temperatures, *Food Chem. Toxicol.* **32**:55-60.

Knize, M. G., Cunningham, P. L., Griffin, E. A., Jr., Jones, A. L., and Felton, J. S., 1994b, Characterization of mutagenic activity in cooked-grain-food products, *Food Chem. Toxicol.* **32**:15-21.

Lee, K. R., Kozukue, N., Han, J. S., Park, J. H., Chang, E. Y., Baek, E. J., Chang, J. S., and Friedman, M., 2004, Glycoakaloids and metabolites inhibit the growth of human colon (HT29) and liver (HepG2) cancer cells, *J. Agric. Food Chem.* **52**:2832-2839.

Liardon, R., Friedman, M., and Philippossian, G., 1991, Racemization kinetics of free and protein-bound lysinoalanine (LAL) in strong acid media. Isomeric composition of bound LAL in processed proteins, *J. Agric. Food Chem.* **39**:531-537.

Lindenmeier, M., Faist, V., and Hofmann, T., 2002, Structural and functional characterization of pronyl-lysine, a novel protein modification in bread crust melanoidins showing in vitro antioxidative and phase I/II enzyme modulating activity, *J. Agric. Food Chem.* **50**:6997-7006.

MacGregor, J. T., Tucker, J. D., Ziderman, II, Wehr, C. M., Wilson, R. E., and Friedman, M., 1989, Non-clastogenicity in mouse bone marrow of fructose/lysine and other sugar/amino acid browning products with *in vitro* genotoxicity, *Food Chem. Toxicol.* **27**:715-721.

Marko, D., Habermeyer, M., Kemeny, M., Weyand, U., Niederberger, E., Frank, O., and Hofmann, T., 2003, Maillard reaction products modulating the growth of human tumor cells in vitro, *Chem. Res. Toxicol.* **16**:48-55.

Matsuda, T., Ishiguro, H., Ohkubo, I., Sasaki, M., and Nakamura, R., 1992, Carbohydrate binding specificity of monoclonal antibodies raised against lactose-protein Maillard adducts, *J. Biochem. (Tokyo)* **111**:383-387.

Miller, A. G., Meade, S. J., and Gerrard, J. A., 2003, New insights into protein crosslinking via the Maillard reaction: structural requirements, the effect on enzyme function, and predicted efficacy of crosslinking inhibitors as anti-ageing therapeutics, *Bioorg. Med. Chem.* **11**:843-852.

Molnar Perl, I., and Friedman, M., 1990a, Inhibition of browning by sulfur amino acids. Part 3. Apples and potatoes, *J. Agric. Food Chem.* **38**:1652-1656.

Molnar Perl, I., and Friedman, M., 1990b, Inhibition of browning of sulfur amino acids. Part 2. Fruit juices and protein-containing foods, *J. Agric. Food Chem.* **38**:1648-1651.

Morales, F. J., and Babbel, M. B., 2002, Antiradical efficiency of Maillard reaction mixtures in a hydrophilic media, *J. Agric. Food Chem.* **50**:2788-2792.

Nagasawa, T., Tabata, N., Ito, Y., Nishizawa, N., Aiba, Y., and Kitts, D. D., 2003, Inhibition of glycation reaction in tissue protein incubations by water soluble rutin derivative, *Mol. Cell. Biochem.* **249**:3-10.

Nakamura, S., Gohya, Y., Losso, J. N., Nakai, S., and Kato, A., 1996, Protective effect of lysozyme-galactomannan or lysozyme-palmitic acid conjugates against *Edwardsiella tarda* infection in carp, *Cyprinus carpio* L, *FEBS Lett.* **383**:251-254.

Nakayama, M., Izumi, G., Nemoto, Y., Shibata, K., Hasegawa, T., Numata, M., Wang, K., Kawaguchi, Y., and Hosoya, T., 1999, Suppression of N(epsilon)-(carboxymethyl)lysine generation by the antioxidant N-acetylcysteine, *Perit. Dial. Int.* **19**:207-210.

Ortwerth, B. J., and Olesen, P. R., 1988, Glutathione inhibits the glycation and crosslinking of lens proteins by ascorbic acid, *Exp. Eye Res.* **47**:737-750.

Öste, R. E., Brandon, D. L., Bates, A. H., and Friedman, M., 1990, Effect of Maillard browning reactions of the Kunitz soybean trypsin inhibitor on its interaction with monoclonal antibodies, *J. Agric. Food Chem.* **38**:258-261.

Öste, R. E., and Friedman, M., 1990, Nutritional value and safety of heated amino acid-sodium ascorbate mixtures, *J. Agric. Food Chem.* **38**:1687-1690.

Pearce, K. N., and Friedman, M., 1988, Binding of copper(II) and other metal ions by lysinoalanine and related compounds and its significance for food safety, *J. Agric. Food Chem.* **36**:707-717.

Pearce, K. N., Karahalios, D., and Friedman, M., 1988, Ninhydrin assay for proteolysis of ripening cheese, *J. Food Sci.* **53**:432-435.

Requena, J. R., Vidal, P., and Cabezas-Cerrato, J., 1993, Aminoguanidine inhibits protein browning without extensive Amadori carbonyl blocking, *Diabetes Res. Clin. Pract.* **19**:23-30.

Ryu, K., Ide, N., Matsuura, H., and Itakura, Y., 2001, N-alpha-(1-deoxy-D-fructos-1-yl)-L-arginine, an antioxidant compound identified in aged garlic extract, *J. Nutr.* **131**:972S-976S.

Sarwar, G., L'Abbe, M. R., Trick, K., Botting, H. G., and Ma, C. Y., 1999, Influence of feeding alkaline treated processed proteins on growth and protein and mineral status of rats, *Adv. Exp. Med. Biol.* **459**:161-177.

Schneider, M., Klotzsche, M., Werzinger, C., Hegele, J., Waibel, R., and Pischetsrieder, M., 2002, Reaction of folic acid with reducing sugars and sugar degradation products, *J. Agric. Food Chem.* **50**:1647-1651.

Schwimmer, S., 1981, *Source Book of Food Enzymology*, The Avi Publishing Company, Westport, CT, pp. 967.

Schwimmer, S., and Friedman, M., 1972, Enzymatic and non-enzymatic genesis of volatile sulfur-containing food flavors, *Flavour Industry* **9**:137-145.

Seidel, W., and Pischetsrieder, M., 1998, Immunochemical detection of N^2-[1-(1-carboxy)ethyl]guanosine, an advanced glycation end product formed by the reaction of

DNA and reducing sugars or L-ascorbic acid in vitro, *Biochim. Biophys. Acta* **1425**:478-484.
Seifert, S. T., Krause, R., Gloe, K., and Henle, T., 2004, Metal complexation by the peptide-bound maillard reaction products *N*(epsilon)-fructoselysine and *N*(epsilon)-carboxymethyllysine, *J. Agric. Food Chem.* **52**:2347-2350.
Sheikh-Zeinoddin, M., Perehinec, T. M., Hill, S. E., and Rees, C. E. D., 2000, Maillard reaction causes suppression of virulence gene expression in *Listeria monocytogenes*, *Int. J. Food Microbiol.* **61**:41-49.
Shoda, H., Miyata, S., Liu, B. F., Yamada, H., Ohara, T., Suzuki, K., Oimomi, M., and Kasuga, M., 1997, Inhibitory effects of tenilsetam on the Maillard reaction, *Endocrinology* **138**:1886-1892.
Song, D. U., Jung, Y. D., Chay, K. O., Chung, M. A., Lee, K. H., Yang, S. Y., Shin, B. A., and Ahn, B. W., 2002, Effect of drinking green tea on age-associated accumulation of Maillard-type fluorescence and carbonyl groups in rat aortic and skin collagen, *Arch. Biochem. Biophys.* **397**:424-429.
Stecchini, M. L., Giavedoni, P., Sarais, I., and Lerici, C. R., 1993, Antimicrobial activity of Maillard reaction products against *Aeromonas hydrophilla*, *Ital. J. Food Sci.* **2**:147-150.
Surh, Y. J., Liem, A., Miller, J. A., and Tannenbaum, S. R., 1994, 5-Sulfooxymethylfurfural as a possible ultimate mutagenic and carcinogenic metabolite of the Maillard reaction product, 5-hydroxymethylfurfural, *Carcinogenesis* **15**:2375-2377.
Taylor, J. L., Demyttenaere, J. C., Abbaspour Tehrani, K., Olave, C. A., Regniers, L., Verschaeve, L., Maes, A., Elgorashi, E. E., van Staden, J., and de Kimpe, N., 2004, Genotoxicity of melanoidin fractions derived from a standard glucose/glycine model, *J. Agric. Food Chem.* **52**:318-323.
Tritscher, A. M., 2004, Human health risk assessment of processing-related compounds in food, *Toxcol. Lett.* **149**:177-186.
Tseng, T. H., Chang, M. C., Hsu, J. D., Lee, M. J., Hsu, C. L., Lan, K. P., and Wang, C. J., 1998, Tumor promoting effect of *N*-nitroso-*N*-(2-hexanonyl)-3'-nitrotyramine (a nitrosated Maillard reaction product) in benzo(a)pyrene-initiated mouse skin carcinogenesis, *Chem.-Biol. Interact.* **115**:23-38.
Usui, T., Shizuuchi, S., Watanabe, H., and Hayase, F., 2004, Cytotoxicity and oxidative stress induced by the glyceraldehyde-related Maillard reaction products for HL-60 cells, *Biosci., Biotechnol., Biochem.* **68**:333-340.
Wang, C. J., Huang, H. P., Lee, M. J., Lin, Y. L., Lin, W. L., and Chang, W. C., 1998, Promotional effect of N-nitroso-N-(3-keto-1,2-butanediol)-3'-nitrotyramine (a nitrosated Maillard reaction product) in mouse fibroblast cells, *Food Chem. Toxicol.* **36**:631-636.
Wijewickreme, A. N., and Kitts, D. D., 1998, Modulation of metal-induced cytotoxicity by Maillard reaction products isolated from coffee brew, *J. Toxicol. Environ. Health A* **55**:151-168.
Yanagimoto, K., Lee, K. G., Ochi, H., and Shibamoto, T., 2002, Antioxidative activity of heterocyclic compounds found in coffee volatiles produced by Maillard reaction, *J. Agric. Food Chem.* **50**:5480-5484.
Yen, G. C., Liao, C. M., and Wu, S. C., 2002, Influence of Maillard reaction products on DNA damage in human lymphocytes, *J. Agric. Food Chem.* **50**:2970-2976.
Yen, G. C., Tsai, L. C., and Lii, J. D., 1992, Antimutagenic effect of Maillard browning products obtained from amino acids and sugars, *Food Chem. Toxicol.* **30**:127-132.
Ziderman, I. I., and Friedman, M., 1985, Thermal and compositional changes of dry wheat gluten-carbohydrate mixtures during simulated crust baking, *J. Agric. Food Chem.* **33**:1096-1102.

Ziderman, I. I., Gregorski, K. S., and Friedman, M., 1987, Thermal analysis of protein-carbohydrate mixtures in oxygen, *Thermochimica Acta* **114**:109-114.

Ziderman, I. I., Gregorski, K. S., Lopez, S. V., and Friedman, M., 1989, Thermal interaction of vitamin C with proteins in relation to nonenzymatic browning of foods and Maillard reactions, *J. Agric. Food Chem.* **37**:1480-1486.

Zou, L., and Atkinson, S., 2003, Characteristic vehicle emissions from the burning biodiesel made from vegetable oils, *Environ. Technol.* **24**:1253-1260.

ACRYLAMIDE FORMATION IN DIFFERENT FOODS AND POTENTIAL STRATEGIES FOR REDUCTION

Richard H. Stadler
Nestlé Product Technology Centre, CH-1350 Orbe, Switzerland; e-mail: richard.stadler@rdls.nestle.com

Abstract: This paper summarizes the progress made to date on acrylamide research pertaining to analytical methods, mechanisms of formation, and mitigation research in the major food categories. Initial difficulties with the establishment of reliable analytical methods have today in most cases been overcome, but challenges still remain in terms of the needs to develop simple and rapid test methods. Several researchers have identified that the main pathway of formation of acrylamide in foods is linked to the Maillard reaction and in particular the amino acid asparagine. Decarboxylation of the resulting Schiff base is a key step, and the reaction product may either furnish acrylamide directly or via 3-aminopropionamide. An alternative proposal is that the corresponding decarboxylated Amadori compound may release acrylamide by a beta-elimination reaction. Many experimental trials have been conducted in different foods, and a number of possible measures identified to relatively lower the amounts of acrylamide in food. The validity of laboratory trials must, however, be assessed under actual food processing conditions. Some progress in relatively lowering acrylamide in certain food categories has been achieved, but can at this stage be considered marginal. However, any options that are chosen to reduce acrylamide must be technologically feasible and also not negatively impact the quality and safety of the final product.

Key words: Acrylamide, food, analysis, mechanisms of formation, mitigation

1. INTRODUCTION

The announcement by the Swedish National Food Authority in April 2002 of the presence of acrylamide predominantly in carbohydrate-rich foods

(Tareke et al., 2002) sparked intensive investigations into acrylamide, encompassing the analysis, occurrence, chemistry, toxicology, and potential health risk of this contaminant in the human diet.

Several research groups have developed methods to quantify reliably acrylamide at relatively low levels in a large variety of different foodstuffs. Most of the methods published so far are based on either GC-MS or LC-MS techniques, with comparable performance of the two approaches. Based on the conclusions of a recent inter-laboratory trial (Wenzl et al., 2003) and the EC / JRC task force group held in April 2003 (Joint European Commission Workshop, 2003), many of the methods did not perform well in difficult matrices such as cocoa and coffee. Consequently, laboratories have adapted their methods to achieve the required precision and sensitivity for those foods in which gaps in the analytical science were initially identified. A fully validated method with adequate performance and that can be applied also to "difficult" matrices such as coffee and cocoa is briefly described in this report.

Several months after the Swedish announcement, a number of research groups simultaneously discovered that acrylamide is formed during the Maillard reaction, and the major reactants leading to the formation of acrylamide are sugars and the amino acid asparagine (Mottram et al., 2002, Sanders et al., 2002, Becalski et al., 2003, Stadler et al., 2002). Fundamental mechanistic studies published in 2003 have revealed a feasible route to acrylamide (Zyzak et al., 2003, Yaylayan et al., 2003), and a number of possible minor pathways have also been described (Stadler et al., 2003, Lingnert et al., 2002). This fundamental knowledge opens the way to concrete studies on kinetic modeling (formation over temperature/time, competitive reaction kinetics with amino acids and sugars), and identifying the rate limiting steps under actual food processing conditions. Measures could then be devised to attempt to reduce acrylamide in food products. This, however, necessitates extensive individual trials for each food category, and in most cases entails a combination of measures that on a case-by-case basis must be applied at the raw material stage, during storage, processing, and final preparation of the product in the home.

This report summarizes the current state of knowledge focusing on the analytical, mechanistic, and mitigation aspects. Furthermore, the progress and complexity of the research is highlighted in a number of different products, so indicating as well the challenges and constraints faced by industry in finding appropriate and practical solutions to this concern.

2. PROGRESS MADE TO DATE IN ANALYTICS

Only recently has a single method been reported that can achieve good sensitivity and selectivity of acrylamide in practically all of the relevant food matrices (Roach et al., 2003), and has been improved for coffee that was found particularly troublesome (Andrzejewski et al., 2004).

Our laboratory has recently reported a method by isotope-dilution liquid chromatography-electrospray ionization tandem mass spectrometry (LC-MS/MS) that achieves good precision, accuracy, and certainty of the analyte in a wide range of foodstuffs. However, acrylamide could not be quantified reliably in difficult matrices such as cocoa powder and coffee, mainly due to considerable loss of the analyte throughout the sample preparation steps (Riediker and Stadler, 2003). Improvements were thus made to the existing method (Figure 1), and sample pre-treatment essentially encompasses (a) protein precipitation with Carrez I and II solutions, (b) extraction of the analyte into ethyl acetate, and (c) solid phase extraction on a Multimode cartridge. This approach provided good performance in terms of linearity, accuracy and precision. Full validation was conducted in soluble chocolate powder, with good precision (Table 1).

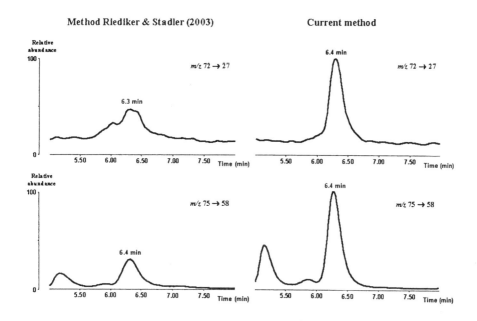

Figure 1. Improvement of the LC-MS/MS methods previously developed in our laboratory to determine acrylamide in "difficult" matrices. Shown is an excerpt of a LC-MS chromatogram of a cocoa powder analyzed using the initial method (left panels) and the improved method (right panels).

Table 1. Precision of the LC-MS/MS method for the quantification of acrylamide in soluble chocolate powder

Spiking level (µg/kg)	Precision (CV in%)	
	Repeatability	Intermediate precision
13	15.8	22.8
305	6.1	9.8
2504	5.4	7.7

The method achieves a limit of determination <10 µg/kg in all food matrices tested, and recovery (not corrected for loss of analyte by isotope dilution) of 43 – 51 % over three concentration ranges. The method was extended to the analysis of acrylamide in various foodstuffs such as mashed potatoes, crisp bread, and butter biscuit and cookies. Furthermore, the accuracy of the method is demonstrated by the results obtained in three inter-laboratory proficiency tests (Delatour et al., in press).

The availability of validated robust and reliable methods to determine acrylamide at low levels in foods are of paramount importance especially for accurate intake assessments. Our laboratory has conducted several hundred analyses of acrylamide on such matrices and can with confidence propose this method as a reference method using isotope dilution LC-MS /MS.

2.1 Stability of acrylamide in food products

Acrylamide may react with inherent food constituents, and we therefore re-analyzed selected dry food products after a certain period of their initial analysis (products kept in their original package at room temperature). As shown in Table 2, acrylamide is stable in certain foods (e.g. breakfast cereals) over prolonged storage periods of up to 12 months.

Table 2. Time-dependent stability of acrylamide in various foodstuffs

Foodstuff	Interval (months)	Difference* (µg/kg)	% Change
Breakfast cereal	12	0	0
Soluble coffee powder	12	515	67
Roasted barley	9	40	15
Roasted coffee	7	56	28
Dried chicory	5	40	19
Roasted chicory	5	620	15
Cocoa	3	3	2

*Initial versus second analysis after the given interval.

On the other hand, loss of acrylamide was appreciable in coffee (roast and ground and soluble) and chicory (dried and roasted) after 5 – 12 months storage. Similar observations for coffee were also recently shown by

Andrzejewski et al. (2004), and it is possible that acrylamide interacts over time with inherent nucleophiles. Further evidence for the reactivity of acrylamide with coffee constituents was demonstrated in an experiment where roast and ground coffee was incubated for 36h at 60°C in closed or open jars. Acrylamide content was reduced by > 20% in the closed jars. The possibility of interaction of acrylamide with volatile nucleophiles (e.g. furfuryl-mercaptan) cannot be excluded and warrants further study.

3. MECHANISMS OF FORMATION

In 2003, several research groups reported more detailed chemical pathways to acrylamide (see Table 3 for a summary). Those published by Yaylayan et al. and Zyzak et al. can be considered the most likely routes by the Maillard reaction. Both groups have shown evidence for the importance of the Schiff base of asparagine, which corresponds to the dehydrated N-glucosyl compound. Decarboxylation of the Schiff base is a key step, and the reaction product may furnish acrylamide either directly or via 3-aminopropionamide (Zyzak et al., 2003).

Table 3. Examples of work published in 2003 on the mechanisms of acrylamide formation in foods

Postulated Mechanism(s)	Reference
Decarboxylation of the Schiff base via a oxazolidin-5-one intermediate, tautomer-ization to the decarboxylated Amadori product and subsequent beta-elimination	Yaylayan et al., 2003
Decarboxylation of the Schiff base, heterocyclic cleavage of the imine	Zyzak et al., 2003
Decarboxylation of the Schiff base, hydrolysis of the imine to afford to afford 3-aminopropionamide that subsequently deaminates	Zyzak et al., 2003
Acrylic acid + NH3 (ammonia from thermal degradation of amino acids) (only approx. 5% of the yield compared to asparagine)	Stadler et al., 2003
Acrolein (from triolein) + asparagine Acrolein + NH3 Acrylic acid + NH3 (amino dehydroxylation)	Yasuhara et al., 2003
Acrylic acid from 2-propenal and subsequent reaction of acrylic acid with NH3	Vattem and Shetty, 2003

Alternatively, the corresponding decarboxylated Amadori product, procured by tautomerization of the decarboxylated Schiff base, may release acrylamide by a beta-elimination reaction (Yaylayan et al., 2003). However, so far the key intermediates in food have not been characterized, and therefore the chemical reactions leading to acrylamide remain largely hypothetical.

To provide further evidence for the route supporting a β–elimination as the rate limiting step, we synthesized the decarboxylated Amadori products and reacted these under low moisture conditions, measuring the formation and yield of the corresponding vinylogous or Strecker degradation products.

The corresponding vinylogous compounds were only generated if a β-proton was available, e.g. styrene from the decarboxylated Amadori compound of phenylalanine (Figure 2) (Blank et al., 2004). Therefore, it is suggested that this thermal pathway may be common to other amino acids resulting under certain conditions in their respective vinylogous reaction products (Stadler et al., 2003).

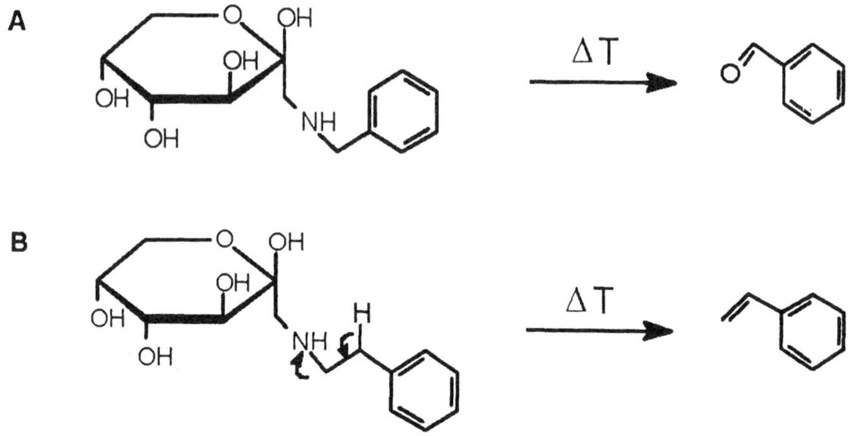

Figure 2. Formation of benzaldehyde (pathway A) and styrene (pathway B) via beta-elimination and Strecker-type degradation, respectively (Blank et al., 2004).

4. FORMATION IN DIFFERENT FOODS AND POTENTIAL MEASURES OF CONTROL

4.1 Potato products

Potato-based foods that are baked, fried, or cooked define a wide range of different products on which much investigative work has been done to date to determine the formation and potential control of acrylamide. Many possible avenues of reduction of acrylamide in potato products, in particular French fries, have been discussed in several recent reports (Biedermann et al., 2002a, 2002b; Noti et al., 2003; Jung et al., 2003;, Haase et al., 2003; Grob et al., 2003). These entail controlling the temperature of storage of the raw potato, selection of certain varieties, and modifying processing (frying) conditions. However, any modifications performed on the raw material constituents will inevitably impact the Maillard reaction and its products, and concomitantly the organoleptic properties (taste and color) of the cooked food.

Table 4. Impact of different parameters and conditions on acrylamide formation in French fries

Study/Topic	Key Findings	Reference
Frying oil	Type of oil not important	
Max. frying temp. for finish frying	175°C	Taeymans et al., in press
Reduction of added sugar	Significant decrease in acrylamide	
Surface/volume ratio (SVR)	Lower SVR decreases acrylamide	
Variation between batches of prefabricated fries (deep frozen)	Considerable variation in acrylamide under standard frying conditions	Franke et al., 2003
Addition of citric acid	Optimal at 0.75%, decrease factor 2 (also considering loss sugars and Asn during the washing step); but strong impact on organoleptic quality	Biedermann et al., 2002a, Biedermann et al., 2002b
Stability of AA / Elimination reactions	Final amount of AA in potato products is dependent on the balance of formation and elimination	Gama-Baumgartner et al., 2004
Addition of citric acid (1 – 2%, w/w)	Significant reduction, impact on sensory at 2%	Jung et al., 2003
Moisture bound by adsorbents added to frying oil	Reduction of acrylamide by > 40%	Gertz and Klostermann, 2002, Gertz et al., 2003
Addition of citric acid to frying oil	No or only slight reduction	
Silicone added to frying oil	Slight increase in acrylamide formation	
Potato cultivar	Impact on acrylamide formation	
Field site	Impact on acrylamide formation	
Frying : load of the fries	Optimal conditions (10%, i.e. 100g/liter oil)	
Frying: size of the fries, larger fries	Slightly lower amounts of acrylamide (higher area/surface ratio)	Haase et al., 2003, Grob et al., 2003, Taeymans et al., in press
Oil type, additives, to improve heat transfer	No significant differences to the control	
Blanching, soaking	Reduction of acrylamide content	
Endpoint of frying	Recommended at max. 170°C (no general browning, but crispy and with adequate flavor)	

However, even though small scale and laboratory trials have shown that products such as French fries can be prepared with acrylamide amounts below 100 µg/kg (Grob et al., 2003), all these measures must be placed in the perspective of consumer acceptance, not forgetting those related to the supply chain management and logistics of harvesting, storage, and transport of the raw potatoes. Table 4 summarizes the key findings in mitigation research around French fries.

4.1.1 Impact of Raw Material Variability

The experimental trials that have been conducted so far on acrylamide in potatoes have shown that the major determinants of acrylamide formation are reducing sugars (mainly glucose and fructose) as well as the (free) amino acid asparagine. The content of sugar (glucose/fructose) in the raw potato is well correlated ($r^2 = 0.85$) to the amount of acrylamide formed upon heating (Biedermann et al., 2003b). A wide range of potatoes were analyzed for free amino acids and sugars (glucose, fructose and sucrose). As already documented in several reports, among the various free amino acids measured, the content of free asparagine was the highest. Widely varying concentrations of asparagine, glucose, fructose and sucrose were observed, as also shown recently by Amrein et al. (2003).

This variability may be one important explanation for the difference in the amounts of acrylamide that may be formed in the products during processing. The reasons for this large spread in asparagine and reducing sugars is probably due to multiple factors, such as potato cultivar, farming systems, field site, fertilization, pesticide/herbicide application, time of harvest, storage time and temperature. Clearly more studies will need to be conducted in the future to enable a better understanding of how these many factors may affect the variability of raw material composition.

4.2 Bread and Bakery wares

Only a few reports on the formation of acrylamide in bread and bakery products have been published (Taeymans et al., in press, Springer et al., 2003, Amrein et al., 2004, Surdyk et al., 2004). In crisp bread, acrylamide concentration could be reduced by decreasing the average longitudinal oven baking temperature and increasing the baking time (Taeymans et al., in press). A similar empirical trial approach has been applied to biscuits. Acrylamide is not present in uncooked dough, but the acrylamide level rises rapidly with time. Temperature and cooking time are closely related in the baking process, as is final moisture content, which in some trials has been shown to be inversely proportional to the acrylamide content in the final product. Acrylamide formation has also been studied with regard to ingredients and formulations. The addition of whole wheat flour and bran to biscuit formulas tended to increase acrylamide in comparison with plain counterparts. Reducing the amount of the raising agent ammonium bicarbonate in formulas lowered acrylamide in plain flour matrices. The addition of lactic acid also lowered acrylamide content in plain flour matrices (Taeymans et al., in press). Experiments are ongoing to determine the relative impact of baking temperature, baking time, final moisture content and biscuit thickness on acrylamide formation. An important observation in general is the large batch-to-batch variation in acrylamide content in biscuits manufactured under industrial conditions and sampled from the line.

A recent study on gingerbread demonstrated that acrylamide is formed evenly over the whole baking process (Amrein et al., 2004). The acrylamide concentration could be considerably lowered when replacing ammonium hydrogencarbonate as raising agent. The same study showed that acidulants such as citric acid also contributed to decreasing the acrylamide content in the final product. In yeast-leavened wheat bread, acrylamide is mainly formed in the crust (99%), and progressively increases with temperature showing a good correlation to color (Surdyk et al., 2004).

4.3 Breakfast Cereals

A marked feature of breakfast cereals is that they possess a clear product identity. They are produced by different and distinct processes that essentially entail a cooking and toasting step. The Maillard reaction develops flavors and color in both the cooking and the toasting steps of the process. Results of trials to date show that most (> 90%) of the acrylamide present in cereals is formed in the toasting step (Taeymans et al., in press). Cereals are characterized by a wide range of acrylamide values observed within and between batches of the same product, processed under the same conditions. This variability creates a difficulty for experimental design and to date no modification to process has had a beneficial effect as large as this variation. Hence, an understanding of what is driving this within-batch and between-batch variation is needed (Taeymans et al., in press).

In model systems, acrylamide begins to form at temperatures >120°C (Stadler et al., 2002). Experimental studies on a wheat biscuit cereal have shown that acrylamide is present in both the surface (270 µg/kg) and cooler centre (60-80 °C, 128 µg/kg) of the biscuit. When biscuits are toasted to the lowest degree compatible with edibility the acrylamide concentration is increased by 15 to 45%. When biscuits are toasted to a near burnt state the acrylamide concentration is decreased by 40 to 50%. Similar results have been seen for several other forms of cereal and during flash frying of potatoes (Taeymans et al., in press), suggesting a balance between formation and elimination with the latter being more rapid at higher temperature.

However, if the conditions of cooking the cereal are varied and the toasting kept constant, then variations in the acrylamide content of the cereal may be seen. This infers that a precursor formed during the cooking stage may be procured in variable amount and converted to acrylamide at toasting. The wet cooking stage may offer at least as much potential for control of acrylamide content, and as emphasized in the "Mechanisms of Formation" section, it would be important to determine the key intermediates/precursors during food processing.

For cereal as for other model systems those spiked with asparagine generate more acrylamide than controls. In terms of the real process knowledge of the "normal" range of free asparagine content for cereals is important. Further investigations are underway to assess the variability of the amount of free asparagine in different wheat varieties.

4.4 Coffee

Compared to the many other fried, roasted and baked food products, roast and ground coffee has been reported to contain relatively low concentrations (170 – 351 µg/kg on a powder basis) of acrylamide (Friedman, 2003). There are no significant differences in acrylamide concentrations in caffeinated versus decaffeinated coffees. Roast and ground coffee is not consumed as such, but prepared as a beverage. Coffee is prepared by the addition of hot water and subsequent filtration. Hence, calculation of the acrylamide content per cup is an important term of exposure levels (Andrzejewski et al., 2004).

Coffee is typically roasted at temperatures in the range of 220 – 250°C, and the roasting time and speed of roast have an important impact on the sensorial properties (aroma/taste). These are carefully fine tuned to a characteristic profile leading to a clear identity of the coffee product.

Experiments have shown that acrylamide is degraded/eliminated during roasting, and the profile of acrylamide formation during the roasting of coffee reflects this effect very clearly (Taeymans et al., in press). In coffee, acrylamide is formed at the beginning of the roasting step, and toward the end of the roasting cycle a loss of acrylamide seems to dominate. Therefore, light roasted coffees may contain relatively higher amounts of acrylamide than very dark roasted beans. The temperature *per se*, however, does not show a significant difference in the formation of acrylamide. Towards the commercial roasting (color) range, the acrylamide level was reduced by a factor of approximately 10 compared to the highest level recorded during the complete roasting cycle (Taeymans et al., in press). However, higher roasting as a potential option to reduce acrylamide could generate other undesirable compounds and negatively impact the taste/aroma of the product. Consequently, no practical solutions are today at hand that would reduce acrylamide levels and concomitantly retain the quality characteristics of coffee, since the roasting step cannot be fundamentally changed.

5. CONCLUSION

Over the past two years, researchers from academia, industry, and national authorities/enforcement laboratories, have gained increasing insight in understanding the presence, formation and potential risk to public health posed by the unexpected discovery of acrylamide in some foods. Major progress has been made in the analytical methodology, with good performance of the methods as judged by inter-laboratory trials. The next steps have been defined and entail the development of rapid and cheaper methods to determine acrylamide with adequate accuracy and precision. A further important task is to prepare certified reference materials that can be used as quality control samples by laboratories. The EC-JRC/IRMM has commenced such a study, choosing the most active laboratories to participate in this exercise.

Several reports have been published on the mechanisms of formation of acrylamide, and there is overall consensus on the key role of asparagine Schiff base in the reaction. Concrete evidence of certain intermediates is, however, so far lacking in foods. More than 20 research papers have been published so far on mitigation research in different foods using mainly experimental and pilot-scale conditions. These have provided avenues that may be pursued to reduce acrylamide levels, and as highlighted in this report industry has achieved moderate success in some selected products. However, since acrylamide formation is directly linked to the desired Maillard reaction that generates important flavor and aroma compounds, any measures taken must assess the impact on overall quality and consumer acceptance.

Finally, a concern that needs to be addressed is the lack of knowledge about the effects of final preparation in food service and domestic situations on acrylamide formation. As recommended by WHO/FAO, SCF, and the U.S. FDA, people should not change their dietary habits and continue to eat a balanced diet rich in fruit and vegetables and moderate their consumption of fried and fatty foods.

ACKNOWLEDGEMENTS

The author thanks the members of the CIAA Acrylamide Expert Group for their valuable contributions and in-depth discussions.

REFERENCES

Amrein, T.M., Bachmann, S., Noti, A., Biedermann, M., Barbosa, M.F., Biedermann-Brem, S., Grob, K., Keiser, A., Realini, P., Escher, F., and Amado, R., 2003, Potential of acrylamide formation, sugars, and free asparagine in potatoes: a comparison of cultivars and farming systems, *J. Agric. Food Chem.* **51**: 5556-5560.

Amrein, T. M. Schoenbaechler, B., Escher, F., and Amado, R., 2004, Acrylamide in gingerbread: critical factors for formation and possible ways for reduction, *J. Agric. Food Chem.* **52**: 4282-4288.

Andrzejewski, D., Roach, J.A.G., Gay, M.L., and Musser, S.M., 2004, Analysis of coffee for the presence of acrylamide by LC-MS/MS, *J. Agric. Food Chem.* **52**: 1996-2002.

Becalski, A, Lau, B. P-Y, Lewis, D and Seaman, S., 2003, Acrylamide in foods: Occurrence, sources and modelling, *J. Agric. Food Chem.* **51**: 802-808.

Biedermann, M. Biedermann-Brem, S. Noti, A. and Grob, K., 2002a, Methods for determining the potential of acrylamide formation and its elimination in raw materials for food preparation, such as potatoes, *Mitt. Lebensm. Hyg.* **93**: 653-667.

Biedermann, M., Noti, A., Biedermann-Brem, S., Mozzetti, V. and Grob, K., 2002b, Experiments on acrylamide formation and possibilities to decrease the potential of acrylamide formation in potatoes, *Mitt. Lebensm. Hyg.* **93**: 668-687.

Blank, I., Robert, F., Vuataz, G., Pollien, P., Saucy, F., and Stadler, R. H., 2004, A new insight into the formation of acrylamide via the early Maillard reaction. Abstracts of Papers, 227th ACS National Meeting, Anaheim, CA, United States, March 28-April 1.

Delatour, T., Perisset, A., Goldmann, T., Riediker, S., and Stadler, R.H. Improved sample preparation to determine acrylamide in difficult matrices such as chocolate powder, cocoa, coffee, and surrogates by liquid chromatography tandem mass spectroscopy. *J. Agric. Food Chem.* **52**: 4625-4631.

Franke, K., Kreyenmeier, F., and Reimerdes, E. H., 2003, Acrylamide - taking a comprehensive look at the events surrounding its formation is critical! *Lebensmitteltechnik*; **35 (3)**: 60-62

Friedman, M., 2003, Chemistry, biochemistry, and safety of acrylamide, *J. Agric. Food Chem.* **51**: 4504-4526.

Gama-Baumgartner, F., Grob, K., and Biedermann, M., 2004, Citric acid to reduce acrylamide formation in French fries and roasted potatoes? *Mitt. Lebensm. Hyg.* **95(1)**: 110-117.

Gertz, C., Klostermann, S., and Kochhar, S.P., 2003, Deep frying: the role of water from food being fried and acrylamide formation, *Lipides*, 2003, **10 (4)**: 297-303.

Gertz, C., and Klostermann, S., 2002, Analysis of acrylamide and mechanisms of its formation in deep-fried products, *Eur. J. Lipid Sci. Technol.* **104**: 762-771.

Grob, K., Biedermann, M., Biedermann-Brem, S., Noti, A., Imhof, D., Amrein, T., Pfefferle, A., and Bazzocco, D., 2003, French fries with less than 100 µg/kg acrylamide. A collaboration between cooks and analysts. *Eur. Food Res. Technol.* **217**: 185-194.

Haase, N.U., Matthaeus, B., and Vosmann, K., 2003, Acrylamide formation in foodstuffs – minimizing strategies for potato crisps. *Deutsche Lebensm. Rund.* **99**: 87-90.

Joint European Commission Workshop "Analytical methods for acrylamide determination in food", Joint European Commission (DG Sanco & DG JRC)/Swedish National Food Administration, 28-29th April, 2003.

Jung, M.Y., Choi, D.S., and Ju, J.W., 2003, A novel technique for limitation of acrylamide formation in fried and baked corn chips and french fries, *J. Food. Sci.* **68**: 1287-1290.

Lingnert, H., Grivas, S., Jagerstad, M., Skog, K., Tornqvist, M., and Aman, P., 2002, Acrylamide in food: mechanisms of formation and influencing factors during heating of food, *Scand. J. Nutr.* **46 (4)**: 159-172.

Mottram, D S, Wedzicha, B I, and Dodson, A T. 2002, Acrylamide is formed in the Maillard reaction, *Nature* **419**: 448.

Noti, A., Biedermann-Brem, S., Biedermann, M., Grob, K., Albisser, P., and Realini, P., 2003, Storage of potatoes at low temperature should be avoided to prevent increased acrylamide during frying or roasting, *Mitt. Lebensm. Hyg.* **94**:167-180.

Roach, J.A.G.; Andrzejewski, D.; Gay, M.L.; Nortrup, D.; Musser, S.M. 2003, Rugged LC-MS/MS survey analysis for acrylamide in foods. *J. Agric. Food Chem.*, **51**: 7547-7554.

Riediker, S., and Stadler, R.H., 2003, Analysis of Acrylamide in Food Using Isotope-Dilution Liquid Chromatography Coupled with Electrospray Ionization Tandem Mass Spectrometry, *J. Chrom. A.* **1020**: 121-130.

Sanders, R. A., Zyzak, D. V., Stojanovic, M., Tallmadge, D. H., Eberhart, B. L., and Ewald, D. K., 2002, An LC/MS acrylamide method and it's use in investigating the role of Asparagine. Presentation at the Annual AOAC International Meeting, Los Angeles, CA, September 22 – 26.

Springer, M., Fischer, T., Lehrack, A., and Freund, W., 2003, Acrylamide formation in baked products, *Getreide, Mehl und Brot*, **57(5)**: 274-278.

Stadler, R. H., Blank, I., Varga, N., Robert, F., Hau, J., Guy, P. A., Robert, M-C. and Riediker, S. Acrylamide from Maillard reaction products. *Nature* **419**, 449 (2002).

Stadler, R.H., Verzegnassi, L., Varga, N., Grigirov, M., Studer, A., Riediker, S., and Schilter, B., 2003, Formation of vinylogous compounds in model Maillard reaction systems. *Chem. Res. Tox.* **16**: 1242-1250.

Surdyk, N., Rosen, J., Andersson, R., and Aaman, P., 2004, Effects of Asparagine, Fructose, and Baking Conditions on Acrylamide Content in Yeast-Leavened Wheat Bread, *J. Agric. Food Chem.* **52(7)**: 2047-2051.

Tareke, E. P., Rydberg, P., Karlsson, S., Erikson, M., and Törnqvist, M., 2002, Analysis of acrylamide, a carcinogen formed in heated foodstuffs, *J. Agric. Food Chem.* **50**: 4998-5006.

Taeymans, D., Ashby, P., Blank, I., Gondé, P., van Eijck, P., Lalljie, S., Lingnert, H., Lindblom, M., Matissek, R., Müller, D., O'Brien, J., Thompson, S., Studer, A., Silvani, D., Tallmadge, D., Whitmore, T., Wood, J., and Stadler, R.H., 2004, A review of acrylamide : an industry perspective on research, analysis, formation and control, *Crit. Rev. Food Sci & Nutr.*, in press.

Vattem, D.A., and Shetty, K., 2003, Acrylamide in food: a model for mechanism of formation and its reduction, *Inn. Food Sci. and Emerging Technol.* **4**: 331-338.

Wenzl, T., de la Calle, B., and Anklam, E., 2003, Analytical methods for the determination of acrylamide in food products : a review, *Food Addit. Contam.* **20**: 885-902.

Yasuhara, A., Tanaka, Y., Hengel, M., and Shibamoto, T., 2003, Gas chromatographic investigation of acrylamide formation in browning model systems, *J. Agric. Food Chem.* **51**: 3999-4003.

Yaylayan, V.A., Wnorowski, A., and Locas, C.P., 2003, Why asparagine needs carbohydrates to generate acrylamide, *J. Agric. Food Chem.* **51**: 1753-1757.

Zyzak, D., Sanders, R.A., Stojanovic. M., Tallmadge, D., Eberhart, B.L., Ewald, D.K., Gruber, D.C., Morsch, T.R., Strothers, M.A., Rizzi, G.P., Villagran, M.D., 2003, Acrylamide formation mechanism in heated foods, *J. Agric. Food Chem.* **51**: 4782-4787.

MECHANISMS OF ACRYLAMIDE FORMATION
Maillard-induced transformation of asparagine

I. Blank[1], F. Robert[1], T. Goldmann[1], P. Pollien[1], N. Varga[1], S. Devaud[1], F. Saucy, T. Huynh-Ba[1], and R. H. Stadler[2]
[1]*Nestlé Research Center, 1000 Lausanne 26, Switzerland;* [2]*Product Technology Center Orbe, CH-1350 Orbe, Switzerland; e-mail : imre.blank@rdls.nestle.com*

Abstract: The formation of acrylamide (AA) from L-asparagine was studied in Maillard model systems under pyrolysis conditions. While the early Maillard intermediate *N*-glucosylasparagine generated ~2.4 mmol/mol AA, the Amadori compound was a less efficient precursor (0.1 mmol/mol). Reaction with α-dicarbonyls resulted in relatively low AA amounts (0.2-0.5 mmol/mol), suggesting that the Strecker aldehyde pathway is of limited relevance. Similarly, the Strecker alcohol 3-hydroxypropanamide generated low amounts of AA (0.2 mmol/mol). On the other hand, hydroxyacetone afforded more than 4 mmol/mol AA, indicating that α-hydroxycarbonyls are more efficient than α-dicarbonyls in transforming asparagine into AA. The experimental results are consistent with the reaction mechanism proposed, *i.e.* (i) Strecker-type degradation of the Schiff base leading to azomethine ylides, followed by (ii) β-elimination of the decarboxylated Amadori compound to release AA. The functional group in β-position on both sides of the nitrogen atom is crucial. Rearrangement of the azomethine ylide to the decarboxylated Amadori compound is the key step, which is favored if the carbonyl moiety contains a hydroxyl group in β-position to the N-atom. The β-elimination step in the amino acid moiety was demonstrated by reacting under pyrolysis conditions decarboxylated model Amadori compounds obtained by synthesis.

Key words: Maillard reaction; acrylamide; asparagine; carbonyls; β-elimination; Strecker degradation; pyrolysis; low moisture conditions.

1. INTRODUCTION

The recent discovery of relatively high amounts of acrylamide (AA) in carbohydrate-rich foods obtained by thermal processing (review by Friedmann, 2003) has led to numerous studies to help understand how AA is formed. Maillard-type reactions have been shown as one major reaction pathway, in particular in the presence of asparagine, which directly provides the backbone of the acrylamide molecule (Mottram et al., 2002; Stadler et al., 2002; Weisshaar and Gutsche, 2002; Biedermann et al., 2002; Becalski et al., 2003; Zyzak et al., 2003). However, other reaction pathways and precursors have also been suggested, such as acrolein formed by oxidative lipid degradation leading to acrylic acid, which can react with ammonia to give AA (Gertz and Klostermann, 2002; Yasuhara et al., 2003). Acrylic acid can also be formed from aspartic acid, in analogy to the formation of AA from asparagine (Stadler et al., 2003).

There are basically two major hypotheses published thus far pertaining to the formation of AA from asparagine in foods by Maillard-type reactions. Mottram et al. (2003) have suggested that α-dicarbonyls are necessary co-reactants in the Strecker degradation reaction affording the Strecker aldehyde as precursor of AA. Glycoconjugates, such as *N*-glycosides and related compounds formed in the early phase of the Maillard reaction, have been proposed as key intermediates leading to AA (Stadler et al., 2002). This hypothesis is supported by the work recently published by Yaylayan et al. (2003) and Zyzak et al. (2003). Both groups have shown some evidence for the importance of the Schiff base of asparagine, which corresponds to the dehydrated *N*-glucosyl compound. The key mechanistic step is decarboxylation of the Schiff base leading to Maillard intermediates that can directly release AA. However, as the key intermediates were not or only partially characterized, the chemical reactions leading to AA remained largely hypothetical.

The objective of this study was to further clarify the mechanism of AA formation from asparagine and to study the role of reaction conditions such as temperature, time, moisture, and pH.

2. EXPERIMENTAL SECTION

2.1 Materials

L-Asparagine, D-fructose, D-glucose, butane-2,3-dione, hydroxyacetone, methylglyoxal, glyoxal, glyoxalic acid, acrylamide (AA), 1-butanal, 1,2-dihydroxybutane, tetrabutylammonium chloride, potassium carbonate, 2,2,6,6-tetramethylpiperidine 1-oxyl (free radical, TEMPO), and *N*-chlorosuccinimide were from Fluka/Aldrich (Buchs, Switzerland). Methanol, formic acid, dichloromethane, water for HPLC, and Silica gel 60 were from Merck (Darmstadt, Germany). 3-Hydroxypropanamide was custom synthesized by Toronto Research Chemicals (Toronto, Canada). 2,3,3-$^2H_3$-

Acrylamide (isotopic purity 98 %) was purchased from Cambridge Isotope Laboratories (Andover, MA, USA). All other reagents were of analytical grade and were used without further purification.
Caution: *Acrylamide (CAS 79-06-1) is classified as toxic and may cause cancer. Wear suitable protective clothing, gloves and eye/face protection when handling this chemical.*

2.2 Synthesis

The synthesis of several Maillard intermediates used in this study, such as potassium N-(D-glucos-1-yl)-L-asparaginate (PGA), N-(1-deoxy-D-fructos-1-yl)-L-asparagine (DFA), N-(1-deoxy-D-fructos-1-yl)-benzylamine (DFB), and N-(1-deoxy-D-fructos-1-yl)-2-phenylethylamine (DFP), has been reported elsewhere (Stadler et al., 2004).

2-Hydroxy-1-butanal was prepared as follows. 1,2-Dihydroxybutane (5 g, 0.056 mol), N-chlorosuccinimide (8.16 g, 0.061 mol), and tetrabutylammonium chloride (1.55 g) were added to dichloromethane (110 mL) in a round flask (0.5 L) at 0°C. Then, an aqueous buffer solution (K_2CO_3/$NaHCO_3$, pH 8.6, 110 mL) and TEMPO (0.87 g) were added to the mixture, which was stirred with a mechanical system for 16 h at 0°C. The organic layer was extracted and the aqueous phase washed with dichloromethane (3 x 100 mL). The organic layers were collected and dried above magnesium sulfate. The solvent was finally filtered and evaporated under vacuum. Distillation of the crude product (55 °C, 0.02 mbar) led to the target compound composed of 2-hydroxy-1-butanal and the corresponding hydroxyketone in a 3:2 ratio. The two tautomers were characterized by mass spectrometry (MS/EI, m/z, rel-%): 2-hydroxy-1-butanal: 88 (M^+), 59 (100), 31 (55); 1-hydroxy-2-butanone: 88 (M^+), 57 (100), 31 (55).

It should be noted that 2-hydroxy-1-butanal is very unstable and easily undergoes a time-dependent tautomerization to its hydroxyketone form under acidic condition (*e.g.* silica gel), as observed by NMR (data not shown). GC-MS clearly indicated the complete conversion of hydroxyaldehyde to hydoxyketone observed during separation of the 2 tautomers by column chromatography.

2.3 Pyrolysis procedure

The chemicals of interest (0.2 mmol each, with 20 µL water added) were heated in a temperature controlled heating module (Brouwer) at 180°C in closed 6 mL Pyrex vacuum hydrolysis tubes (16 cm x 0.9 mm) that were immersed in silicone oil. After a defined heating period, the tubes were cooled in ice. For quantification of acrylamide, the pyrolysates were spiked with $^2H_3$-acrylamide (250 ng, 125 µL of 5 µg/mL), suspended in water (2.325 mL), vortexed (1 min), and sonicated in an ultrasonic water bath (5 min). After suspension, the extracts were centrifuged (3.5 min, 9000 rpm). SPE extraction cartridges (Isolute Multimode, 3 mL, 500 mg, IST Hengoed Mid Glamorgan, UK) were preconditioned with one-bead volume of methanol and two-bead volumes of water. A portion of the clear supernatant (1 mL) was loaded onto the SPE cartridge and water (1 mL) was added, both fraction being collected. The aqueous extract (2 mL) was then reduced to 500 µL under N_2 at 60°C and filtered (0.2 µm syringe filters) prior to analysis of an aliquot (0.06 mL) by LC-MS/MS, as described below.

Samples for on-line measurement of headspace volatiles obtained in pyrolysis experiments were analyzed by proton transfer reaction mass spectrometry (PTR-MS) as previously described (Pollien et al., 2003). The precursors (each 0.35 mol) were ground, mixed, and heated from room temperature to 190°C at a 5°C/min heating rate. Acrylamide (*m/z* 72) was monitored in the scan mode (*m/z* 21-200, 0.2 s/mass).

2.4   Quantification of acrylamide

This was performed by liquid chromatography tandem mass spectrometry (LC-MS/MS) as recently described (Stadler et al., 2004; Riediker and Stadler, 2003) using some modifications as follows. Analytical separation was achieved by using a Shodex RSpack DE-413L HPLC column (polymethacrylate gel, 250 x 6 mm i.d., Showa Denko K.K., Japan). The elution mode was a gradient, beginning with 0.01% aqueous formic acid/methanol 90:10 (v/v) and ramping linearly to 0.01% aqueous formic acid/methanol 60:40 (v/v) within 12 min. The initial flow rate was set at 0.6 mL/min, and reduced by post-column splitting after the LC column to 0.3 mL/min.

The different fragment ion transitions at t_R = 7.5 min were *m/z* 72→55, *m/z* 72→54, and *m/z* 72→27 for acrylamide as well as *m/z* 75→58 and 75→29 for the internal standard. The MS settings were 3.2 kV capillary voltage and for acrylamide 22 V for the cone voltage. The collision energy was set at –20 eV for all acrylamide transition reactions, except the fragmentation transitions *m/z* 72→55 and 72→58 that were set at -11 eV. External calibration curves (6 points) were established in the concentration range from 15 pg/µL to 20000 pg/µL acrylamide, containing a defined amount of the internal standard (250 pg/µL). Good linearity was obtained for all standard curves.

3. RESULTS AND DISCUSSION

3.1 The Strecker aldehyde route

To clarify the role of the Strecker aldehyde route compared to that employing glycoconjugates, the formation of AA from the Strecker alcohol of asparagine (3-hydroxypropanamide) was studied under pyrolytic conditions. This Strecker alcohol is the direct precursor of AA as it can be formed by a one-step dehydration process. As shown in Figure 1, the lower amounts of AA were generated from the Strecker alcohol compared to the binary mixture of glucose/asparagine, in particular at milder temperatures.

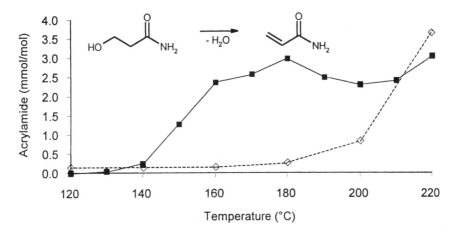

Figure 1. Formation of acrylamide from equimolar fructose/asparagine mixtures (■) and the Strecker alcohol 3-hydroxypropanamide (◇) upon pyrolysis for 5 min. (Adapted from Stadler et al., 2004).

Considering the fact that the Strecker aldehyde first needs to be generated through a cascade of reactions and then reduced to the alcohol, it seems unlikely that the Strecker aldehyde route plays a major role in AA formation. Furthermore, it is questionable if the reducing potential of Maillard systems (Ledl and Schleicher, 1990) is sufficiently strong to reduce the Strecker aldehyde to the corresponding alcohol.

All attempts thus far failed to identify substantial amounts of the Strecker aldehyde 3-oxopropanamide.. As shown in Figure 2, even on-line measuring tools based on PTR-MS indicated only traces of a compound with m/z 88 ($C_3H_5NO_2$, protonated). Possible fragments pointing to the Strecker aldehyde could neither be found, *i.e.* $[M+1-NH_3]^+$ (m/z 71) and $[M+1-H_2O]^+$ (m/z 70). However, the signal representing acrylamide (m/z 72) formed under the same reaction conditions could easily be monitored.

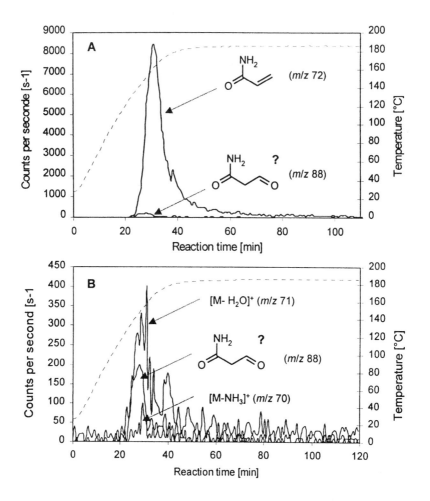

Figure 2. (A) Formation of acrylamide (*m/z* 72) and putative 3-oxopropanamide (*m/z* 88) from binary mixtures of fructose and asparagine monohydrate monitored by PTR-MS. (B) Traces of potential fragments of putative 3-oxopropanamide.

These data suggest that acrylamide is preferably generated compared to the Strecker aldehyde. Interestingly, no published data could be found on 3-oxopropanamide as Strecker aldehyde of asparagine using SciFinder®. In general, it has only been mentioned in a few papers, mainly dealing with computational chemistry related to intramolecular N-H···O resonance-assisted hydrogen bonding in β-enaminones (Gilli et al., 2000; Rios and Rodriguez, 1991). In fact, 3-oxopropanamide may preferably occur as 3-amino-3-hydroxy-2-propenal stabilized by intramolecular N-H···O bonding.

In agreement with this suggestion, only relatively low amounts of AA were formed in the presence of α-dicarbonyls, which are known to promote the formation of Strecker aldehydes from amino acids (Schönberg and Moubacher, 1952; Yaylayan, 2003). As shown in Table 1, the typical amounts of AA generated from asparagine in the presence of α-dicarbonyls were 0.2-0.5 mmol/mol (samples A-C). However, the α-dicarbonyl moiety is not a prerequisite for generating AA, as shown for glyoxylic acid (sample D) and 1-butanal (sample E). Surprisingly, hydroxyacetone gave rise to ca. 4 mmol/mol AA (sample G), ~10 times higher than in the samples containing α-dicarbonyls or the Strecker alcohol (sample F). However, it should be noted that the reproducibility of data obtained by pyrolysis is limited. As an example, values of 0.2-0.7 mmol/mol were obtained with methylglyoxal in the course of this study. The data are therefore only indicative, but show the tendency in reactivity.

Table 1. Acrylamide generated from asparagine in the presence of carbonyls upon pyrolysis (180 °C, 5 min, 20 μL water).

Sample[a]	Carbonyl reactant	Acrylamide[b]	CV (%)[c]
A	Butane-2,3-dione (diacetyl)	0.26	6.1
B	2-Oxopropanal (methylglyoxal)	0.52	5.7
C	Glyoxal	0.38	20.2
D	Glyoxylic acid	0.08	3.2
E	1-Butanal	0.01	2.9
F	3-Hydroxypropanamide	0.24[d]	<20
G	Hydroxyacetone (acetol)	3.97	23.4
H	2-Hydroxy-1-butanal[e]	15.8	13.3
I	Glucose	2.22	11.8

[a] Sample heated at 180°C for 5 min. [b] Concentration in mmol/mol asparagine. [c] Coefficient of variation. [d] Taken from Stadler et al., 2004. [e] Mixture of both tautomers (see text).

The structural features of the carbonyl reactants were further studied, in particular the role of the functional group in α-position to the carbonyl group, by reacting asparagine with various functionalized carbonyls. 2-Hydroxy-1-propanal was commercially not available to compare it with hydroxyacetone. The higher homologue, 2-hydroxy-1-butanal, was synthesized and turned out to be very unstable, easily tautomerizing to the corresponding hydroxyketone as shown in Figure 3.

Figure 3. Tautomeriztion of 2-hydroxy-1-aldehydes to 1-hydroxy-2-ketones.

Thus, the sample synthesized was composed of both tautomers, with the equilibrium favoring the hydroxyketone. Consequently, a direct comparison with acetol was not possible. 2-Hydroxy-1-butanal (Table 1, sample H) was much more efficient in forming AA (15.8 mmol/mol) than was glucose (sample I, 2.2 mmol/mol) or any of the dicarbonyls. These obsrvations demonstrate the importance of the α-hydroxyl group of the carbonyl for acrylamide formation.

3.2 The glycoconjugate route

The type of early Maillard intermediate was studied to identify the key mechanistic step transforming asparagine into AA. In dry model systems, N-glucosyl asparagine was found to be a more efficient precursor (2.4 mmol/mol) than the binary mixture of glucose and asparagine (0.7 mmol/mol). Moreover, the corresponding Amadori compound N-(1-deoxy-D-fructos-1-yl)-L-asparagine (DFA) formed only traces of AA (0.1 mmol/mol), suggesting the key precursor to be a Maillard intermediate prior to the Amadori rearrangement. As shown in Figure 4, AA can be formed from N-glucosyl asparagine under mild reaction conditions.

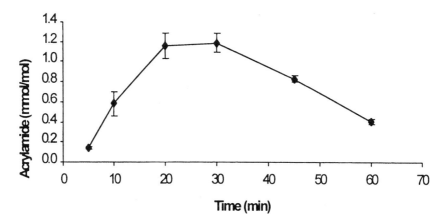

Figure 4. Formation of acrylamide from potassium N-(D-glucos-1-yl)-L-asparaginate over time at 100°C. (Adapted from Stadler et al., 2004).

The Schiff base, which is relatively stable under low moisture conditions (Zyzak et al., 2003), and the decarboxylated Amadori compound (Yaylayan et al., 2003) have been suggested as a key intermediates . We have prepared model decarboxylated Amadori compounds, *i.e.* N-(1-deoxy-D-fructos-1-yl)-

Mechanisms of Acrylamide Formation

benzylamine (DFB) and N-(1-deoxy-D-fructos-1-yl)-2-phenylethylamine (DFP) (Stadler et al., 2004), to study the validity of the β-elimination step suggested by Yaylayan et al. (2003). Only the latter Maillard intermediate shows a β-proton susceptible to a Hofmann-type β-elimination reaction. Indeed, GC-MS analysis of the reaction products indicated that styrene was only generated from DFP, while its lower homologue DFB (no β-proton) resulted in the Strecker degradation product benzaldehyde (Figure 5). This observation supports the hypothesis that AA formation may proceed via β-elimination of early Maillard intermediates based on asparagine. Consequently, the Strecker aldehyde route is likely to be a minor source of acrylamide compared to asparagine glycoconjugates.

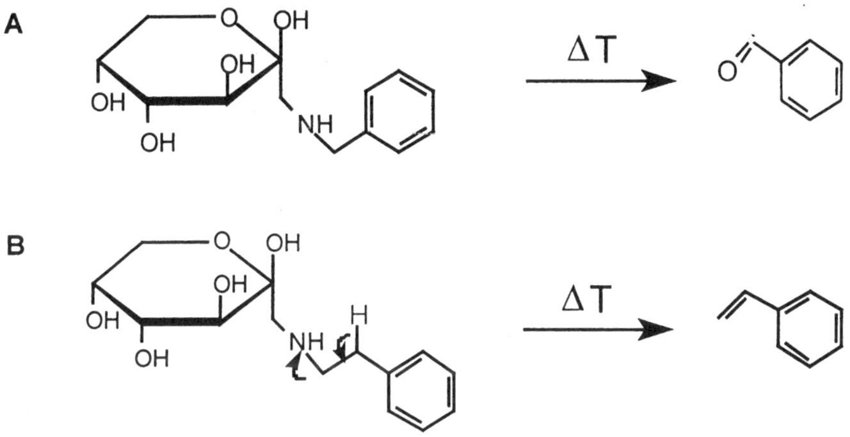

Figure 5. Formation of (A) benzaldehyde and (B) styrene from model Amadori compounds. (Adapted from Stadler et al., 2004).

3.3 Mechanism of acrylamide formation from asparagine

The experimental data shown above confirm the role of early Maillard intermediates as suggested earlier, *i.e.* N-glycosyl asparagine (Stadler et al., 2002) and the corresponding decarboxylated Schiff base (Zyzak et al., 2003) and decarboxylated Amadori compound (Yaylayan et al., 2003). The role of the β-elimination step was substantiated to release AA from the Maillard intermediate. Furthermore, some critical features of the carbonyl reactant were found, *i.e.* α-hydroxycarbonyls (*e.g.* acetol) are more efficient than α-dicabonyls (*e.g.* 2,3-butandione). On the basis of these observations, the following mechanistic scheme is proposed, which is in line with the experimental data obtained (Figure 6).

The first critical step is the amino-carbonyl reaction of asparagine and a carbonyl compound, preferably an α-hydroxycarbonyl, resulting in the corresponding conjugate, which under elevated temperatures dehydrates forming the Schiff base. Under low moisture conditions, both the N-glycosyl compound and Schiff base are relatively stable. However, in aqueous systems, the Schiff base may hydrolyze to the precursors or rearrange to the Amadori compound (pathway I), not an efficient precursor in AA formation. Even under low moisture conditions, this reaction is the main pathway initiating the early Maillard reaction cascade that leads to 1- and 3-deoxyosones. The deoxosones then further decompose to generate color and flavor (Ledl and Schleicher, 1990). This is in agreement with the relatively low transformation yield of asparagine to AA, typically below 1 mol%.

Alternatively, the Schiff base may decarboxylate to the intermediary azomethine ylide (pathway II), which after tautomerization leads to the decarboxylated Amadori compound (pathway III). The prerequisite for this reaction is the presence of an OH-group in β-position to the N-atom. As α-hydroxycarbonyls are proper precursors to yield such azomethine ylides, in contrast to α-dicarbonyls, reactants such as 1-hydroxy-2-ketones (*e.g.* fructose, acetol) and 2-hydroxyaldehydes (glucose, 2-hydroxy-1-butanal) generate more AA than do α-dicarbonyls (2,3-butanedione, methylglyoxal).

Decarboxylation of the Schiff base to the azomethine ylide may proceed *via* the zwitterionic form (pathway IIa), claimed as more probable (Grigg et al., 1988, Grigg and Thianpatanagul, 1984) compared to the classical Strecker degradation mechanism (Schönberg and Moubacher, 1952). Alternatively, the Schiff base may undergo an intramolecular cyclization to the oxazolidine-5-one derivative (pathway IIb). Such compounds have been reported to easily decarboxylate (Manini et al., 2001), thus giving rise to stable azomethine ylides, which after tautomerization lead to the decarboxylated Amadori compound (pathway III).

Acrylamide is then released, along with the corresponding aminoketone, via a β-elimination reaction and cleavage of the carbon-nitrogen covalent bond. Our model decarboxylated Amadori compounds confirmed the β-elimination step. This mechanistic pathway is supported by the fact that co-pyrolysis of a reducing sugar with aspartic acid, glutamine, and phenylalanine also leads to the corresponding vinylogous compounds, *i.e.* acrylic acid (Stadler et al., 2003), 3-butenamide (Weisshaar and Gutsche, 2002), and styrene (Keyhani and Yaylayan, 1996), respectively.

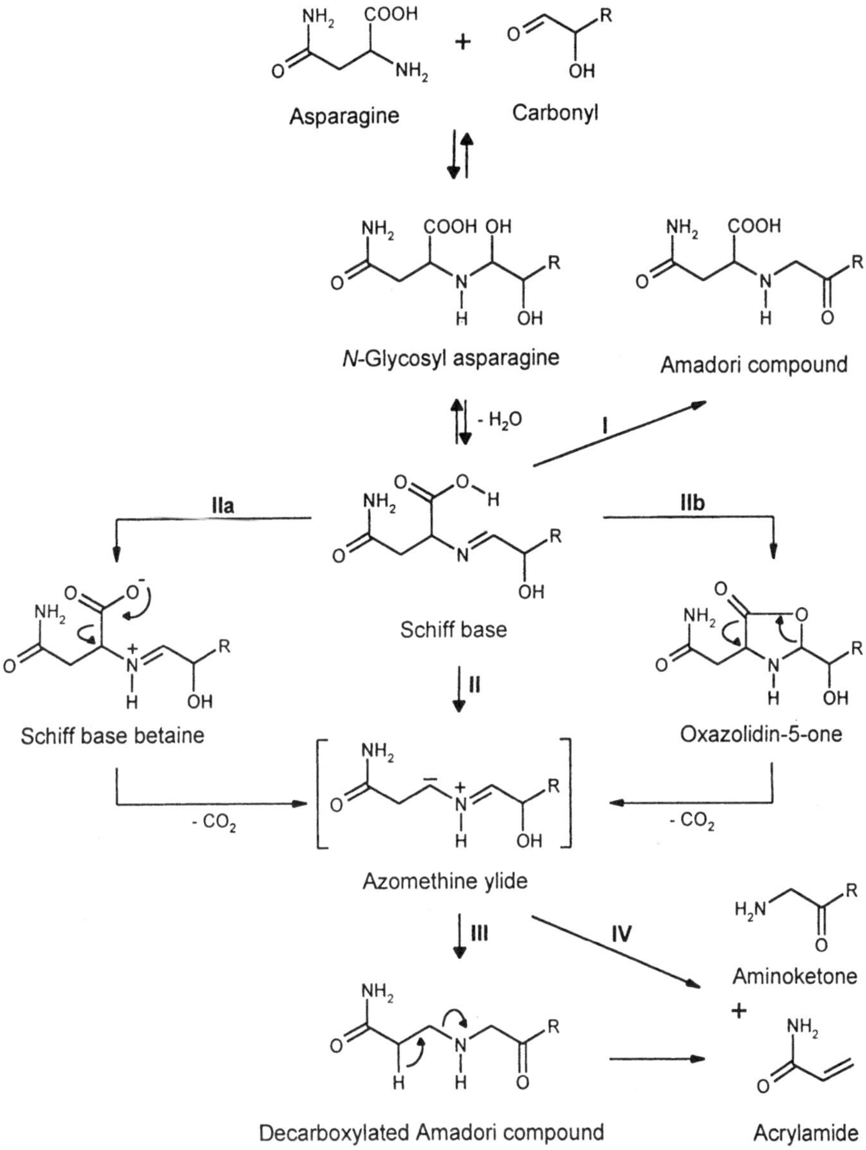

Figure 6. Formation of acrylamide from asparagine in the presence of α-hydroxycarbonyls. R represents the rest of the carbonyl moiety. (Adapted from Stadler et al., 2004).

Zyzak and coworkers (2003) have reported some evidence for the decarboxylated Schiff base of asparagine in model systems containing an excess of reducing sugar by MS(ESI+) measurement of *m/z* 251, which however may also represent the decarboxylated Amadori compound. They suggested AA to be formed directly from the azomethine ylide (pathway IV) and claimed that the decarboxylation of the Schiff base to be the limiting step. In the oxazolidine-5-one pathway (Yaylayan et al., 2003), it is not the decarboxylation step that is thought to be the limiting step, but rather the cleavage of the strong carbon-nitrogen covalent bond in the β-elimination reaction.

The formation of AA in Maillard reaction systems containing reducing sugars, *i.e.* polyhydroxy carbonyls, may alternatively proceed as shown in Figure 7. This scheme employs the classical Strecker degradation mechanism leading to a concomitant release of CO_2 and water. H-transfer in the intermediary imine gives rise to a secondary amine, which undergoes the β-elimination reaction to afford AA. The driving force for this reaction is the carbonyl group of the sugar moiety in γ-position to the nitrogen atom. However, none of the intermediates were thus far isolated to substantiate this hypothesis.

Results obtained with binary mixtures composed of carbonyls and asparagine (Table 1) clearly indicate that the α-dicarbonyl function is not a prerequisite for AA formation. Also Becalski et al. (2003) found low AA levels in dry systems containing diacetyl (0.2 mmol/mol). Mottram et al. (2002) reported 0.5 mmol/mol. However this concentration was almost 10-times lower compared to the glucose system. On the other hand, we found α-hydroxycarbonyls to be very efficient precursors for AA formation.

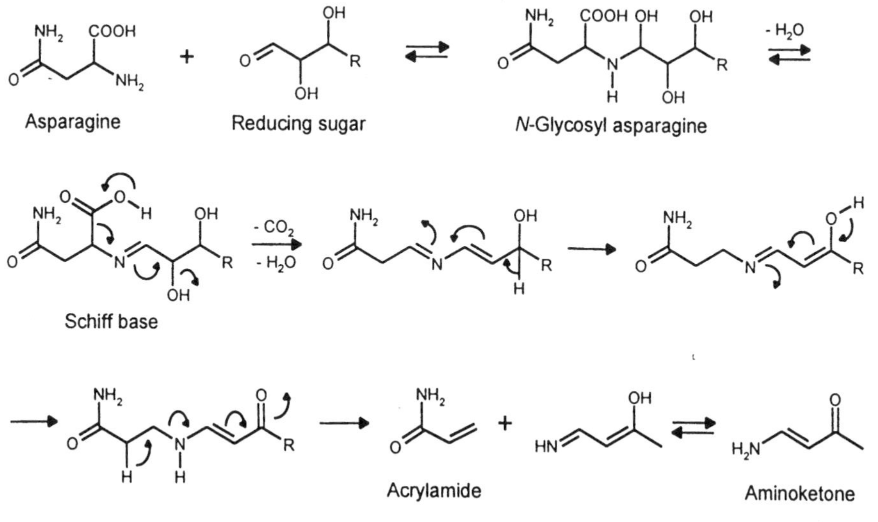

Figure 7. Hypothesis of the formation of acrylamide from asparagine in the presence of reducing sugars.

This observation can be explained by the type of azomethine ylide preferably formed upon decarboxylation of the Schiff base. As shown in

Figure 8, the Schiff base leads after decarboxylation to the azomethine ylide with a 1,3-dipole structure shown in two resonance-stabilized forms. In general, the final location of the proton in the neutral imines depends on the kinetically controlled proton transfer to the site of the dipole with the greatest electron density (Grigg and Thianpatanagul, 1984). However, as the negative charge in **1b** can delocalize at the carbonyl group in β-position to the nitrogen atom, the 1,2-prototropic H-shift may preferably lead to imine **2b**, which upon hydrolysis furnishes the Strecker aldehyde.

Alternatively, the azomethine ylide **1a** may react to imine **2a**, which hydrolyzes to the decarboxylated amino acid, *i.e.* 3-aminopropionamide from asparagine. This compound has been reported to release acrylamide (Zyzak et al., 2003, Granvogl et al., 2004). Thus, the relatively low amounts of acrylamide reported in Table 1 (samples A-C) is most likely due to preferred formation of imine **2b**, resulting in a Strecker aldehyde that does not release high amounts of AA.

In the presence of α-hydroxycarbonyls, imines **4a** and **4b** can be formed (Figure 9). Due to the higher electron density in the azomethine ylide **3b** (hydroxy group in β-position to the N-atom), the 1,2-prototropic H-shift preferably leads to imine **4b**, which upon hydrolysis furnishes the Strecker aldehyde. However, imine **4a** formed in a side reaction *via* the azomethine ylide **3a** has the required structural features to rearrange to the decarboxylated Amadori compound, which can then undergo the β-elimination reaction, directly affording the vinylogous compound, *i.e.* acrylamide from asparagine. This reaction is less probable with imine **2a** obtained in the presence of α-dicarbonyls (Figure 8).

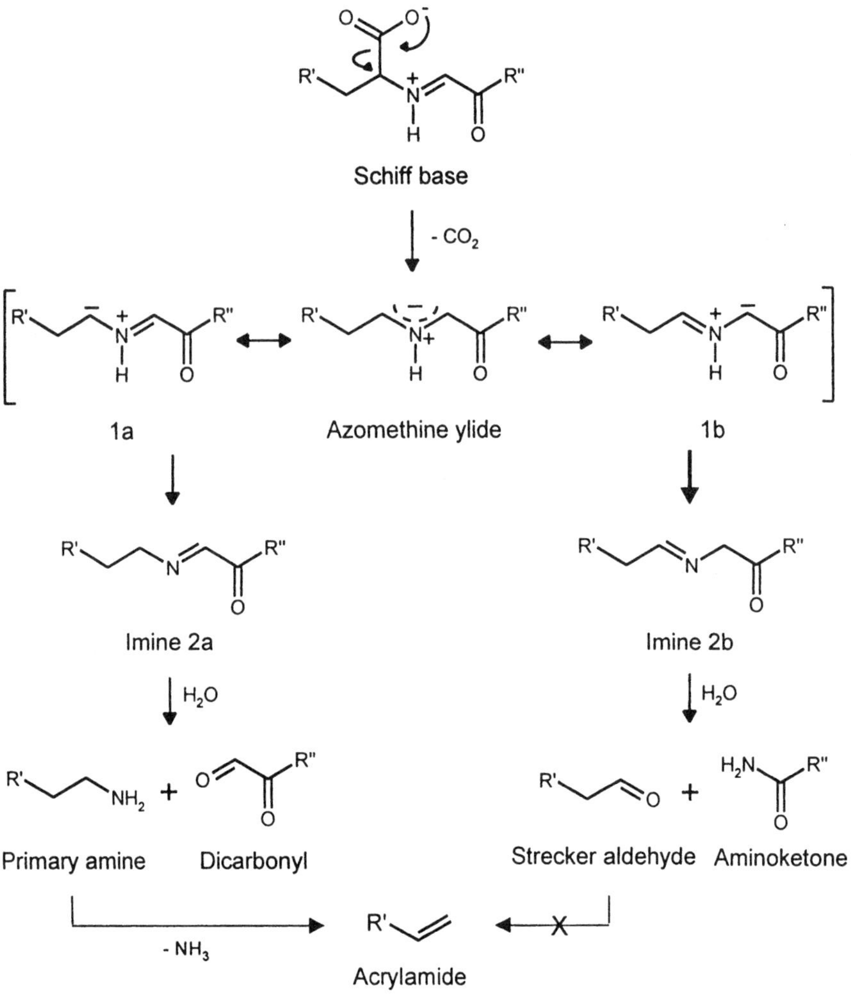

Figure 8. Formation of acrylamide (R' = $CONH_2$), 3-aminopropanamide (primary amine), and the 3-oxopropanamide (Strecker aldehyde) from asparagine (R' = $CONH_2$) in the presence of an α-dicarbonyl.

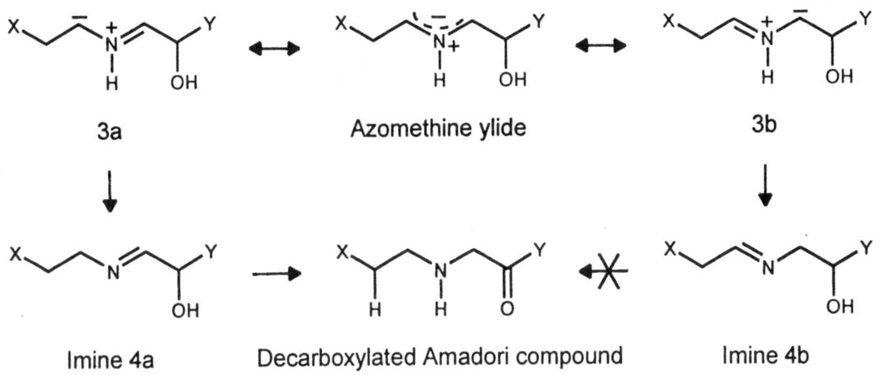

Figure 9. Formation of the azomethine ylide shown as 1,3-dipole and resonance-stabilized structures **3a** and **3b** leading *via* 1,2-prototropy to the neutral imines **4a** and **4b**. (Adapted from Stadler et al., 2004).

Thus, the relatively low transformation yield (< 1 mol%) of asparagine to AA might be explained as described below:
- The major Maillard reaction flux follows 1,2- and 2,3-enolization pathways (Amadori pathway I in Figure 6) forming deoxyosones;
- Imines **2b** and **4b** are preferably formed, leading to Strecker aldehydes with low efficiency for acrylamide release;
- Imines **2a** and **4a** are minor reaction products, of which only **4a** shows proper structural features needed to rearrange to the decarboxylated Amadori compound, which can then release AA;
- Reaction conditions such as temperature, moisture, pH, and time may markedly influence the various steps of the reaction cascade until AA is released.

3.4 Effect of reaction parameters on the formation of acrylamide from asparagine

It is known that AA formation is favored under low moisture conditions. The water content may affect both the chemical reaction and the physical state of the reaction system, which is linked to reaction temperature and heat transfer. However, the role of water in AA formation has not been studied in great detail. As shown in Figure 10, the amounts of AA do indeed depend on the water content. In the reaction system based on fructose, the AA increases with increasing water content. They accumulate in the 200 µL water sample and decline at higher water content. The glucose system showed a broad maximum in the 20-200 µL water samples. The physical state of the reaction systems changes from solid to suspension (50-200 µL water) with increasing solubility to become a solution (500 µL water).

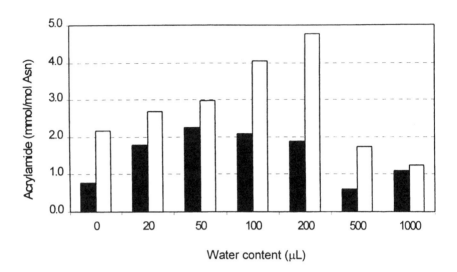

Figure 10. Formation of acrylamide in binary mixtures of asparagine and glucose (black bars) or fructose (white bars) as function of moisture (180 °C, 5 min).

Low pH has been reported as a means to control AA amounts in certain food products (Jung et al., 2003). This idea refers to the fact that the initial amino-carbonyl reaction is hampered due to protonation of the amino group at low pH. Our data indicate only in a relatively weak influence of the pH on AA formation from asparagine (Figure 11). However, the fructose reaction series formed significantly less AA at pH 3 (2.2 mmol/mol), while pH 8 represented optimum reaction conditions forming more AA (3.3 mmol/mol). Also Rydberg et al. (2003) reported optimal AA formation at around pH 8.

Interestingly, all samples containing fructose (Figure 10) gave rise to higher amounts of AA compared to glucose. Even in the sample with no water, fructose was more efficient, despite the fact that glucose, as an aldohexose, should be chemically more reactive, since the aldehyde group is not hydrated. Therefore, other parameters than chemical reactivity may play a role to explain this phenomenon, which is subject of ongoing studies dealing with the physical state and its effect on AA formation (Robert et al., 2004).

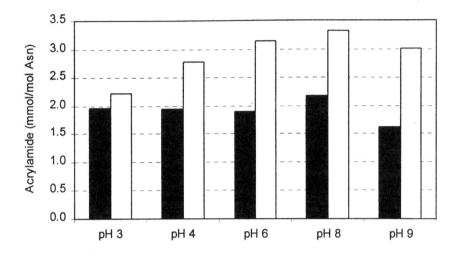

Figure 11. Formation of acrylamide in binary mixtures of asparagine and glucose (black bars) or fructose (white bars) as function of pH (180°C, 5 min, 20 µL water).

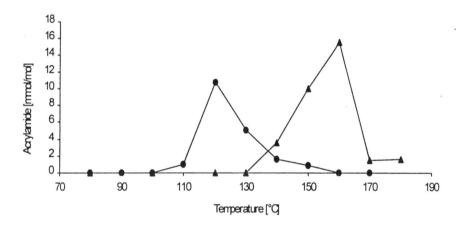

Figure 12. Formation of acrylamide by heating a binary equimolar mixture of asparagine and glucose for 5 (▲) and 60 (●) minutes.

In another series of experiments, the role of reaction time and temperature (heat load) was investigated. As shown in Figure 12, the interplay of these parameters strongly influenced AA formation. Pyrolysis of glucose and asparagine at different temperatures for either 5 min or 60 min resulted in different curves, indicating high amounts of AA formed already at 120°C at long reaction times, whereas 160°C was required to obtain highest amounts of AA at short pyrolysis times. These two parameters are covariant and represent a means for controlling AA formation under food processing conditions. The decline of the curves is most likely due to polymerization as recently shown (Stadler et al., 2004).

3.5 Conclusions

Our studies have confirmed that glycoconjugates of asparagine are the major source of AA in foods under low moisture conditions at elevated temperatures in the presence of reducing sugars or a suitable carbonyl source. Moisture content and heat load (reaction time *versus* temperature) may be suitable parameters to minimize AA formation. Occurrence and chemical reactivity of the precursors are also important factors to consider. However, the physical state of the reaction system may be equally essential to achieve reduced AA amounts, while keeping desirable product attributes such as flavor and color generated by similar Maillard reaction pathways. This aspect is subject of further studies and will be published elsewhere.

REFERENCES

Becalski, A.; Lau, B. P.-Y.; Lewis, D.; Seaman, S., 2003, Acrylamide in food: Occurrence, sources, and modeling, *J. Agric. Food Chem.* **51**: 802-808.

Biedermann, M.; Noti, A.; Biederman-Brem, S.; Mozzetti, V.; Grob, K., 2002, Experiments on acrylamide formation and possibilities to decrease the potential of acrylamide formation in potatoes, *Mitt. Lebensm. Hyg.* **93**: 668-687.

Friedman, M., 2003, Chemistry, biochemistry, and safety of acrylamide. A review, *J. Agric. Food Chem.* **51**: 4504-4526.

Gertz, C.; Klostermann, S., 2002, Analysis of acrylamide and mechanisms of its formation in deep-fried products, *Eur. J. Lipid Sci. Technol.* **104**: 762-771.

Gilli, P.; Bertolasi, V.; Ferretti, V.; Gilli, G., 2000, Evidence for intramolecular N-H···O resonance-assisted hydrogen bonding in enaminones and related heterodienes. A combined crystal-structure, IR, NMR spectroscopic, and quantum-mechanical investigation, *J. Am. Chem. Soc.* **122**: 10405-10417.

Granvogl, M.; Koehler, P.; Schieberle, P., 2004, An alternative pathway in the generation of acrylamide from asparagine, *J. Agric. Food Chem.* **52**: in press.

Grigg, R.; Thianpatanagul, S., 1984, Decarboxylative transamination. Mechanism and application to the synthesis of heterocyclic compounds, *J. Chem. Soc.,Chem. Comm*, 180-181.

Jung, M. Y.; Choi, D. S.; Ju, J. W., 2003, A novel technique for limitation of acrylamide formation in fried and baked corn chips and in French fries, *J. Food Sci.* **68**: 1287-1290.

Keyhani, A.; Yaylayan, V. A., 1996, Pyrolysis/GC/MS analysis of *N*-(1-deoxy-D-fructos-1-yl)-L-phenylalanine: Identification of novel pyridine and naphthalene derivatives, *J. Agric. Food Chem.* **44**: 223-229.

Ledl F.; Schleicher E., 1990, New aspects of the Maillard reaction in foods and in the human body, *Angew. Chem. Int. Ed. Engl.* **29**: 565-594.

Manini, P.; d'Ischia, M.; Prota, G., 2001, An unusual decarboxylative Maillard reaction between L-DOPA and D-glucose under biomimetic conditions: Factors governing competition with Pictet-Spengler condensation, *J. Org. Chem.* **66**: 5048-5053.

Mottram, D. S.; Wedzicha, B. L.; Dodson, A. T., 2002, Food chemistry: Acrylamide is formed in the Maillard reaction, *Nature* **419**: 448-449.

Riediker, S.; Stadler, R. H., 2003, Analysis of acrylamide in food using isotope-dilution liquid chromatography coupled with electrospray ionisation tandem mass spectrometry, *J. Chromatogr.* **1020**: 121-130.

Rios, M. A.; Rodriguez, J., 1991, Analysis of the effect of substitution on the intramolecular hydrogen bond of malonaldehyde by ab initio calculation at the 3-21G level, *Theochem.* **74**: 149-158.

Robert, F.; Vuataz, G.; Pollien, P.; Saucy, F.; Alonso, M.-I.; Bauwens, I., Blank, I., 2004, Acrylamide formation from asparagine under low-moisture Maillard reaction conditions. 1. Physical and chemical aspects in crystalline model systems, *J. Agric. Food Chem.* **52**: submitted.

Rydberg, P.; Eriksson, S.; Tareke, E.; Karlsson, P.; Ehrenberg, L.; Törnqvist, M, 2003, Investigations of factors that influence the acrylamide content of heated foodstuffs, *J. Agric. Food Chem.* **51**: 7012-7018.

Schönberg, A.; Moubacher, R., 1952, The Strecker degradation of α-amino acids, *Chem. Rev.* **50**: 261-277

Stadler, R. H.; Robert, F.; Riediker, S.; Varga, N.; Davidek, T.; Devaud, S.; Goldmann, T.; Blank, I., 2004, In-depth mechanistic study on the formation of acrylamide and other vinylogous compounds by the Maillard reaction, *J. Agric. Food Chem.* **52**: 5550-5558.

Stadler, R. H.; Verzegnassi, L.; Varga, N.; Grigorov, M.; Studer, A.; Riediker, S.; Schilter, B., 2003, Formation of vinylogous compounds in model Maillard reaction systems, *Chem. Res. Tox.* **16**: 1242-1250.

Stadler, R. H.; Blank, I.; Varga, N.; Robert, F.; Hau, J.; Guy, Ph. A.; Robert, M.-C.; Riediker, S., 2002, Food chemistry: Acrylamide from Maillard reaction products,, *Nature* **419**: 449-450.

Weisshaar R.; Gutsche B., 2002, Formation of acrylamide in heated potato products - model experiments pointing to asparagine as precursor, *Deutsche Lebensm. Rundsch.*. **98**: 397-400.

Yasuhara, A.; Tanaka, Y.; Hengel, M.; Shibamoto, T., 2003, Gas chromatographic investigation of acrylamide formation in browning model systems, *J. Agric. Food Chem.* **51**: 3999-4003.

Yaylayan, V. A., 2003, Recent advances in the chemistry of Strecker degradation and Amadori rearrangement: Implications to aroma and color formation, *Food Sci. Technol. Res.* **9**: 1-6.

Yaylayan V. A.; Wnorowski, A., Perez Locas C., 2003, Why asparagine needs carbohydrates to generate acrylamide, *J. Agric. Food Chem.* **51**: 1753-1757.

Zyzak, D. V.; Sanders, R. A.; Stojanovic, M.; Tallmadge, D. H; Eberhart, B. L.; Ewald, D. K.; Gruber, D. C.; Morsch, T. R.; Strothers, M. A.; Rizzi, G. P.; Villagran, M. D, 2003, Acrylamide formation mechanism in heated foods, *J. Agric. Food Chem.* **51**: 4782-4787.

MECHANISTIC PATHWAYS OF FORMATION OF ACRYLAMIDE FROM DIFFERENT AMINO ACIDS

Varoujan A. Yaylayan[1], Carolina Perez Locas[1], Andrzej Wnorowski[1] and John O'Brien[2]
[1]*McGill University, Department of Food Science and Agricultural Chemistry, 21, 111 Lakeshore, Ste. Anne de Bellevue, Quebec, Canada, H9X 3V9;* [2]*Danone Group, 17 Rue du Heldor, 75439, Paris, France; e-mail: varoujan.yaylayan@mcgill.ca*

Abstract: Studies on model systems of amino acids and sugars have indicated that acrylamide can be generated from asparagine or from amino acids that can produce acrylic acid either directly such as β-alanine, aspartic acid and carnosine or indirectly such as cysteine and serine. The main pathway specifically involves asparagine and produces acrylamide directly after a sugar-assisted decarboxylation and 1,2-elimination steps and the second non-specific pathway involves the initial formation of acrylic acid from different sources and its subsequent interaction with ammonia to produce acrylamide. Aspartic acid, β-alanine and carnosine were found to follow acrylic acid pathway. Labeling studies with [^{13}C-4]aspartic acid have confirmed the occurrence in aspartic acid model system, of a previously proposed sugar-assisted decarboxylation mechanism identified in asparagine model systems. In addition, creatine was found to be a good source of methylamine and was responsible for the formation of N-methylacrylamide in model systems through acrylic acid pathway. Furthermore, certain amino acids such as serine and cysteine were found to generate pyruvic acid that can be converted into acrylic acid and generate acrylamide when reacted with ammonia.

Key words: asparagine, acrylic acid, pyruvic acid, aspartic acid, creatine, carnosine, acrylamide, N-methylacrylamide, mechanisms of acrylamide formation.

1. INTRODUCTION

Preliminary studies (Stadler et al., 2002; Mottram et al., 2002) that followed the initial discovery of acrylamide in cooked food have lead not only to the unambiguous identification of asparagine as the main amino acid precursor of acrylamide, but also confirmed the origin of its carbon atoms and the amide nitrogen through labeling studies. Although thermal decarboxylation and deamination reactions (Yaylayan et al., 2003) of asparagine alone, in principle, can produce acrylamide, the presence of sugars was necessary to effect the conversion of asparagine into acrylamide. Subsequent studies (Becalski et al., 2003; Zyzak et al, 2003) have indicated that any carbonyl containing moiety can perform a similar transformation and that asparagine alone prefers to undergo intramolecular cyclization and form an imide rather than decarboxylate to form acrylamide. Studies related to the detailed mechanism (Yaylayan et al. 2003) of this transformation in model systems have indicated that decarboxylated Amadori product of asparagine is the key precursor of acrylamide. Furthermore, the decarboxylated Amadori product was shown to be formed under relatively mild conditions through the intramolecular cyclization of the initial Schiff base and formation of oxazolidin-5-one intermediate (Manini et al., 2001) and subsequent generation of a stable azomethine ylide which is prone to undergo an irreversible 1,2-prototropic shift (Grigg 1989) to produce decarboxylated Schiff base and eventually decarboxylated Amadori product. Similar conclusions, using model food systems, were drawn by Zyzak et al. (2003) depicting direct decarboxylation of the Schiff base, but without invoking, oxazolidin-5-one as an intermediate. As part of our investigation of other sources of acrylamide in food and using Py-GC/MS as in integrated reaction, separation and identification system (Yaylayan, 1999) we have studied, in addition to selected α-amino acids, β-alanine and the dipeptide carnosine (N-β-alanyl-L-histidine) as potential sources of acrylamide.

2. MATERIALS AND METHODS

All reagents, chemicals and $^{15}NH_4Cl$ were purchased from Aldrich Chemical company (Milwaukee, WI) and used without further purification. The labeled [^{13}C-4]aspartic acid was purchased from Cambridge Isotope Laboratories (Andover, MA).

2.1 Pyrolysis-GC/MS analysis

A Hewlett-Packard GC with Mass selective detector (5890 GC/5971B MSD) interfaced to a CDS pyroprobe 2000 unit was used for the Py-GC/MS analysis. One mg samples of pure reactants was introduced inside a quartz tube (0.3mm thickness), plugged with quartz wool, and inserted inside the coil probe with a total heating time of 20s. The column was a fused silica

DB-5 column (50m length x 0.2mm i.d. x 0.33 μm film thickness; J&W Scientific). The pyroprobe interface temperature was set at 250°C. Capillary direct MS interface temperature was 280°C; ion source temperature was 180°C. The ionization voltage was 70 eV, and the electron multiplier was 2471 V. All samples were injected in splitless mode. Three methods of analysis were used.

- **Method 1** had a delayed pulse of 65 psi followed by constant flow of 0.775mL/min, and a septum purge of 2 mL/min. The initial temperature of the column was set at 40°C for 2 minutes and was increased to 100°C at a rate of 30°C/min, immediately the temperature was further increased to 250°C at a rate of 8°C/min and kept at 250°C for 5 min.
- **Method 2** had a delayed pulse of 65 psi followed by constant flow of 1 mL/min, and a septum purge of 2 mL/min. The initial temperature of the column was set at -5°C for 2 minutes and was increased to 50°C at a rate of 30°C/min, immediately the temperature was further increased to 250°C at a rate of 8°C/min and kept at 250°C for 5 min.
- **Method 3** had a delayed pulse of 65 psi followed by constant flow of 1mL/min, and a septum purge of 2 mL/min. The initial temperature of the column was set at -5°C for 2 minutes and was increased to 50°C at a rate of 8°C/min, immediately the temperature was further increased to 100°C at a rate of 3°C/min followed by a increase to 250°C at a rate of 20°C/min and kept at 250°C for 5 min.

The identity and purity of the chromatographic peaks were determined using NIST AMDIS version 2.1 software. The reported percent label incorporation values (corrected for natural abundance and for % enrichment) are the average of duplicate analyses and are rounded to the nearest multiple of 5%. Fig. 4, shows examples of the three methods.

3. RESULTS AND DISCUSSION

Studies with model systems containing selected amino acids and glucose (see Table 1) have indicated that there are two general pathways of acrylamide formation; a major pathway that generates acrylamide directly from asparagine and the second minor pathway that generates acrylamide though reaction of ammonia with acrylic acid (see Fig. 1). Furthermore, studies have also indicated that acrylic acid itself can be generated either directly from certain amino acids or dipeptides such as carnosine, β-alanine and aspartic acid or indirectly from amino acids such as serine and cysteine, through reduction of pyruvic acid into lactic acid and its subsequent dehydration into acrylic acid.

Table 1. Efficiency (area/mole of amino acid) of acrylic acid & acrylamide generation from 1 mg samples of either amino acid or amino acid/glucose (3:1) mixtures pyrolyzed at 350° C (data generated using method 1)

Model System	Acrylic acid x 10^{12}	Acrylamide x 10^{12}
Asparagine	not detected	trace
Asparagine/glucose	1.20 ± 0.18	5.18 ± 0.6=57
β-alanine	98.30 ± 0.35	14.4 ± 0.5
β-alanine/glucose	108.00 ± 7.10	13.9 ± 0.7
Carnosine	27.5 ± 6.9	12.86 ± 4.15
Carnosine/glucose	18.46 ± 1.08	5.11 ± 0.55
Aspartic acid	2.0 ± 0.08	0.18 ± 0.08
Aspartic acid/glucose	22.53 ± 2.95	0.53 ± 0.04
Cysteine	1.7 ± 0.1	trace
Cysteine/glucose	1.5 ± 0.1	trace
Serine	0.38 ± 0.01	not detected
Serine/glucose	0.68 ± 0.01	not detected

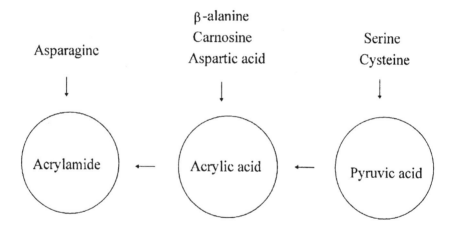

Figure 1. Precursors of acrylamide.

3.1 Direct formation of acrylamide from asparagine

Asparagine is the only amino acid capable of directly generating acrylamide. Consequently it is considered the main source of acrylamide in food. The studies related to the detailed mechanism of this transformation have indicated that sugars and other carbonyl compounds play a specific role in the decarboxylation process of asparagine – a necessary step in the generation of acrylamide. It has been proposed (Yaylayan et al., 2003) that

Schiff base intermediate formed between asparagine and the sugar provides a low energy alternative to the decarboxylation from the intact Amadori product through generation and decomposition of oxazolidin-5-one intermediate (Manini et al., 2001) leading to the formation of a relatively stable azomethine ylide (see Fig. 2).

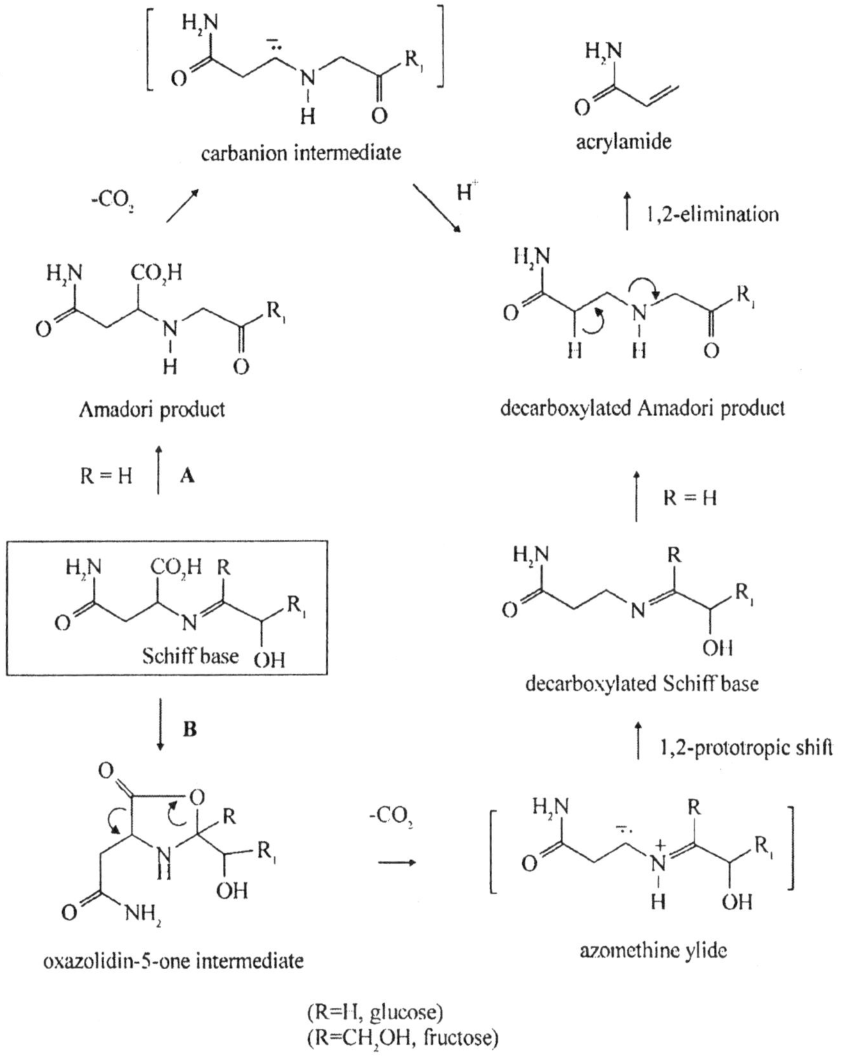

Figure 2. Mechanism of formation of acrylamide through thermally-induced decarboxylation of intact Amadori products (pathway A) and through sugar-assisted pre-Amadori decarboxylation (pathway B).

Literature data indicates the propensity of such protonated ylides to undergo irreversible 1,2-prototropic shift (Grigg et al., 1989) and produce, in this case, decarboxylated Schiff base which can easily rearrange into corresponding Amadori product. Decarboxylated Amadori products can either undergo the well known β-elimination process initiated by the sugar moiety to produce 3-aminopropanamide and 1-deoxyglucosone or undergo 1,2-elimination initiated by the amino acid moiety to directly generate acrylamide. On the other hand the decarboxylated Schiff intermediate can either hydrolyze and release 3-aminopropanamide or similarly undergo amino acid initiated 1,2-elimination to directly form acrylamide (Yaylayan and Stadler, 2004). However, their relative contribution to acrylamide formation is still under investigation.

3.2 Direct formation of acrylic acid from β-alanine, carnosine and aspartic acid

Some amino acids can generate acrylic acid directly during their thermal decomposition. Such amino acids require the presence of ammonia to convert acrylic acid into acrylamide. One of the main sources of ammonia in food is the free amino acids. Sohn and Ho (1995) have identified asparagine, glutamine, cysteine and aspartic acid as the most efficient ammonia generating amino acids under thermal treatment.

3.2.1 Formation of acrylamide from β-alanine

The mechanism of decomposition of β-alanine to generate both reactants required for the formation of acrylamide, ammonia and acrylic acid, is shown in Fig. 3. Pyrolysis of β-alanine alone generated mainly acrylic acid and acrylamide, indicating deamination as a major pathway of thermal decomposition of β-alanine. The resulting acid can then interact with the available ammonia to form acrylamide (Fig. 4b). When β-alanine was pyrolyzed in the presence of excess $^{15}NH_4Cl$ the resulting acrylamide incorporated both the labeled (added) and unlabeled (generated from β-alanine) ammonia. Similarly, pyrolysis of commercial acrylic acid in the presence of an ammonia source (NH_4Cl, $(NH_4)_2CO_3$, etc.) also generated acrylamide (Fig. 4a). Comparison of figures 4a and 4b indicates the efficiency of conversion of β-alanine into acrylic acid and ammonia. No significant change in the efficiency of β-alanine conversion into acrylamide was observed in the presence of glucose (see Table 1).

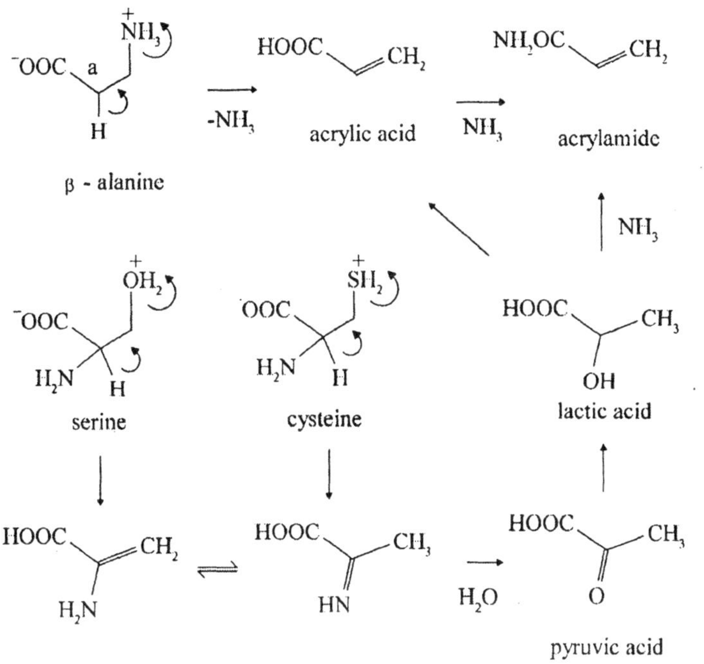

Figure 3. Proposed mechanisms of formation of acrylamide from β-alanine, serine and cysteine.

3.2.2 Formation of acrylamide from aspartic acid

Aspartic acid, on the other hand, can also form acrylic acid and subsequently acrylamide (Stadler et al., 2003; Yaylayan et al., 2004; Becalski et al., 2003) (Fig. 4c), but unlike β-alanine and similar to asparagine, it produces more acrylic acid in the presence of glucose (see Table 1). In order to identify the mechanism of acrylic acid formation from aspartic acid, [^{13}C-4]-aspartic acid was pyrolyzed alone and in the presence of glucose. According to Fig. 5, aspartic acid can undergo decarboxylation of either C-1 or C-4 carboxylate moieties. C-1 decarboxylation can generate β-alanine and C-4 decarboxylation can generate α-alanine as shown in Fig. 5. Unlike α-alanine, β-alanine is known to produce acrylic acid and consequently it was expected to observe 100% label retention in the acrylic acid mass spectrum when [^{13}C-4]-aspartic acid was pyrolyzed alone. However, analysis of the data showed the formation of 65% of labeled acrylamide and 35% unlabelled product (Fig. 6c) indicating existence of a third pathway capable of formation of acrylamide with C-4 decarboxylation.

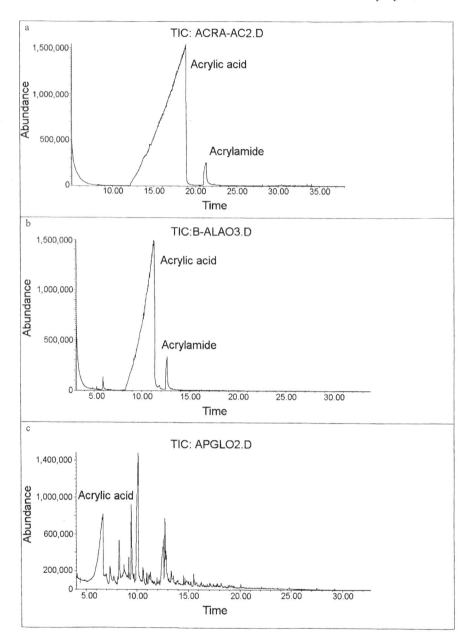

Figure 4. Pyrograms of model systems consisting of equimolar mixtures of (a) acrylic acid/ammonium carbonate (using method 3) (b) β-alanine alone (using method 2) and (c) aspartic acid/glucose (using method 1).

A concerted mechanism where decarboxylation occurs simultaneously with deamination can explain the formation of unlabelled acrylic acid as shown in Fig. 5. Interestingly, when [^{13}C-4]-aspartic acid was pyrolyzed in the presence of glucose only 100% labeled acrylic acid was observed (Fig. 6b), indicating preferential decarboxylation of C-1 carboxylate moiety consistent with the mechanism of sugar-assisted decarboxylation shown in Fig. 2. This observation, along with increased ability of aspartic acid to generate acrylamide in the presence of glucose (see Table 1), provides evidence for the ability of the Schiff base to provide a low energy pathway for decarboxylation of amino acids relative to decraboxylation from intact Amadori products that passes through a carbanion intermediate rather than the more stable azomethine ylide as shown in Fig. 2. Furthermore, similar to asparagine, reaction with sugar and formation of oxazolidine intermediate can prevent cyclization to form maleic anhydride (equivalent to succinimide in the case of asparagine) and enhance acrylic acid generation as observed.

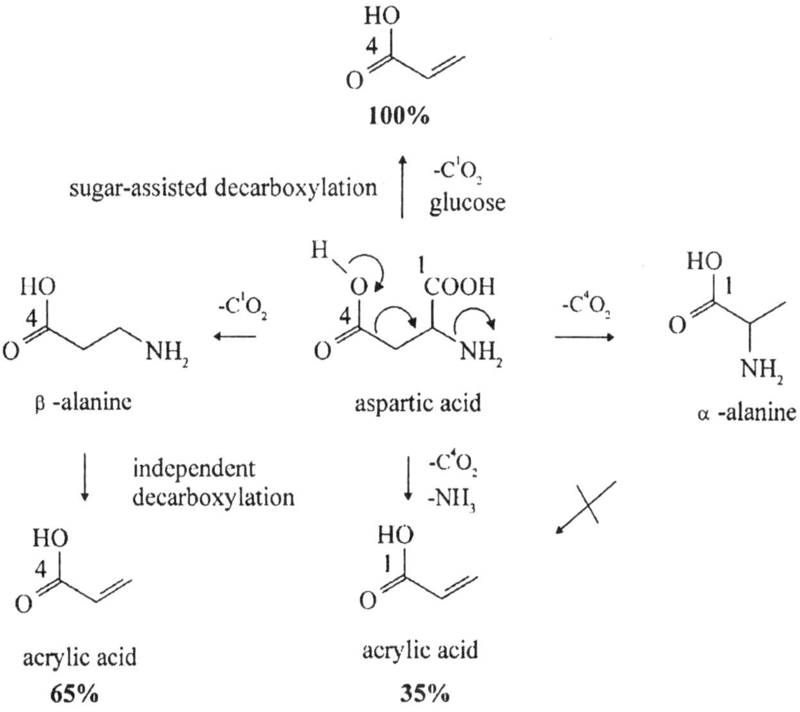

Figure 5. Decarboxylation pathways of aspartic acid based on labeling studies.

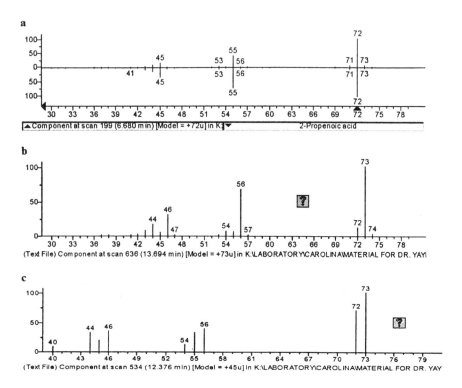

Figure 6. Mass spectrum of (a) acrylic acid generated from unlabeled aspartic acid compared with authentic NIST library spectrum in head to tail fashion. (b) acrylic acid generated from [^{13}C-4]aspartic acid/glucose mixture. (c) acrylic acid generated from [^{13}C-4]aspartic acid alone.

3.2.3 Formation of acrylamide from carnosine

The dipeptide carnosine (N-β-alanyl-L-histidine) when pyrolyzed alone produced acrylic acid and acrylamide in amounts higher than asparagine/glucose model system. However, in the presence of glucose the amounts became comparable (see Table 1) due to the interaction of carnosine with reducing sugars (Chen and Ho, 2002). Fig. 7 depicts two possible pathways of formation of acrylamide from carnosine, one through hydrolysis of the peptide bond and release of β-alanine and its subsequent deamination, the second through release of 3-aminopropanamdie and its deamination. However, the conspicuous absence of acrylamide in meat products at the scale expected to that of potatoes (Friedman, 2003) has lead us to investigate its possible fate in meat products using carnosine containing model systems. Carnosine was reacted in the presence of lysine (a reactive amino acid) and creatine (a major constituent of meat) and their effect on the amounts of acrylamide and its precursor acrylic acid was calculated. Lysine did not exert any significant effect on the formation efficiencies of acrylamide and acrylic acid. Creatine on the other hand, not only significantly reduced the acrylic acid content but also gave rise to two new potentially toxic (Hashimoto et al., 1981 & WHO, 1985) acrylamide derivatives; N-methylacrylamide and N,N-dimethylacrylamide. The decrease in acrylic acid formation can be explained by its accelerated conversion into acrylamide derivatives due to the efficient generation of ammonia and methylamines from added creatine (Yaylayan et al., 2004).

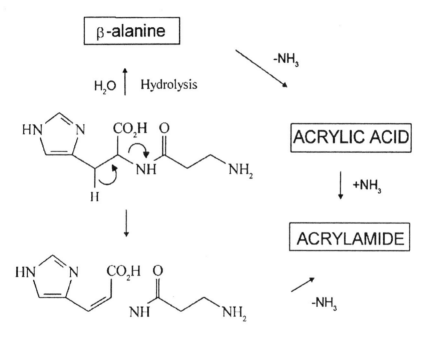

Figure 7. Proposed mechanism of acrylamide formation from carnosine.

3.3 Indirect formation of acrylic acid from serine & cysteine

Dehydration of serine alone (see Fig. 3) and in the presence of sugars has been shown to generate pyruvic acid (Wnorowski and Yaylayan, 2003). Conversion of β-alanine into acrylic acid and release of acrylamide from decarboxylated Amadori product follow a similar mechanism of 1,2-elimination (see Figs 2 & 3). Cysteine can also lose a hydrogen sulfide molecule to generate acrylic acid as shown in Fig. 3. Acrylic acid was detected along with pyruvic acid when serine was pyrolyzed at 350°C. This observation can be justified by proposing the reduction of pyruvic acid into lactic acid and its subsequent dehydration into acrylic acid. Model studies with lactic acid have indicated that such transformations are possible in the presence of ammonia; mixtures of lactic acid and ammonium salts produced lactamide, acrylic acid and acrylamide when pyrolyzed at 350°C.

4. CONCLUSION

Although, in theory, there are more than one amino acid that can generate acrylamide, however, the efficiency of the conversion of acrylic acid into acrylamide is limited by the availability of free ammonia in the vicinity of its production in the food matrix, in addition, this limitation is further compounded by the extreme volatility of ammonia at temperatures that are conducive to acrylamide formation. Recent studies (Stadler et al., 2003) have indicated that aspartic acid/fructose mixtures generated acrylamide at levels 1000-fold below the levels measured for asparagines/fructose mixtures.

ACKNOWLEDGMENTS

The authors acknowledge funding for this research by the Natural Sciences and Engineering Research Council of Canada (NSERC) and by Danone (France).

REFERENCES

Becalski, A., Lau, B. P.-Y., Lewis, D., and Seaman, S. W., 2003, Acrylamide in Foods: Occurrence, sources and modeling, *J. Agric. Food Chem,* **51**:802-808.

Chen, Y., Ho, Chi-T. 2002, Effects of Carnosine on Volatile Generation from Maillard Reaction of Ribose and Cysteine, *J. Agric. Food Chem.* **50**:2372-2376.

Friedman, M., 2003, Chemistry, biochemistry, and safety of acrylamide. A review, *J. Agric. Food Chem.* **51**:4504-4526.

Grigg, R., Malone, J. F. Mongkolaussavaratana, T., and Thianpatanagul, S., 1989, X=Y-ZH Compounds as Potential 1,3-dipoles: Part 23 mechanisms of the reactions of ninhydrin and phenalene trione with amino acids. X-ray crystal structure of protonated Ruhemnan's purple, a stable azomethine ylide, *Tetrahedron,* **45**(12):3849-3862.

Hashimoto, K., Sahamoto, J., Tanii, H., 1981, Neurotoxicity of acrylamide and related compounds and their effect on male gonads in mice, *Arch. Toxicol.* 47:179-189.

Manini, P., d'Ischia, M., and Prota, G., 2001, An unusual decarboxylative Maillard reaction between L-DOPA and D-glucose under biomimetic conditions: factors governing competition with Pictet-Spengler condensation, *J. Org. Chem.* 66:5048-5053.

Mottram, D. S., Wedzicha, B. L., and Dodson, A. T., 2002, Acrylamide is formed in the Maillard reaction. *Nature* 419:448-449.

Sohn, M., and Ho, C.-T., 1995, Ammonia generation during thermal degradation of amino acids, *J. Agric. Food Chem.* 43:3001-3003.

Stadler, R. H., Blank, I., Varga, N., Robert, F., Hau, J., Guy, P., Robert, M-C, and Riediker, S., 2002, Acrylamide from Maillard reaction products, *Nature* 419:49-450.

Stadler, R. H., Verzegnassi, L., Varga, N., Grigorov, M., Studer, A., Reidiker, S., Schilter, B., 2003, Formation of vinylogous compounds in model Maillard reaction systems, *Chem. Res. Toxicol.* 16:242-1250.

Wnorowski, A., and Yaylayan, V., 2003, Monitoring carbonyl-amine reaction between pyruvic acid and α-amino alcohols by FTIR spectroscopy – A possible route to Amadori products, *J. Agric. Food Chem.* 51:6537-6543.

World Health Organization (WHO). International Programme on Chemical Safety. Environmental Health Criteria No 49, 1985.

Yaylayan, V. A., 1999, Analysis of complex reaction mixtures: novel applications of Py-GC/MS and microwave assisted synthesis (MAS), *Am. Lab.* 31:30-31.

Yaylayan, V. A., Wnorowski, A and Perez, L. C., 2003, Why asparagine needs carbohydrates to generate acrylamide, *J. Agric. Food Chem.* 51:1753-1757

Yaylayan, V. A., Perez, L. C., Wnorowski, A., and O'Brien, J., 2004, The role of creatine in the generation of N-methylacrylamide: A New toxicant in cooked meat, *J. Agric. Food Chem.*, **52**: 5559-5565.

Yaylayan, V., and Stadler, R., 2004, Formation of acrylamide in food: A mechanistic perspective, *J. Assoc. Off. Ana. Chem.* in press.

Zyzak, D., Sanders, R. A., Stojanovic, M., Tallmade, D. H., Eberhart, B. L., Ewald, D. K., Gruber, D. C., Morsch, T. R., Strothers, M. A., Rizzi, G. P., and Villagran, M. D., 2003, Acrylamide formation mechanism in Heated Foods, *J. Agric. Food Chem.* 51:4782-4787.

NEW ASPECTS ON THE FORMATION AND ANALYSIS OF ACRYLAMIDE

Peter Schieberle[1], Peter Köhler[2] and Michael Granvogl[1]
[1]*Lehrstuhl für Lebensmittelchemie der Technischen Universität München, Lichtenbergstraße 4, D-85748 Garching, Germany;* [2]*Deutsche Forschungsanstalt für Lebensmittelchemie, Lichtenbergstraße 4, D-85748 Garching, Germany; e-mail: Peter.Schieberle@Lrz.tum.de*

Abstract: The effectiveness of different compounds in the generation of acrylamide (AA) from asparagine, was determined by reacting asparagine with mono-, di- and polysaccharides, as well as four different oxo-compounds known to be involved in carbohydrate metabolism/degradation. Quantitation of AA formed either under aqueous conditions or in low water model systems revealed glucose and 2-oxopropionic acid as the most effective compounds in AA generation, when reacted in model systems with a low water content (about 1 mol-% yield). Interestingly, heating of asparagine in the presence of 2-oxopropionic acid generated quite high amounts of 3-aminopropionamide (3-APA), which itself effectively generated AA upon heating in aqueous solution, as well as in low water systems. Because this is the first report on amounts of 3-APA generated by Maillard-type reactions, the general role of 3-APA as key intermediate in AA formation is discussed in detail. In addition, first results on the development and application of an HPLC/fluorescence method for AA quantitation are presented.

Key words: Acrylamide, 3-aminopropionamide, formation, HPLC/fluorescence; analysis

1. INTRODUCTION

Shortly after the characterization of acrylamide (AA) as a constituent of processed foods (Tareke et al., 2002), the thermal degradation of free asparagine in the presence of carbohydrates following Maillard-type reactions was proposed as the major route in AA formation (Mottram et al., 2002; Stadler et al., 2002; Weisshaar and Gutsche, 2002). Labeling experiments have also showed that the carbon skeleton in AA, as well as the nitrogen in the amide group, stem from asparagine (Stadler et al., 2002). Thus, the stoichiometry of generatation of AA from asparagine requires the loss of CO_2 and NH_3.

Mottram et al. (2002) have suggested the Strecker degradation of asparagine induced by α-dicarbonyl compounds as an important pathway in AA formation, because significant amounts were formed when 2,3-butandione and asparagine were reacted in aqueous solution at high temperatures. From a Strecker reaction of asparagine, after decarboxylation and transamination, the aldehyde 3-oxopropionamide will be generated. However, to form AA from this intermediate, a reduction into 3-hydroxypropionamide followed by a β-elimination of water has to be assumed. However, recently, Stadler et al. (2004) showed that the latter intermediate was not very effective as AA precursor, thereby suggesting only a minor importance of the Strecker reaction in AA formation.

Instead, Stadler et al. (2002) had previously proposed the N-glycoside of asparagine as a direct precursor of AA. This group recently established that the potassium salt of N-(D-glucos-1-yl)-L-asparaginate was more effective by a factor of twenty-three in AA formation than the respective Amadori product of asparagine, N-(1-deoxy-D-fructos-1-yl)-L-asparagine (Stadler et al., 2004). In agreement with mechanisms proposed earlier by Yaylayan et al. (2003) and Becalski et al. (2003), the authors (Stadler et al., 2004) suggested a thermally induced decarboxylation of the Schiff base formed from asparagine and several aldehydes as the key step in AA formation. From the azomethine ylide formed as transient intermediate, the generation of AA is suggested to occur via a β-elimination. This reaction was found to be significantly favored when α-hydroxy aldehydes, such as 1-hydroxypropan-2-one (acetol) were reacted with asparagine, whereas α-oxo aldehydes were shown to be much less effective (Stadler et al., 2004). In addition, because the Amadori product was a less effective precursor as compared to the respective N-glucosides, the authors also ruled out the importance of the Amadori product as the key intermediate in AA generation. Overall, these studies indicated that the degradation of asparagine, or the respective Schiff base, into AA occurs without going through any measurable transient intermediate.

In a recent publication, Zyzak et al. (2003) were able to identify 3-aminopropionamide (3-APA) as a transient intermediate formed during AA generation from asparagine. However, because only a slight increase in the yields of AA was observed as compared to asparagine itself, the role of 3-APA in AA formation was not considered important. In a more recent study, however, we showed that 3-APA may also be formed in foods by an enzymatic decarboxylation of asparagine (Granvogl et al., 2004) and that 3-APA is a very effective precursor of AA under certain reaction conditions.

Because (i) discrepancies on the role of the Strecker degradation in AA formation still exist and (ii) the importance of 3-APA in AA formation still needs to be clarified, one of the aims of the present study was to shed more light on the role of this amide as a precursor of AA.

Several procedures have been developed for AA quantitation. However, nearly all of them use mass spectrometry. Because smaller companies may not have this equipment, there is still a demand for simpler and cheaper methods for AA quantitation. Thus, another aim of this study was to develop a method for AA quantitation by HPLC with fluorescence detection.

2. MATERIALS AND METHODS

2.1 Chemicals

3-Aminopropionamide (β-alaninamide hydrochloride) was obtained from Chemos (Regenstauf, Germany), and $[^2H_3]$-acrylamide (98%) from CIL (Andover, MA, USA). Acrylamide (99.9 %), erythrose, arabinose, ribose, fructose, glucose, lactose, sucrose, cellulose, starch, asparagine monohydrate, hydroxyacetaldehyde (glyoxal), 2-oxopropanal (methylglyoxal), 1-hydroxypropan-2-one (acetol), 2-oxopropionic acid (pyruvic acid), cystamine dihydrochloride, and silica gel KG 60 (0.063 – 0.200 mm) were from VWR International (Darmstadt, Germany). 5-Dimethylamino-1-naphthalene sulfonylchloride (dansyl chloride), glycinamide hydrochloride, and N,N-dimethylacrylamide were from Aldrich (Sigma-Aldrich, Steinheim, Germany) and 5-(4,6-dichloro-s-triazine–2-ylamino)-fluoresceine hydrochloride (DTAF) was from Fluka (Sigma-Aldrich Chemie, Taufkirchen, Germany). All other reagents were of analytical grade.

CAUTION: Acrylamide as well as $[^2H_3]$-acrylamide are hazardous and must be handled carefully.

2.2 Model studies

2.2.1 Model I

Equimolar amounts of asparagine (Asn) and several carbohydrates or carbonyl compounds, respectively (0.1 mmol each), were singly homogenized with silica gel (3 g, KG 60, 0.063-0.200 mm, containing 10 % water; VWR International, Darmstadt, Germany) and heated at 170°C in sealed glass vials in a metal bloc for the times given in the tables. 3-

Aminopropionamide hydrochloride (3-APA) was treated in the same way, but without added carbohydrates.

2.2.2 Model II

Equimolar amounts of Asn and the respective carbohydrate (0.25 mmol each) were singly dissolved in the respective buffer (10 mL; 0.07 mol/L; pH 5.0 or 7.0) and were heated at different temperatures (100 to 180 °C) for 20 min in sealed glass vials. 3-APA hydrochloride was treated in the same way, but without added carbohydrates.

2.3 Quantitation of 3-aminopropionamide

The determination of 3-APA was carried out as follows: After cooling the reaction mixture to RT, water (total volume: 25 mL) and the internal standard glycinamide hydrochloride (148 µg) were added. After ultrasonification (3 min), aqueous sodium hydrogen carbonate (25 mL, 0.5 mol/L) was added and the pH was adjusted to 10 ± 0.2 with sodium hydroxide (2.5 mol/L). 5-Dimethylamino-1-naphthalenesulfonylchloride (dansyl chloride; 32.4 mg dissolved in 15 mL acetone) was added and the mixture was stirred at room temperature for 3 h in the dark. The sulfonamides formed were extracted with dichloromethane (total volume: 60 mL), washed with aqueous sodium hydrogen carbonate (10 mL, 0.25 mol/L, adjusted to pH 10 ± 0.2), dried over anhydrous sodium sulfate and, finally, evaporated to dryness at about 20 kPa and 35°C. The residue was taken up in acetonitrile (4.0 mL), filtered (0.45 µm; Spartan®13/0.45RC, Schleicher & Schuell, Dassel, Germany) and diluted 1 in 10 with acetonitrile/formic acid (0.1 %, v/v in water) 30/70 (v/v) prior to LC/MS/MS analysis, which was performed as described recently (Granvogl et al., 2004).

The quantitation of acrylamide was carried out by GC/MS as described previously (Jezussek and Schieberle, 2003).

2.4 Quantitation of acrylamide by HPLC/fluorescence detection

2.4.1 Preparation of the derivatization reagent

A solution of 5-(4,6-dichloro-s-triazine-2-ylamino)-fluoresceine hydrochloride (DTAF; 188 µmol) in acetone (40 mL) was added to a solution of cystamine (Cysa-Cysa) dihydrochloride (300 µmol) in sodium bicarbonate (60 mL; 0.5 mol/L; pH 10 ± 0.2) and the mixture was stirred for

12 h at RT. The solvent was evaporated in vacuum and the aqueous phase was extracted with ethyl acetate (total volume: 80 mL). After adjusting the pH to 2.0 with hydrochloric acid (2.5 mol/L; 15 mL), the target compound (DTAF-Cysa-Cysa-DTAF) was extracted with ethyl acetate (total volume: 160 mL). Its structure was confirmed by LC/MS/ESI$^+$: *m/z* (M+1) = 1069 (100 %; no further signal above 10 %).

For reduction, the dimeric DTAF-Cysa-Cysa-DTAF (42 µmol) was dissolved in n-propanol/methanol (10 mL; 1+1, v+v), aqueous sodium hydrogen carbonate (10 mL; 0.5 mol/L) and tributylphosphine (60 µmol) were added and the mixture was stirred under argon atmosphere at room temperature for 30 min in the dark. The solvent was evaporated at about 20 kPa and 35°C and the remaining aqueous phase was extracted with ethyl acetate (total volume: 80 mL). After adjusting the pH to 2.0 ± 0.2 with hydrochloric acid (1.0 mol/L), the target compound was isolated by extraction with ethyl acetate (total volume: 160 mL), dried over anhydrous sodium sulfate and evaporated to dryness at about 20 kPa and 35°C. The solid residue was dissolved in methanol (10.0 mL) and stored under argon atmosphere to prevent re-oxidation. The structure of the reagent was confirmed by LC/MS/ESI$^+$: *m/z* (M+1) = 536 (100 %, no further signal above 10 %).

2.4.2 Preparation of reference compounds

Acrylamide (AA) or N,N-dimethylacrylamide (DMA, each 365 nmol), respectively, were singly dissolved in ethanol (10 mL) and the pH was adjusted to 10 ± 0.2 with sodium hydroxide (2.5 mol/L). DTAF-Cysa (0.9 µmol dissolved in 300 µL of methanol) was added; the mixture was purged with argon for 30 s and stirred at room temperature for 3 h in the dark. An aliquot of this solution (200 µL) was made up to 4 mL with acetonitrile/formic acid (0.1 %, v/v in water) 40/60 (v/v) for HPLC/ fluorescence as well as HPLC/MS analysis.

2.5 HPLC/fluorescence analysis

Analyses were performed by means of a high performance liquid chromatograph (Kontron, Eching, Germany) equipped with a Luna Phenyl-Hexyl 100 Å HPLC column (250 x 4.6 mm i.d., 5 µm, Phenomenex, Aschaffenburg, Germany) and a C18 precolumn (4 x 3.0 mm i.d., Phenomenex). The sample (10 µL) was analyzed at a flow rate of 0.8 mL/min using a fluorescence detector SFM 25 operating at an excitation wavelength of 480 nm and an emission wavelength of 520 nm. The solvents used were aqueous formic acid (0.1 %, by vol.) and acetonitrile. A gradient was applied by increasing the concentration of acetonitrile from 40 to 50 % within 15 min, and then, to 100 % in 2 min. Prior to fluorescence detection, an aqueous solution of sodium hydroxide (0.1 mL/min, 1 mol/L) was automatically added to the eluent.

2.6 Quantitation of acrylamide in rusk

Powdered rusk (10 g) was dispersed in tap water (10 mL) and the internal standard N,N-dimethylacrylamide (DMA, 36.4 µg), dissolved in ethanol, was added. For equilibration, the sample was stirred for 60 s and then homogenized using an Ultraturrax (Jahnke & Kunkel, IKA-Labortechnik, Staufen, Germany) for 2 min. After ultrasonification for 2 min, the suspension was centrifuged twice (first run: 4000 rpm, 10 min at 10°C; centrifuge GR 412, Jouan, Unterhaching, Germany; second run: 15000 rpm, 15 min at 10°C; Beckman J2-HS, München, Germany). The supernatant (about 40 mL) was defatted with n-hexane (total volume: 40 mL) and, after precipitation of proteins using acetone (20 mL), the sample was centrifuged again (4000 rpm, 10 min at 10°C; centrifuge GR 412, Jouan). The solvent was evaporated at about 20 kPa and 35°C. To an aliquot of the remaining aqueous solution (5 mL), sodium hydroxide (2.5 mol/L) was added and the pH was adjusted to 10 ± 0.2. After the addition of DTAF-Cysa in methanol (350 µg in 100 µL), the reaction mixture was purged with argon for 30 s and stirred at room temperature for 3 h in the dark. An aliquot of this solution (100 µL) was made up to 1.0 mL with acetonitrile/formic acid (0.1 %, v/v in water) 40/60 (v/v) for HPLC/fluorescence analysis.

3. RESULTS AND DISCUSSION

3.1 Model studies on acrylamide formation

In most of the model studies recently reported in the literature, glucose and fructose have been used to generate AA from asparagine. To obtain a more detailed view on the effectiveness of the carbohydrate moiety in generating AA from asparagine, the amino acid was reacted for 30 min at 170°C in the presence of five monosaccharides, two disaccharides and two polysaccharides (model I). Among the carbohydrates, glucose, followed by fructose and lactose, was the most effective in AA generation (Table 1). In the presence of these carbohydrates, 0.8 to 1.0 mol-% AA were formed. Also the non-reducing carbohydrate sucrose was quite effective, whereas cellulose and starch only led to a small amount of AA. Thus, the non-reducing glycosidic bonds in starch or cellulose do not deliver significant amounts of intermediates able to degrade asparagine into AA.

Table 1. Influence of the carbohydrate moiety on the formation of acrylamide (AA) from asparagine (Asn)[a]

Carbohydrate	AA (mmol/mol Asn)
Erythrose	6.4
Arabinose	6.7
Ribose	7.6
Fructose	8.9
Glucose	9.6
Lactose	8.1
Sucrose	5.7
Cellulose	0.5
Starch	0.07

[a] Asparagine (0.1 mmol) and the respective carbohydrate (0.1 mmol) were mixed with silica gel (3 g, containing 10 % water) and reacted for 30 min at 170°C in a closed glass vial.

The amount of water present in the matrix had a very significant effect on the yields of AA. Thus, when the reaction of asparagine and glucose or fructose, respectively, was performed in a completely dry system (Table 2), the yields dropped to about one third of those formed in a matrix with 10 % water. Interestingly, the yields of AA were further increased, when the reaction was performed at a water content of 25 %. The reaction of sucrose and asparagine showed the same behavior indicating that the formation of acrylamide needs at least a certain amount of water to generate distinct transient intermediates able to release AA on thermal treatment.

Table 2. Influence of the water content on the formation of acrylamide (AA) from asparagine (Asn) in the presence of different carbohydrates[a]

Expt.	water content (%)	AA (mmol/mol Asn) formed in the presence of		
		glucose	fructose	sucrose
1	0	1.9	1.5	1.5
2	2.5	2.3	1.7	1.8
3	5	2.8	1.9	1.8
4	10	6.2	4.8	3.7
5	25	9.4	7.6	5.5

[a] Asparagine (0.1 mmol) and the respective carbohydrate (0.1 mmol) were mixed with silica gel (3 g, containing 10 %containing the amount of water given in table) and reacted for 30 min at 170°C in a closed glass vial.

To further evaluate the influence of water on AA formation, asparagine and glucose were reacted in an aqueous buffer at increasing temperatures. The results (Table 3) showed that at temperatures below 160°C, heating of the aqueous solution at pH 7.0 for 20 min only generated low amounts of AA. However, at 160°C or 180°C, respectively, much higher amounts of AA were formed. However, as compared to the same model mixture reacted in a low water system (25 %; Table 2), the yields in the aqueous solution were clearly lower. At pH 5.0, the yields of AA were further lowered as compared to pH 7.0, thereby establishing a significant effect of the pH on AA formation. In general, these data suggest that the presence of water and a neutral pH facilitate formation of primary precursors for AA formation. However, in an aqueous solution these are less effectively converted into AA as compared to a low moisture system.

Table 3. Influence of temperature and pH on the formation of acrylamide (AA) from asparagine (Asn) and glucose in aqueous solution[a]

Temperature (%)	AA (µmol/mol Asn) formed at pH	
	7.0	5.0
100	12	6
120	12	8
140	73	9
160	754	24
180	1108	347

[a a] Asparagine monohydrate (0.5 mmol = 75 mg) and glucose (0.5 mmol = 90 mg) were dissolved in phosphate buffer (10 mL) and reacted for 20 min in a closed glass vial.

In further model studies, the two α-dicarbonyl compounds ethanedial (glyoxal) and 2-oxopropanal (methylglyoxal), known as carbohydrate degradation products, as well as 1-hydroxypropan-2-one (hydroxyacetone) were reacted with asparagine. As shown in Table 4, all intermediates were able to generate AA in significant yields. However, both α-dicarbonyl compounds were less effective than hydroxyacetone, glucose or fructose, respectively (cf. Tables 1 and 4). These results are in good agreement with data published by Stadler et al. (2004) and suggested that the Strecker reaction may not be the most predominant pathway in AA formation, because, e.g., 1-hydroxypropan-2-one, the effective "catalyst" of asparagine degradation (cf. Table 4) is not likely to initiate a Strecker reaction.

Table 4. Influence of different oxo-compounds on the formation of acrylamide (AA) from asparagine (Asn)[a]

Oxo compound	AA (mmol/mol Asn)
Ethanedial (glyoxal)	2.2
2-Oxopropanal (methylglyoxal)	2.1
1-Hydroxypropan-2-one (hydroxyacetone)	6.4
2-Oxopropionic acid (pyruvic acid)	9.9

[a a] Asparagine (0.1 mmol) and the respective oxo-compound (0.1 mmol) were mixed with silica gel (3 g; 10 % water content) and reacted for 30 min at 170°C in a closed glass vial.

Stadler et al. (2004) suggested that an electron donating group, α to the carbonyl group in the "catalyzing" oxo compound, such as a hydroxyl group, should facilitate AA formation from asparagine. For the same reason, a carboxyl group α to the aldehyde function should enhance this electron donating effect, and thus 2-oxopropionic acid (pyruvic acid) should also be an effective "catalyst" in asparagine degradation. To check this assumption, the α-oxo acid was reacted with asparagine (model I). As shown in Table 4, in the presence of this acid, asparagine was very effectively converted into AA. A suggestion for the reaction pathway is given in Figure 1. The formation of a Schiff base between 2-oxopropionic acid and asparagine leads to an intermediate dicarbonic acid (I in Figure 1), which might easily decarboxylate yielding the ylide II (Figure 1). After tautomerization, this intermediate might then directly be hydrolyzed into 3-aminopropionamide (3-APA).

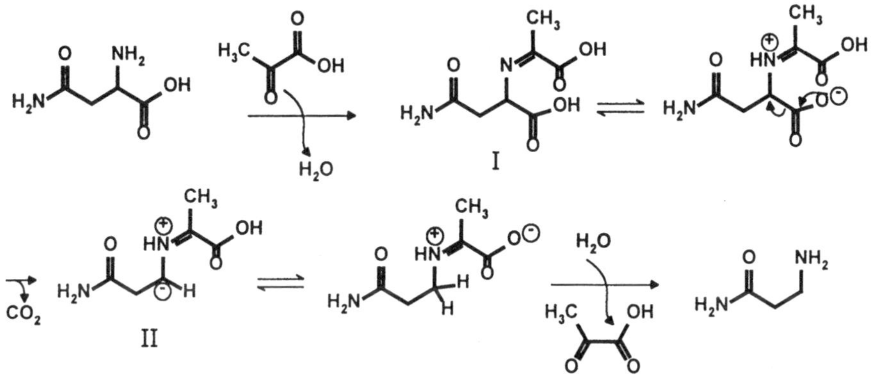

Figure 1. Reaction pathway leading to the formation of 3-aminopropionamide from asparagine and 2-oxopropionic acid.

Recently, we showed that 3-APA is a very effective precursor of acrylamide, either when heated under low water conditions or in aqueous systems (Granvogl et al., 2004). To prove that 3-APA is formed in the asparagine/2-oxopropionic acid model, its concentrations were monitored in parallel with AA formation. The data (Table 5) show that 3-APA is formed in the asparagine/2-oxopropionic acid mixture in quite high amounts in the first minutes of the reaction (expt. 1; Table 5). At this early stage, however, only low amounts of AA were present. The fact that 3-APA is formed prior to AA clearly corroborates the pathway given in Figure 1. Increasing the reaction time yielded 3-APA in similar quantities, whereas AA was present in much higher amounts after 20 and 30 min as compared to a reaction time of 5 min. After 30 min, 3-APA as well as AA were much increased.

However, 3-APA was always present in higher amounts as compared to AA. These results suggest that 3-APA is an important precursor of AA in the thermal degradation of asparagine.

Table 5. Formation of 3-aminopropionamide (3-APA) and acrylamide (AA) from asparagine (Asn) and 2-oxopropionic acid[a]

Expt.	Reaction time (min)	3-APA		AA	
		µg	mmol/mol Asn	µg	mmol/mol Asn
1	5	26.3	2.99	0.6	0.08
2	10	18.7	2.13	2.1	0.30
3	20	24.7	2.80	11.3	1.60
4	30	38.0	4.32	13.3	1.87

[a] Asparagine (0.1 mmol) and 2-oxopropionic acid (0.1 mmol) were mixed with silica gel (3 g; 10 % water content) and heated at 170°C in a closed glass vial.

To establish the key role of 3-APA in AA formation, two further series of model experiments were performed. In the first series, 3-APA was reacted singly in a model with low water content (model I) and the AA formation with time was recorded. As shown in Figure 2, after a short induction period, the amount of AA steadily increased reaching about 27 mol-% after 30 min, based on the amount of 3-APA used. When the same reaction was performed in an aqueous reaction system, significant amounts of AA were formed already at temperatures below 160°C (Table 6). However, at 180°C, the yields reached nearly 65 mol-%, in particular when the buffer was replaced by tap water (cf. expts. 5 and 6; Table 6). These data confirmed the important role of 3-aminopropionamide as a transient intermediate in AA formation.

Table 6. Influence of the temperature on the generation of acrylamide (AA) from 3-aminopropionamide (3-APA) in aqueous solution[a]

Expt.	Temp. (°C)	AA (mmol/mol 3-APA)
1	100	1.2
2	120	7.5
3	140	41
4	160	147
5	180	409
6	180	649[b]

[a] 3-Aminopropionamide (0.001 mmol) was dissolved in phosphate buffer and reacted for 20 min in a closed glass vial (model II; pH 7.0). [b] The buffer was replaced by tap water.

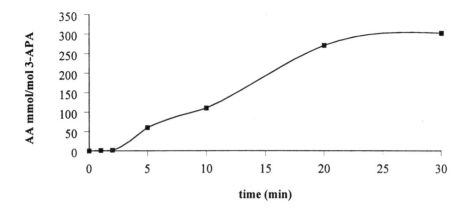

Figure 2. Time course of the formation of acrylamide (AA) from 3-aminopropionamide (3-APA). Reaction conditions: 170°C on silica gel (3g, 10 % water content).

3-APA might also be formed as an intermediate in the Strecker reaction of asparagine as suggested in Figure 3. A decarboxylation of the Schiff base formed from an α-dicarbonyl compound and asparagine may lead to the formation of three different tautomers (I-III in Figure 3). Hydrolysis of tautomer II would lead to 3-oxopropionamide, the typical Strecker aldehyde, which is not able to generate AA directly. In contrast, if tautomer III is hydrolyzed, 3-APA will be formed, which in turn will yield AA after thermal treatment. Based on these suggestions it is obvious why the Strecker reaction is not as effective in AA formation, because only tautomer III will yield a transient intermediate able to generate AA. However, it still needs to be clarified which tautomer is mainly formed in the Strecker reaction.

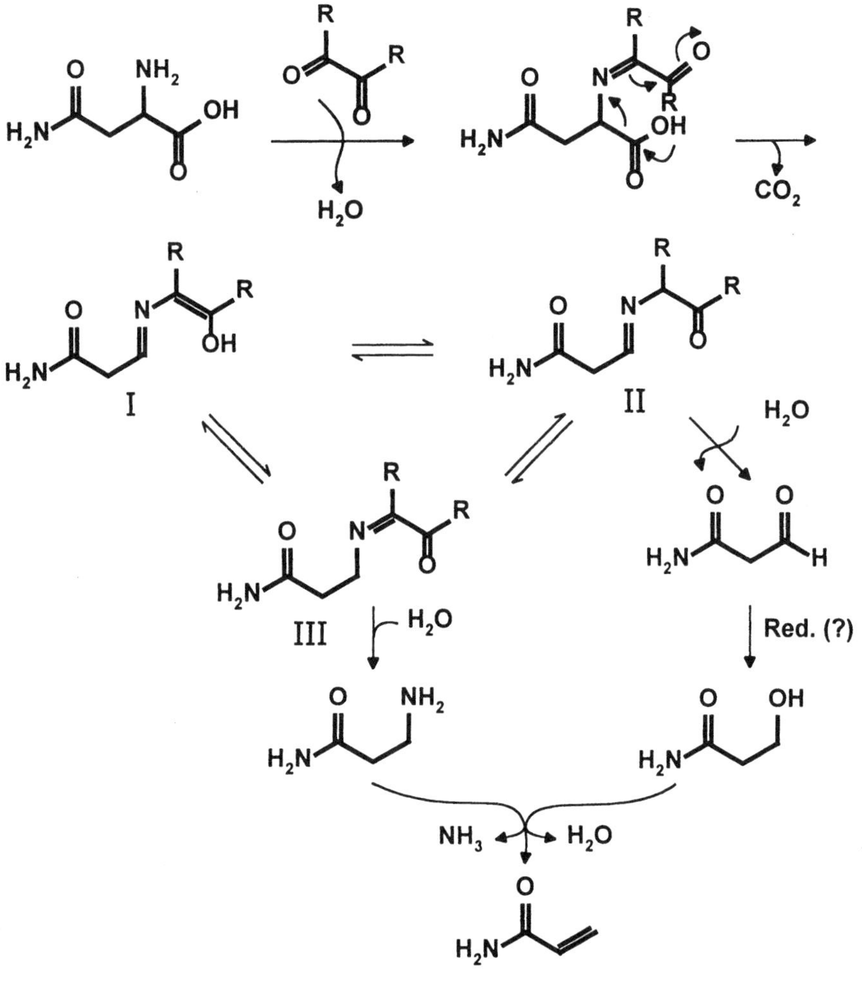

Figure 3. Hypothetical reaction pathway suggesting the formation of 3-aminopropionamide (3-APA) and acrylamide (AA) via a Strecker–type degradation of asparagine initiated by an α-dicarbonyl compound.

3-APA has also recently been identified as transient intermediate in a glucose/asparagine model system (Zyzak et al., 2003). The authors explained its formation via a decarboxylation of the Schiff base and the degradation of an ylide intermediate. However, there is also another possibility to explain the formation of 3-APA and AA from a glycosyl amine structure (Figure 4): Tautomerization of the Schiff base formed from asparagine and an α-hydroxycarbonyl compound, e.g. a carbohydrate or a degradation product such as hydroxyacetone, yields enaminol II. Such compounds have been shown by us to be very susceptible to oxidation

(Hofmann and Schieberle, 2000). The vinylogous β-keto acid formed by an oxidation step will easily decarboxylate yielding an α-oxo-imine, which may then be hydrolyzed into 3-APA and an osone (Figure 4).

In conclusion, the data suggest 3-APA as a key intermediate in the formation of AA from asparagine. Discrepancies in the recent literature on the mechanisms of AA formation may arise from the fact that AA formation may occur via several Maillard-type reaction sequences.

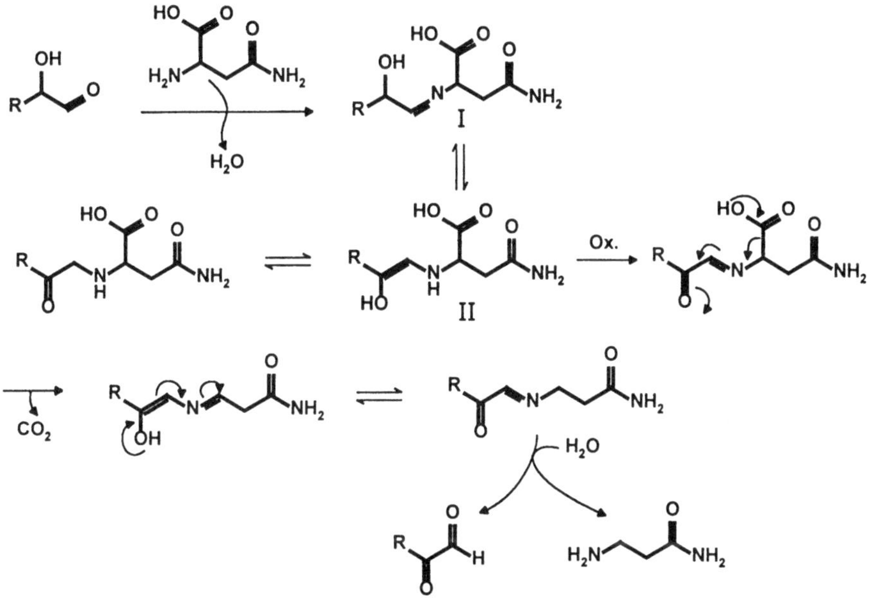

Figure 4. Hypothetical formation pathway leading to 3-aminopropionamide (3-APA) after oxidation of an enaminol formed in the Amadori-type rearrangement of asparagine and its subsequent decarboxylation and hydrolysis.

3.2 Development of the HPLC/fluorescence method

In a previous study we showed that acrylamide effectively reacted with 2-mercaptobenzoic acid forming a stable thioether, which allowed the selective and sensitive detection of acrylamide using a benchtop LC mass spectrometer (Jezussek and Schieberle, 2003). Following the same concept of derivatization, several fluorescent compounds were bound to the amino groups of cystamine (Cysa-Cysa) as exemplified by 5-(4,6-dichloro-s-triazine–2-ylamino)-fluoresceine hydrochloride (DTAF) in Figure 5.

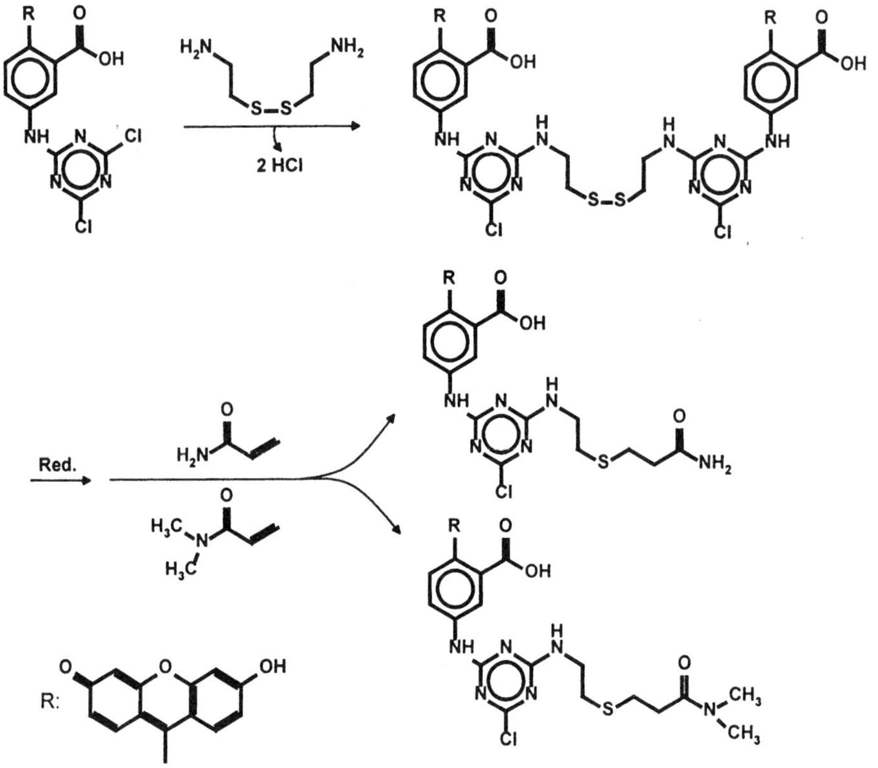

Figure 5. Synthetic route used in the derivatization of acrylamide (AA) and N,N-dimethylacrylamide (DMA). The derivatization reagent was prepared by a reaction of 5-(4,6-dichloro-s-triazine–2-ylamino)-fluoresceine hydrochloride (DTAF) with cystamine ((Cysa-Cysa) followed by reduction into DTAF-Cysa.

The fluorescent disulfide (DTAF-Cysa-Cysa-DTAF) formed was reduced into the corresponding thiol (DTAF-Cysa), which was subsequently reacted with acrylamide or N,N-dimethylacrylamide (DMA), selected as the internal standard. The yields of the DTAF-Cysa-Cysa-DTAF and the DTAF-Cysa, respectively, were determined to be 88% or 75%, respectively. The mass spectra of the synthesized reference compounds AA-DTAF-Cysa and DMA-DTAF-Cysa shown in Figure 6 corroborated the structures given in Figure 5. Four other fluorescent molecules were reacted the same way and showed similar yields (data not shown).

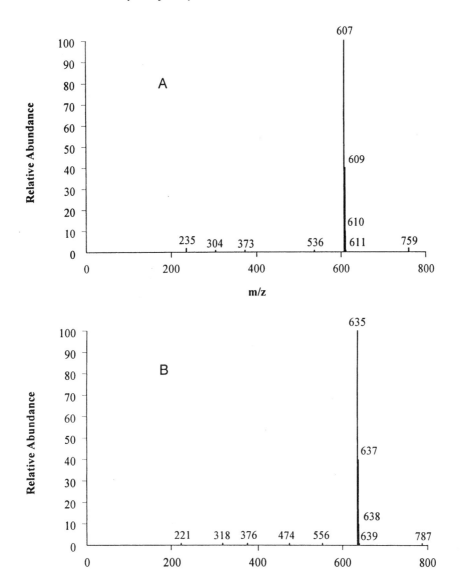

Figure 6. Mass spectra (MS/ESI) of the acrylamide derivative (A) and the N,N-dimethylacrylamide derivative (B) formed by derivatization with DTAF-Cysa.

For calibration, seven model mixtures containing AA and DMA in ratios of 10:1 to 1:10 were reacted with DTAF-Cysa and the HPLC chromatograms were monitored by fluorescent detection as exemplified in Figure 7 for a 1:3 mixture of AA and DMA. The results indicated a good separation of the acrylamide (1 in Fig. 7) and the N,N-dimethylacrylamide derivative (2 in Fig. 7). However, a calculated response factor of 0.25 indicated that the reactivity of DMA vs. DTAF-Cysa was much lower as compared to AA. Other compounds, such as N-methylacrylamide, N-ethylacrylamide, N-isopropylacrylamide, and 2-methylacrylamide showed even less reactivity than DMA (data not shown) and can, therefore, not be recommended as internal standards. Nevertheless, the reproducibility of the derivatization procedure with DTAF-Cysa as well as the limit of quantitation (0,3 µg/L) were very satisfying and, thus, the method was applied to several foods, in particular, cereal products.

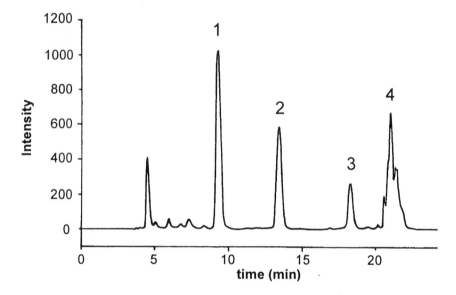

Figure 7. Synthetic route used in the derivatization of acrylamide (AA) and N,N-dimethylacrylamide (DMA). The derivatization reagent was prepared by a reaction of 5-(4,6-dichloro-s-triazine–2-ylamino)-fluoresceine hydrochloride (DTAF) with cystamine ((Cysa-Cysa) followed by reduction into DTAF-Cysa.

In Figure 8, an HPLC chromatogram obtained in the analysis of a commercial rusk sample is shown. The dotted line represents a reference solution of AA-DTAF-Cysa and DMA-DTAF-Cysa. Although the amount of AA was calculated to be 250 µg/kg, this sensitivity of the method is too low to be recommended for AA analysis in foods. There are two main reason for this. First, obviously too many compounds present in food extracts do react with DTAF-Cysa, thereby causing a very high baseline of the fluorescence detector and, secondly, the derivatives seem to form complexes with the food matrix and, thus, the yields of the extraction were very low. Although the application of LC/MS significantly increased the selectivity and sensitivity of the method (data not shown), our purpose still is the development of a method sensitive and selective enough for AA analysis without MS detection. Further studies are underway to solve this challenge. Nevertheless, the method can be proposed for the sensitive and selective quantitation of AA in aqueous matrices, such as drinking water.

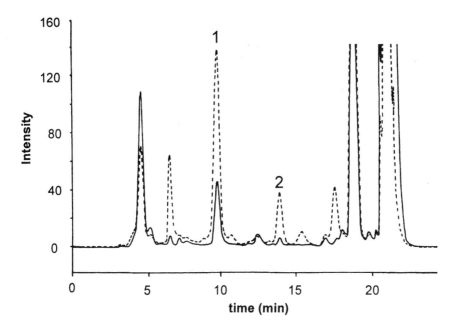

Figure 8. HPLC separation of the DTAF-derivatives of acrylamide (1) and N,N-dimethylacrylamide (2) isolated from a sample of rusk (full line).
Dotted line: reference sample for comparison

REFERENCES

Becalski, A., Lau, B. P.-Y., Lewis, D., and Seaman, S., 2003, Acrylamide in food: Occurrence, sources, and modeling, *J. Agric. Food Chem.* **51**:802-808.

Granvogl, M., Jezussek, M., Koehler, P., and Schieberle, P., 2004, Quantitation of 3-Aminopropionamide in Potatoes - A Minor but Potent Precursor in Acrylamide Formation, *J. Agric. Food Chem.* **52**:4751-4757.

Hofmann, T., and Schieberle, P., 2000, Formation of aroma-active Strecker-aldehydes by a direct oxidative degradation of Amadori compounds, *J. Agric. Food Chem.* **48**:4301-4305.

Jezussek, M., and Schieberle, P., 2003, A new LC/MS-method for the quantitation of acrylamide based on a stable isotope dilution assay and derivatization with 2-mercaptobenzoc acid. Comparison with two GC/MS-methods, *J. Agric. Food Chem.* **51**,:7866-7871.

Mottram, D. S., Wedzicha, B. L., and Dodson, A. T., 2002, Food chemistry: Acrylamide is formed in the Maillard reaction, *Nature* **419**:448-449.

Stadler, R. H., Blank, I., Varga, N., Robert, F., Hau, J., Guy, P. A., Robert, M.-C., and Riediker, S., 2002, Food chemistry: Acrylamide from Maillard reaction products, *Nature* **419**:449-450.

Stadler, R. H., Robert, F., Riediker, S., Varga, N., Davidek, T., Devaud, S., Goldmann, T., Hau, J., and Blank, I., 2004, In-depth mechanistic study on the formation of acrylamide and other vinylogous compounds by the Maillard reaction, *J. Agric. Food Chem.* http://dx.doi.org/10.1021/jf0495486.

Tareke, E., Rydberg, P., Karlsson, P., Eriksson, S., and Törnqvist, M., 2002, Analysis of acrylamide, a carcinogen in heated foodstuffs, *J. Agric. Food Chem.* **50**:4998-5006.

Weisshaar, R., and Gutsche, V., 2002, Formation of acrylamide in heated potato products – model exeperiments pointing to asparagine as precursor, *Deutsche Lebensm. Rund.* **98**:397-400.

Yaylayan, V. A., Wnorowski, A., and C.P., L., 2003, Why asparagine needs carbohydrates to generate acrylamide, *J. Agric. Food Chem.* **51**:1753-1757.

Zyzak, D. V., Sanders, R. A., Stojanovic, M., Tallmadge, D. H., Eberhart, B. L., Ewald, D. K., D.C., G., Morsch, T. R., Strothers, M. A., Rizzi, G. P., and Villagran, M. D., 2003, Acrylamide formation mechanism in heated foods, *J. Agric. Food Chem.* **51**:4782-4787.

FORMATION OF ACRYLAMIDE FROM LIPIDS

Stefan Ehling, Matt Hengel and Takayuki Shibamoto
Department of Environmental Toxicology, University of California Davis, One Shield Ave. Davis CA 95616; e-mail: sehling@ucdavis.edu

Abstract: Heating amino acids with dietary oils or animal fats at elevated temperatures produced various amounts of acrylamide. The amount of acrylamide formation corresponded to the degree of unsaturation of the oils and animal fats. The decreasing order of acrylamide formation from dietary oils or animal fats with asparagine was sardine oil (642 µg/g asparagine) > cod liver oil (435.4 µg/g) > soybean oil (135.8 µg/g) > corn oil (80.7 µg/g) > olive oil (73.6 µg/g) > canola oil (70.7 µg/g) > corn oil (62.1 µg/g) > beef fat (59.3 µg/g) > lard (36.0 µg/g). Three-carbon unit compounds such as acrylic acid and acrolein, which are formed from lipids by oxidation also produced acrylamide by heat treatment with amino acids, in particular with asparagine. The results of the present study suggest that acrylamide forms in asparagine-rich foods during deep fat frying in the absence reducing sugars.

Key words: acrolein, acrylamide, acrylic acid, amino acids, carbonyl compounds, dietary oils, lipid oxidation

1. INTRODUCTION

The widely recognized main acrylamide formation mechanism in foods is the interaction of asparagine with reducing sugars (typically glucose and fructose) at elevated temperatures. Large amounts of acrylamide (over 1,000 ppm with respect to asparagine) have been reported in model systems consisting of asparagine and glucose (Mottram et al., 2002; Yasuhara et al., 2003).

According to the reaction mechanism described by Yaylayan et al. (2003), the Schiff base formed from asparagine and a reducing sugar does not undergo the Amadori rearrangement but it forms a cyclic intermediate, oxazolidin-5-one. This intermediate gives the decarboxylated Amadori product via decarboxylation, and subsequently produces acrylamide and an amino sugar. Since the Amadori rearrangement, which requires an α-dicarbonyl compound such as glucose and fructose, is not involved in the formation of acrylamide, any carbonyl compound can react with asparagine to form acrylamide.

There are several reports on the acrylamide formation from asparagine with miscellaneous carbonyl compounds other than reducing sugars. Heating asparagine with octanal, 2-octanone, or 2,3-butanedione formed various amounts of acrylamide (Becalski et al., 2003). Glycolaldehyde and glyceraldehyde produced about 2 times and 1.75 times more acrylamide than glucose with asparagine, respectively (Yaylayan et al., 2003). The sugar 2-deoxyglucose, which does not undergo the Amadori rearrangement because of the absence of a carbonyl group on the second carbon, produced 0.71 times more acrylamide than glucose with asparagine (Zyzak et al., 2003). Acrolein formed from lipids upon oxidation yielded 114 µg acrylamide/g asparagine. Acrylic acid, which is an oxidation product of acrolein, produced significant amount of acrylamide with ammonia (190,000 µg/g ammonia) (Yasuhara et al., 2003).

In the present study, various lipids and their oxidative breakdown products were heated with asparagine and other amino acids to investigate the role of lipids in acrylamide formation.

2. MATERIALS AND METHODS

2.1 Chemicals and reagents

Asparagine was purchased from CN Biosciences, Inc. (La Jolla, CA). Acrylic acid, ammonium chloride, and linoleic acid were form Fluka Chemical Corporation (Milwaukee, WI). Stearic acid was from Fischer Scientific (Fair Lawn, NJ). L-lysine, L-arginine, methacrylamide, acrolein, oleic acid, and squalene were from Aldrich (Milwaukee, WI). L-glutamine, L-methionine, L-histidine, L-cysteine, DL-alanine, and glycine were purchased from Sigma (St. Louis, MO). Acrylamide was from Bio-Rad Laboratories (Hercules, CA). $^{13}C_3$-Acrylamide was purchased from Cambridge Isotope Laboratories (Andover, MA). Cod liver oil was bought from Arista Industries, Inc. (Wilton, CT). Sardine oil was provided by the Japan Institute for the Control of Aging (Fukuroi, Japan). α–Tocopherol-stripped corn oil was purchased from ICN Biomedicals, Inc. (Aurora, OH). Beef fat was provided by the Meat Lab at the University of California at Davis. Olive oil, canola oil, soybean oil, corn oil, and lard were purchased from a local supermarket.

2.2 Preparation of reaction mixtures

2.2.1 Asparagine and various lipids

The samples were prepared according to the method reported previously with slight modification (Yasuhara et al., 2003). L-Asparagine (1.76 g) and 5 g each of lipid were heated in a 100 mL round bottom flask in an oil bath at 180°C for 30 min. The reaction mixtures were stirred with a magnetic stirrer. After the reaction mixture was allowed to cool to room temperature, 50 mL of dichloromethane was added and the mixture stirred for 20 min. The resulting solution was filtered through a Whatman #1 paper disk and the clear filtrate was loaded onto a glass column (185 mm × 15 mm o.d.) packed with 10 g of 60-200 μm mesh silica gel (Mallinkrodt Baker, Inc., Phillipsburg, NJ). Anhydrous sodium sulfate (1g) was placed at the top of the column prior to loading a sample. The column was eluted with 100 mL of dichloromethane, and then with a 100 mL of ethyl acetate/methanol (95/5) solution. The latter fraction was condensed with a rotary evaporator. After 200 μL of internal standard methacrylamide (1 mg/mL in ethyl acetate) was added, the volume was adjusted to 10 mL with ethyl acetate. The ethyl acetate solution was analyzed for acrylamide using a gas chromatograph equipped with a nitrogen phosphorus detector (GC/NPD).

2.2.2 Asparagine and stearic acid

The reaction mixture was stirred with 50 mL ethyl acetate for 20 min, and then cooled to + 5°C in a refrigerator until the stearic acid crystallized. The content of the flask was filtered through a bed of anhydrous sodium sulfate. The clear ethyl acetate solution was then prepared for GC/NPD analysis as described above.

2.2.3 Amino acids and acrylic acid

Amino acids (0.5 g) and 5 g of acrylic acid were heated in a 100 mL round bottom flask in an oil bath at 120°C for 30 min. The reaction mixture was stirred with a magnetic stirrer. After the reaction mixture was allowed to cool to room temperature, 50 mL of ethyl acetate was added and the mixture stirred for 20 min. The resulting solution was filtered through a Whatman #1 paper disk and the filtrate was subsequently condensed with a rotary evaporator. The condensed sample was passed through a Pasteur pipette packed with anhydrous sodium sulfate and prepared for GC/NPD analysis as described above.

2.2.4 Amino acids and acrolein

The samples were prepared according to the method reported previously (Roach et al., 2003). Amino acid (0.5 g) and 0.18 g acrolein were heated in 10 mL water in a sealed bottle at 180°C for 30 min in an oven. After the reaction mixture cooled down to room temperature, 1 mL of internal standard $^{13}C_3$-acrylamide (200 ng/mL in water) was added, and 1.5 mL of the solution was purified on an OASIS-HLB solid phase extraction column (Waters Corporation, Milford, MA). The purified sample was then analyzed for acrylamide by LC-MS/MS.

2.2.5 Amino acids and triolein or oils

Amino acids (1.76 g) and 5 g of triolein or soybean oil were heated at 180°C for 30 min. After addition of 1 mL of internal standard $^{13}C_3$-acrylamide (200 ng/mL in water) and 20 mL of water, the reaction mixture was stirred for 20 min. The water layer (10 mL) was centrifuged in a polypropylene centrifuge tube at 9,000 rpm for 15 min. A 1.5 mL portion of the centrifugate was purified on an OASIS-HLB solid phase extraction column as described above, and then subjected to LC-MS/MS analysis.

2.3 Instrumentation

A Hewlett Packard model 6890 gas chromatograph equipped with a 30 m × 0.25 mm i.d. DB-WAX fused silica capillary column (D_f = 0.25 μm) was used. The injector and the detector temperatures were 250°C. The column was held at 50°C for 1 min, then programmed to 180°C at 20°C/min and held for 4 min, and then programmed to 230°C at 30°C/min and held for 1 min. Helium carrier gas flow rate was 30 cm/sec. Injection volume was 1 μL in the splitless mode. Under these conditions acrylamide eluted at 10.2 min.

A Hewlett Packard 1100 liquid chromatograph equipped with a 50 × 2.1 mm Hypercarb column (Thermo, San Jose, CA) with a 5 μm particle size was used. Mobile phase was aqueous 0.1% acetic acid with 4% methanol (isocratic) at 400 μL/min. Injection volume was 10 μL. Under these conditions acrylamide eluted at 1.4 min. The LC was interfaced with an Applied Biosystems API 2000 MS/MS instrument using an atmospheric pressure chemical ionization (ACPI) source operating in the positive ion mode. The nebulizer temperature was 475°C. Multiple reaction monitoring was performed for the transition m/z 72→55 for acrylamide and 75→58 for $^{13}C_3$-acrylamide.

3. RESULTS AND DISCUSSION

Recovery efficiency of acrylamide was over 90% from the reaction mixtures consisting of asparagine and various lipids or acrylic acid.

Levels of acrylamide formed from asparagine (1.76 g) and various dietary oils (5 g) heated at 180°C for 30 min are shown in Fig. 1.

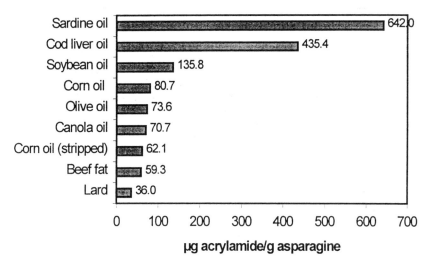

Figure 1. Formation of acrylamide from asparagine (1.76 g) and various lipids (5 g) at 180°C (30 min). Analysis performed by GC/NPD. Values are the average of two experiments.

Heating asparagine with lard or beef fat, which contains low levels of unsaturated lipids, produced the lowest amounts of acrylamide among the lipids tested (36.0 and 59.3 µg/g asparagine, respectively). The amounts of acrylamide formed from asparagine with canola oil or olive oil, which contain mostly monounsaturated fatty acids were 70.7 and 73.6 µg/g asparagine, respectively. The amount of acrylamide formed from asparagine and corn oil, which contains mono-and polyunsaturated lipids, was slightly higher (80.7 µg/g asparagine) than those from asparagine with lard, beef fat, canola oil, or olive oil. Interestingly, when regular corn oil was replaced with α-tocopherol-stripped corn oil, the formation of acrylamide (62.1 µg/g asparagine) was reduced by 23%. The results from the asparagine-monounsaturated oil systems in the present study are comparable with that reported for the asparagine-triolein experiment (88.6 µg/g asparagine) (Yasuhara et al., 2003).

Soybean oil (which contains high levels of linolenic acid) and asparagine produced 135.8 µg/g asparagine of acrylamide. When cod liver oil and sardine oil, which contain highly unsaturated fatty acids, were used, large amounts of acrylamide (435.4 and 642.0 µg/g asparagine, respectively) were formed. The results from the present study suggest that the higher the degree of unsaturation of the lipid, the larger the formation of acrylamide.

Many carbonyl compounds (aldehydes and ketones) produced from lipids upon oxidation react with asparagine to form acrylamide. It was previously reported that potato chips, which are fried in crude, unrefined oils (e.g. virgin olive oil) contained higher levels of acrylamide than potato chips fried in refined or processed oils (Becalski et al., 2003). This may be due to the removal of oxidized products and/or precursors of carbonyl compounds by the refining process. In the present study, beef fat (crude fat) with asparagine produced more acrylamide than lard (refined fat product) with asparagine did.

Fig. 2 shows the results of acrylamide formation from asparagine and simple lipids. The heating asparagine with stearic acid (a saturated fatty acid) produced 91.0 µg/g asparagine, which is a rather high level for a saturated fat, suggesting that the oxidation also occurred in a saturated fatty acid.

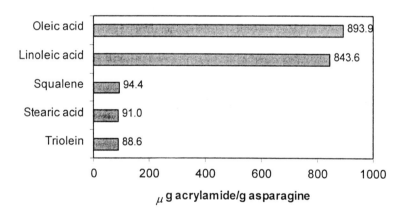

Figure 2. Formation of acrylamide from asparagine (1.76 g) and various simple lipids (5 g) at 180°C (30 min). Analysis performed by GC/NPD. Values are the average of two experiments.

Asparagine and squalene (a polyunsaturated hydrocarbon found in olive oil and certain fish oils) produced 94.4 µg/g asparagine, suggesting that these small hydrocarbon fractions of crude oils also contribute to the generation of acrylamide. When the free fatty acids oleic and linoleic acid were heated with asparagine, high levels of acrylamide formed (893.9 and 843.6 µg/g asparagine, respectively). These values are about 50% of that reported in the asparagine – glucose system. The unsaturation of these compounds, together with the presence of the free carboxyl group, is likely generate many lipid oxidation products and subsequently generate a high levels of acrylamide.

Fig. 3 shows a typical gas chromatogram of an ethyl acetate extract of the reaction mixture of asparagine and canola oil.

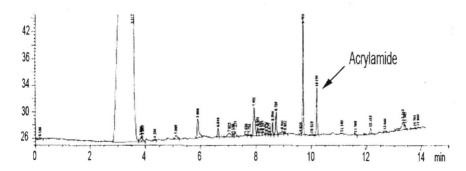

Figure 3. Gas chromatogram of an ethyl acetate extract from the reaction mixture of asparagine (1.76 g) and canola oil (5 g) heated at 180°C.

The amounts of acrylamide formed from asparagine and vegetable oils (olive oil, canola oil, corn oil and soybean oil), which are commonly used for deep fat frying of foods, are less than those formed in the asparagine – glucose system. Reducing sugars, such as glucose, seem to play an important role in acrylamide formation with asparagine. However, the present study indicated that lipids also contribute to the formation of acrylamide in lipid-rich foods and in deep fat frying.

Fig. 4 shows hypothetical pathways for the formation of acrylamide based on three-carbon unit compounds, such as acrolein and acrylic acid, generated from lipids at elevated temperature. Other food constituents, besides lipids, also form these three-carbon unit compounds when they are heated. The amino acid methionine yields methional via Strecker degradation and methional subsequently produces acrolein (Friedman, 2003). Amino acids produce ammonia upon Strecker degradation, which reacts with acrylic acid derived from lipids to form acrylamide (Yasuhara et al., 2003). The backbone of several amino acids can also produce acrylic acid (Gertz et al., 2002; Yaylayan et al., 2004). Acrylic acid is also formed from the pyrolysis of glucose through methyl glyoxal (Gertz et al., 2002). Acrylic acid produced a significantly large amount of acrylamide with ammonia (190,000 µg/g ammonia) at an elevated temperature (Yasuhara et al., 2003). This result suggested that ammonia from a nitrogen source, such as an amino acid, and acrylic acid from a lipid can yield acrylamide at a deep fat frying temperature.

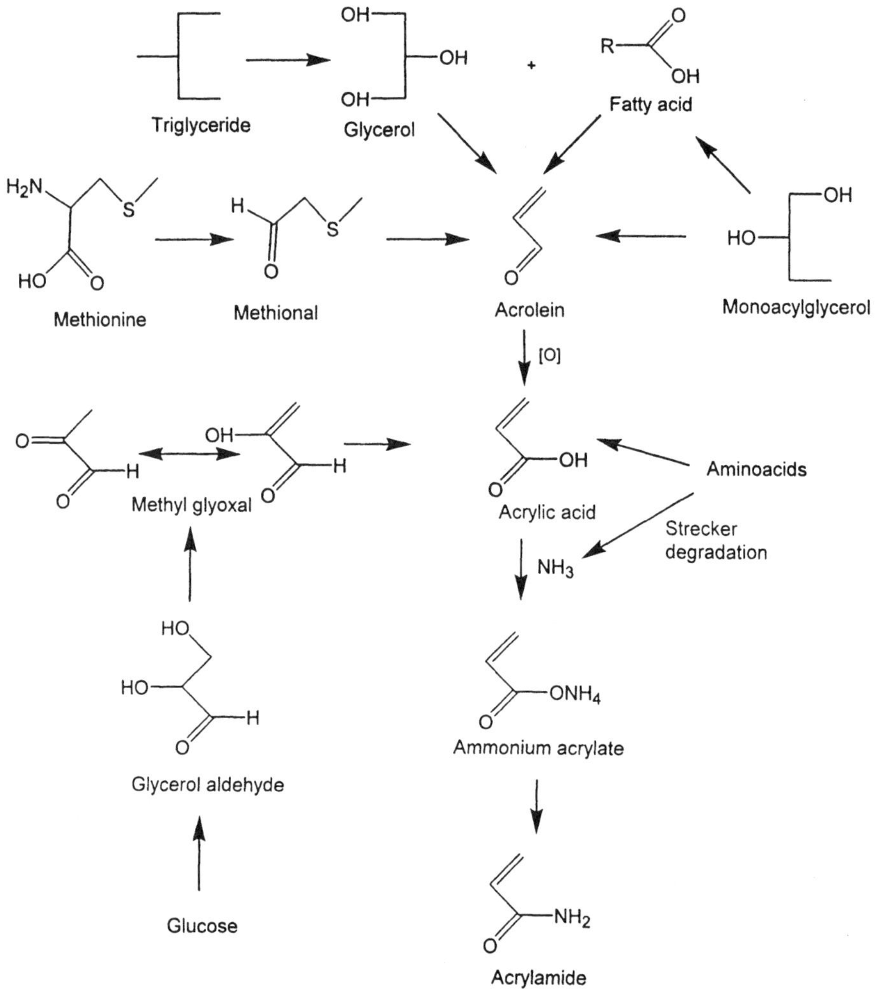

Figure 4. Hypothesized pathways of acrylamide formation *via* acrolein and acrylic acid (Gertz et al., 2002; Stadler et al., 2002; Friedman, 2003; Yasuhara et al., 2003; Yaylayan et al., 2004).

Fig. 5 shows the results of acrylamide formation in the reaction mixtures consisting of acrylic acid and various nitrogen sources. All amino acids tested and also ammonium chloride produced substantial amounts of acrylamide. The highest amount of acrylamide was obtained from the reaction between asparagine and acrylic acid (1675.5 µg/g asparagine). The

Formation of Acrylamide from Lipids

other amino acids containing nitrogen in their side chain—glutamine (1566.8 µg/g amino acid), lysine (1533.6 µg/g amino acid), and arginine (893.9 µg/g amino acid)—also produced large amounts of acrylamide.

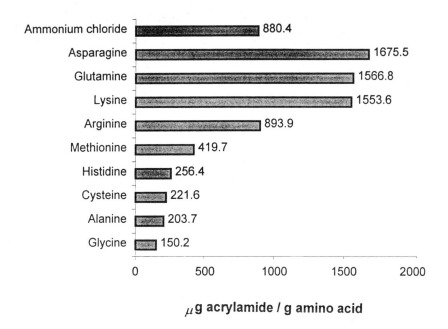

Figure 5. Formation of acrylamide from amino acids (0.5 g) and acrylic acid (5 g) at 120°C (30 min). Analysis performed by GC-NPD. Data represent the average of two experiments.

The strongly acidic medium might promote the release of ammonia from amino acids, consequently the formation of acrylamide was increased. Methionine, cysteine, alanine, and glycine, which contain only the α-amino nitrogen, produced much lower amounts of acrylamide (under 500 µg/g amino acid) than glutamine, lysine, and arginine did.

Acrolein, which gives acrylic acid via oxidation, has been known to form from lipid glycerides in large amounts (Umano and Shibamoto, 1987). When acrylic acid was replaced with acrolein, only asparagine produced a large amount of acrylamide in these reactions (10,372 µg/g acrolein). A very small amount of acrylamide was formed in the reaction between glutamine and acrolein (9 µg/g acrolein). When other amino acids, i.e. arginine, lysine, histidine or glycine were reacted with acrolein under these conditions, no detectable amounts of acrylamide were formed. However, the results suggest that acrolein is one of the precursors of acrylamide.

Since acrylic acid but not acrolein generated large amounts of acrylamide when reacted with a nitrogen source, it appears that the oxidation of acrolein to acrylic acid is a critical step in this pathway. Acrolein may instantaneously react with various food constituents rather than undergoing oxidation to acrylic acid. Common frying temperatures are well above the boiling point of acrolein (51°C) and even that of acrylic acid (140°C), so that these three-carbon unit compounds formed from lipids may readily leave the system and enter the gaseous phase, and consequently do not contribute significantly to the formation of acrylamide. When heating several amino acids besides asparagine with triolein or soybean oil, detectable amounts of acrylamide were not formed.

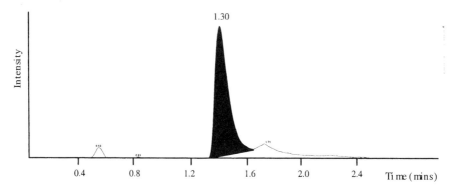

Figure 6. LC-MS/MS chromatogram of the reaction mixture resulted upon the heating of asparagine with acrolein at 180°C for 30 min. Retention time of acrylamide is 1.4 min.

The reaction mixtures obtained from an amino acid and acrolein were analyzed by LC-MS/MS. Fig. 6 shows a typical LC-MS/MS chromatogram of a reaction mixture of asparagine and acrolein.

In the present study, it was observed that lipids play a role in the formation of acrylamide in foods. Acrylamide formed from lipids via the three-carbon unit compounds produced from lipids at elevated temperatures. The overall contribution of lipids to the formation of acrylamide in foods may not be as high as that of reducing sugars. However, it is obvious that acrylamide forms from asparagine-rich foods, such as potatoes; via deep fat frying practices using various dietary oils.

ACKNOWLEDGEMENTS

We thank Paul Kuzmicky for technical assistance with the LC-MS/MS.

REFERENCES

Becalski A., Lau B.P.Y., Lewis D., and Seaman S.W., 2003, Acrylamide in foods: occurrence, sources and modeling, *J. Agric. Food Chem.* **51**:802-808.

Friedman M., 2003, Chemistry, biochemistry and safety of acrylamide. A review, *J. Agric. Food Chem.* **51**:4504-4526.

Gertz C. and Klostermann S., 2002, Analysis of acrylamide and mechanisms of its formation in deep-fried products, *Eur. J. Lipid Sci. Technol.* **104**:762-771.

Mottram D.S., Wedzicha B.L., and Dodson A.T., 2002, Acrylamide is formed in the Maillard reaction, *Nature* **419**:448-449.

Roach J.A.G., Andrzejewski D., Gay M.L., Nortrup D., and Musser S.M., 2003, Rugged LC-MS/MS survey analysis of acrylamide in foods, *J. Agric. Food Chem.* **51**:7457-7554.

Stadler R.H., Blank I., Varga N., Robert F., Hau J., Guy P.A., Robert M.C., and Riediker S., 2002, Acrylamide from Maillard reaction products, *Nature,* **419**:449-450.

Umano, K., and Shibamoto T., 1987, Analysis of acrolein from heated cooking oils and beef fat, *J. Agric. Food Chem.* **35**, 909-912.

Yasuhara, A., Tanaka Y., Hengel M., and Shibamoto T., 2003, Gas chromatographic investigation of acrylamide formation in browning model systems, *J. Agric. Food Chem.* **51**:3999-4003.

Yaylayan V.A., Perez Locas C., and Wnorowski A., 2005, Mechanistic pathways of formation of acrylamide from different amino acids, in: *Chemistry and Safety of Acrylamide in Food,* M Friedman and D.S. Mottram, eds, Springer, New York, pp. 191-204.

Yaylayan V.A., Wnorowski A., and Perez Locas C., 2003, Why asparagine needs carbohydrates to generate acrylamide, *J. Agric. Food Chem.* **51**:1753-1757.

Zyzak D.V., Sanders R.A., Stojanovic M., Tallmadge D.H., Eberhart B.L., Ewald D.K., Gruber D.C., Morsch T.R., Strothers M.A., Rizzi G.P., and Villagran M.D., 2003, Acrylamide formation mechanism in heated foods, *J. Agric. Food Chem.* **51**:4782-4787.

KINETIC MODELS AS A ROUTE TO CONTROL ACRYLAMIDE FORMATION IN FOOD

Bronislaw L. Wedzicha[1], Donald S. Mottram[2], J. Stephen Elmore[2], Georgios Koutsidis[2] and Andrew T. Dodson[2]
[1]*Procter Department of Food Science, University of Leeds, Leeds, LS2 9JT;* [2]*School of Foo Biosciences, The University of Reading, Whiteknights, Reading, RG6 6AP, United Kingdom; e-mail: b.l.wedzicha@leeds.ac.uk*

Abstract: A kinetic model for the formation of acrylamide in potato, rye and wheat products has been derived, and kinetic parameters calculated for potato by multi-response modeling of reducing sugar (glucose and fructose), amino acid, asparagine and acrylamide concentrations with time. The kinetic mechanism shares, with Maillard browning, a rate limiting (probably dicarbonylic) intermediate, and includes reaction steps of this intermediate which are competitive with respect to acrylamide formation. A pathway representing physical and/or chemical losses of acrylamide accounts for the measured reduction of acrylamide yield at long reaction times. A mechanistic hypothesis regarding the competing reactions of Strecker aldehyde formation and tautomerization followed by beta-elimination to give acrylamide, features in the kinetic model and can be used to determine the factors which steer the reaction towards acrylamide. A predictive application of this model is for 'what-if' experiments to explore the conditions which lead to reduced acrylamide yields.

Key words: acrylamide, potato, wheat, rye, kinetics, model, multi-response modeling

1. INTRODUCTION

Acrylamide formation in food is under kinetic control, *i.e.*, the amount which is measured in cooked or processed food depends on the conditions under which the food has been heated and time of heating. Fig. 1 illustrates our suggested mechanism for acrylamide formation involving the reaction of asparagine with reactive carbonyl intermediates, such as 3-deoxyhexosulose (3DH) derived from the Maillard reaction. This mechanism proposes that the reaction proceeds *via* the Strecker degradation of asparagine, and that acrylamide is formed either as a result of β-elimination from a protonated tautomer of the Strecker intermediate or *via* 3-aminopropionamide. Since 3DH and other dicarbonylic species are key intermediates in browning reactions, the kinetics of the formation of acrylamide could be seen as a derivative of the kinetic model for browning. The kinetics of browning have been modeled successfully (Swales & Wedzicha, 1992; Gogus *et al.*, 1998; Leong & Wedzicha, 2000; Mundt & Wedzicha, 2003). This paper will show how the kinetics of acrylamide formation in heat treated potato (and to some extent in heated wheat and rye products) can be modeled accurately on the basis of the scheme shown in Fig. 1 and the known kinetics of browning of aldoses and ketoses.

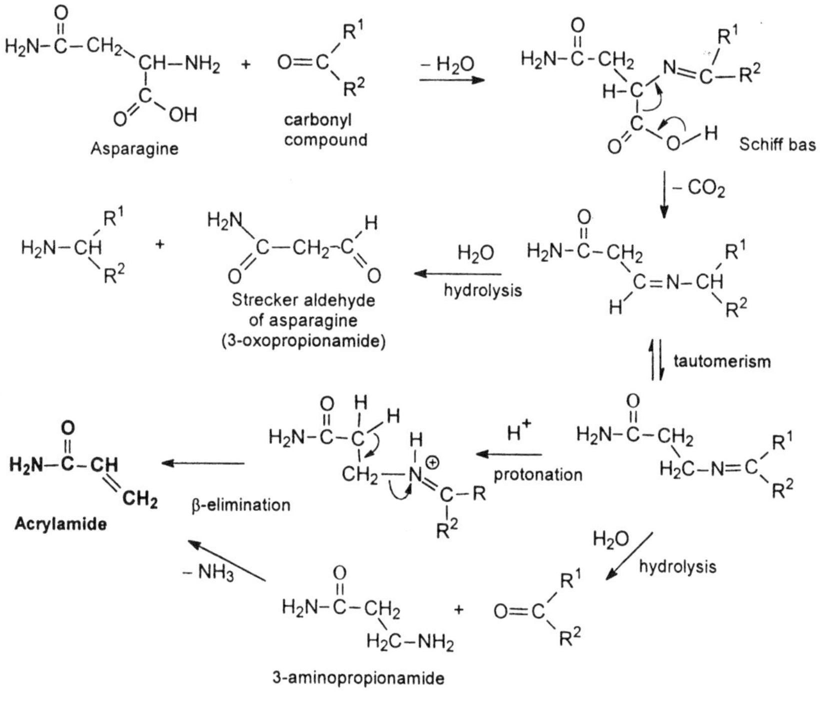

Figure 1. Suggested mechanism for the formation of acrylamide from intermediates formed during the Maillard reaction.

An understanding of kinetics is fundamental to our ability to control acrylamide formation in food. Kinetic models tend to be much simpler than mechanistic models because they describe the rates of formation of substances in terms of the rate limiting processes; these are the control points in chemical reactions. It therefore follows that the kinetic approach may be used to determine how the control points in acrylamide formation are affected by reaction (cooking, thermal processing) conditions. If similar models can be formulated for color and flavor formation, texture change, *etc*, which occur alongside acrylamide formation, we will then have a multi-response model which can be optimized in order to minimize acrylamide formation whilst maximizing (or otherwise optimizing) the other attributes of the food quality.

All the data referred to in this paper are published separately by Elmore *et al.* (2004).

Fig. 2 shows the formation of acrylamide with time as potato, wheat and rye 'cakes' are heated at 180°C. Considering the features of these plots, the model needs to account for the following systems:
- an induction phase during which time no acrylamide is being formed,
- an increase in acrylamide yield with time,
- the presence of at least one limiting component which causes the concentration-time plot to level off,
- loss of acrylamide with time, *i.e.*, possible reactions of formed acrylamide with other food components.

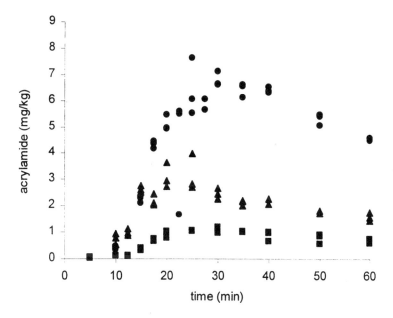

Figure 2. Formation of acrylamide in potato (●), rye (▲) and wheat (■) cakes heated at 180°C. The data for these plots are taken from Elmore et al. (2004).

2. STARTING POINT FOR THE MODEL

Fig. 3 illustrates the basic kinetic model which is based on the following assumptions:

1. The key intermediate and precursor of acrylamide, 'int', is formed from the aldose by a stepwise process, which also involves the amino acid. We propose that this intermediate is common both to acrylamide and 'normal' Maillard products. The Maillard reaction involves a number of steps in which the aldose undergoes amine-assisted dehydration. The actual number of steps involved in the kinetic model depends on model resolution and discussion, but we will assume initially that there are two steps which lead to the formation of 'int'.

2. Intermediate 'int' reacts with all amino acids, including asparagine, to form Maillard products. These may or may not be colored or flavored. Initially, it is assumed that all amino acids are similarly reactive (k_3) towards 'int', and that the reaction is of first order with respect to 'int' and the amino acid. This is a reasonable assumption for a process which is expected to involve a carbonyl-amine reaction. Such a reaction usually proceeds by a bimolecular nucleophilic attack on the carbonyl group by the amino group. Elmore et al. (2004) demonstrated that amino acids are effectively used up by the time acrylamide concentration reaches a maximum. The limiting components in the model could therefore be the amino acids (either as a whole or asparagine in particular).

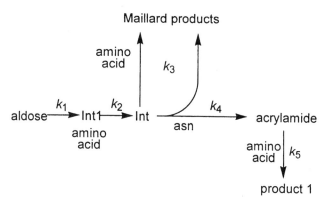

Figure 3. Basic kinetic model for the formation of acrylamide as a result of the reaction of asparagine with key intermediates in the Maillard reaction.

3. When asparagine reacts with the intermediate 'int' it produces acrylamide as a 'side reaction' product. We regard it as a side reaction from the Maillard reaction because the yield of acrylamide is a fraction of

a percent based on the amount of asparagine which has undergone reaction (Mottram *et al.*, 2002). It is assumed that this reaction is also of first order with respect to 'int' and asparagine (rate constant k_4), because the first step in this process is postulated to involve a nucleophilic attack by asparagine on 'int'.

4. Acrylamide undergoes further reaction characterized by k_5. The concentration of acrylamide is likely to be far smaller than that of any food component with which it could react at a significant rate. It is envisaged that two reaction types may exist. The first is where acrylamide reacts spontaneously with another substance whose concentration remains constant throughout the observation period. The second is where acrylamide reacts with a food component whose concentration changes during the observation period. An example of the latter could be the Michael reaction of acrylamide with the amino groups of amino acids, which are themselves being lost. It will be assumed that this is a first order reaction with respect to acrylamide and the substance with which it is reacting.

3. SELF-CONSISTENCY OF THE KINETIC DATA

3.1 Behavior of amino acids

The most elementary test of any kinetic model is to determine whether the data are self-consistent in relation to the model. The model shows explicitly that asparagine and the remaining amino acids compete for the intermediate 'int'. Given the assumptions regarding the kinetics of individual reactions within the model, the rates of loss of asparagine and the other amino acids (aa) are:

$$-\frac{d[\text{asn}]}{dt} = k_4[\text{asn}][\text{int}]$$

$$-\frac{d[\text{aa}]}{dt} = k_3[\text{aa}][\text{int}]$$

Because [int] for both reactions varies in the same way with time, the relative rates of the loss of non-asparagine amino acids and of asparagine should be the same at every instant of the reaction. Thus, the amount of amino acid (non-asparagine) and asparagine which have undergone reaction at any instant should be related linearly to each other. Fig. 4 demonstrates that this indeed is true for potato, rye and wheat. Remarkably, the data for these three food materials overlap and the relationship is thus followed for nearly four orders of magnitude in amino acid and asparagine concentration. From the point of view of reaction kinetics, the fact that a trend is observed over such a large concentration range allows us to conclude that the rate of loss of the amino acid pool occurs at a similar rate as the loss of asparagine. Most importantly, this feature appears to depend only on the amino acids in question, and is independent of the food matrix in which they are contained or the concentrations of other potential reactants or precursors, such as sugars.

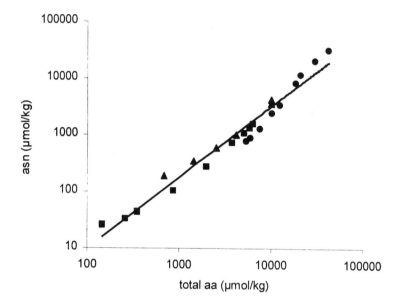

Figure 4. Relationship between asparagine and total amino acid (non asparagine) contents of potato (●), rye (▲) and wheat (■) cakes whilst they are being heated at 180°C. For each food product the data point with the highest concentration is at t=0. Successive data points are at increasing heating time. Data for these plots are taken from Elmore et al. (2004).

Since we postulate that asparagine is converted both to Maillard products and to acrylamide, it is relevant now to test whether the amount of acrylamide formed is dependent on the amount of asparagine which has undergone reaction. Fig. 5 shows the linear relationship between the two measurements for potato, suggesting that there is a constant fraction of asparagine converted to acrylamide at any instant. The slope of the line drawn through the points is 314 (dimensionless), suggesting that the yield of acrylamide, based on asparagine, is ~0.3 %. This result is typical of the results also obtained for rye and wheat. The data shown in Fig. 5, and for the corresponding plots for rye and wheat cakes, were those obtained whilst the concentration of acrylamide was increasing; once the amino acid has been exhausted, the concentration of acrylamide falls slowly, and the nature of the reacting system changes. Fig. 5 also serves to define the rate of loss of asparagine relative to the rate of formation of acrylamide, which will be referred to as the *relative rate*.

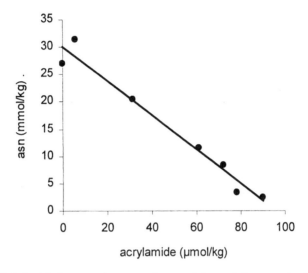

Figure 5. Relationship between the asparagine remaining and the amount of acrylamide formed in potato cakes. Data for these plots are taken from Elmore et al. (2004).

Since the rate of loss of the amino acid pool (non-asparagine) is linearly related to the rate of loss of asparagine (Fig. 4), it is to be expected that the amount of each amino acid lost should similarly be related to the amount of acrylamide formed. Thus, we extend the results demonstrated by Fig. 5 to include excellent linear correlations between the amounts of individual amino acids remaining and the amount of acrylamide formed in potato, rye and wheat cakes as they are heated at 180°C. In each case, we calculate the rate of loss of each amino acid relative to the rate of formation of acrylamide as the *relative rate*.

Figs 6-8 show this relative rate plotted for rye, wheat and potato, respectively, against the initial concentration of each amino acid analyzed. Each point on a particular graph, therefore, indicates a different amino acid.

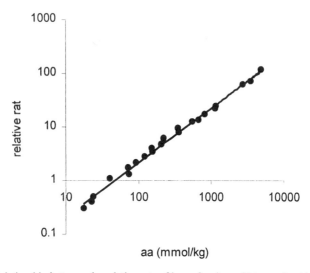

Figure 6. Relationship between the relative rate of loss of amino acid to acrylamide formation and the concentration of the respective amino acid for rye. Data for these plots are taken from Elmore et al. (2004).

The slopes of these graphs are 1.01, 1.01 and 1.11 for rye, wheat and potato respectively, indicating that if plotted on a linear scale, these graphs are excellent straight lines, over nearly a 3 order of magnitude change in concentration. Again, from the point of view of a study of reaction kinetics, the fact that a trend is observed over such a large concentration range allows us to conclude that the relative rate of loss of each amino acid depends only on its initial concentration and not on the nature of that amino acid. Therefore, all the 20 amino acids analyzed by Elmore *et al.* (2004) are similarly related to the formation of acrylamide. One explanation for this surprising observation in terms of the kinetic behavior of the respective reactions is that the rates of reactions of the amino acids are fast, *e.g.*, the amino acids simply mop up the intermediate 'int' as it is formed. The amount of each amino acid undergoing reaction then depends on the probability of that amino acid molecule colliding with 'int' which, in turn, depends on the number of molecules of that amino acid present. One could simply state that if there were double the number of molecules of one amino acid relative to another, then that amino acid would react twice as much as the other.

Similarly, we suggest that the conversion of asparagine to acrylamide can be expressed as a probability that the Maillard reaction of asparagine 'diverts' from 'normal' Maillard products to acrylamide. We find from Fig. 5, and related plots for rye and wheat cakes, that for a given food system, this probability appears to be constant throughout the heating process whilst acrylamide is being formed.

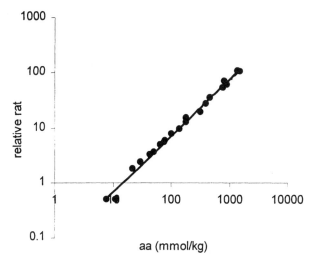

Figure 7. Relationship between the relative rate of loss of amino acid to acrylamide formation and the concentration of the respective amino acid for wheat. Data for these plots are taken from Elmore et al. (2004).

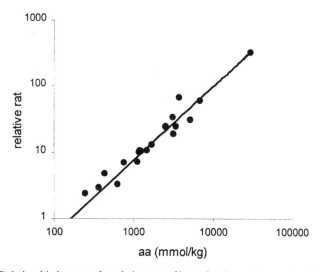

Figure 8. Relationship between the relative rate of loss of amino acid to acrylamide formation and the concentration of the respective amino acid for potato. Data for these plots are taken from Elmore et al. (2004).

3.2 Behavior of sugars

We have now demonstrated that the formation of acrylamide can be used to follow the progress of the reactions derived from intermediate 'int'. Elmore *et al.* (2004) report changes in the concentrations of sugars as acrylamide is formed with time. It is now relevant to consider how the formation of acrylamide, and thus the formation of intermediate 'int', is related kinetically to the loss of sugar (Fig. 3). This is exemplified for potato cakes in Figs 9 and 10 for glucose and fructose, respectively.

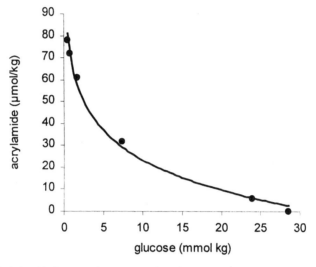

Figure 9. Relationship between the amount of acrylamide formed and the amount of glucose which remains as potato cakes are heated at 180°C. Data for these plots are taken from Elmore et al. (2004).

Fig. 9 shows that glucose is 'lost' significantly before any acrylamide is formed. This observation implies that the conversion of glucose to intermediate 'int' is by way of a two-step or multi-step process, with each step being rate-determining. On the other hand, Fig. 10 shows that the relationship between acrylamide formation and the 'loss' of fructose is linear, suggesting a single step conversion of fructose to 'int'. It is interesting to compare this result with previously published research on the kinetics of the glucose-glycine-sulfite and fructose-glycine-sulfite reactions. Sulfite is used to inhibit the browning of glucose and fructose, and it does this in both cases by reacting with an intermediate derived from 3DH in a fast process (*cf.* the model proposed here). It was found that glucose is indeed converted to this intermediate in a two-stage process (Leong &

Wedzicha, 2000), whereas fructose is converted to the same intermediate in a one-stage process (Swales & Wedzicha, 1992).

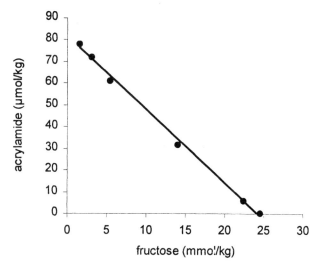

Figure 10. Relationship between the amount of acrylamide formed and the amount of fructose which remains as potato cakes are heated at 180°C. Data for these plots are taken from Elmore et al. (2004).

The different kinetic behaviors of glucose and fructose, and the fact that over the timescale of the formation of acrylamide both sugars (which are present initially in similar concentrations) are lost completely, means that both contribute significantly to the formation of acrylamide and the kinetic model has to be modified to account for their respective contributions. The suggested revised model is illustrated in Fig. 11.

Similarly, the closest correlations between acrylamide formation and loss of sugar in rye and wheat cakes are for maltose. It is found that acrylamide-maltose data are non-linear and the plots resemble closely Fig. 9. Wedzicha and Kedward (1995) and Mundt and Wedzicha (2004) have shown that the kinetics of the maltose-glycine-sulfite reaction are similar to the kinetics those of the glucose-glycine-sulfite reaction. Both reactions are multi-step processes, and maltose behaves as does the aldose in Fig. 11.

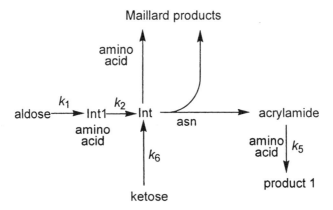

Figure 11. Revised kinetic model to take account of the different kinetic behaviors of glucose and fructose.

4. VALIDATION OF THE KINETIC MODEL

4.1 The rate equations

Thus far, the discussion has concentrated on verifying that the relationships *between* components of the model agree with experiment. The ultimate test of a kinetic model is, of course, that it should be capable of predicting the time-course of the formation of acrylamide. This is obtained first, by formulating the rate equations, and second, by integrating these numerically. The rate equations are formulated as follows:

$$\frac{d[\text{int}1]}{dt} = k_1[\text{aldose}][\text{aa}] - k_2[\text{int}1]$$

$$\frac{d[\text{acryl}]}{dt} = \{k_2[\text{int}1] + k_6[\text{ketose}]\}/k[\text{aa}] - k'_5[\text{acryl}][\text{aa}] - k_5[\text{acryl}]$$

$$-\frac{d[\text{aldose}]}{dt} = k_1[\text{aldose}][\text{aa}]$$

$$-\frac{d[\text{ketose}]}{dt} = k_6[\text{ketose}]$$

$$-\frac{d[\text{aa}]}{dt} = k_2[\text{Int}1] + k_6[\text{ketose}]$$

$$\frac{d[\text{M}]}{dt} = \{k_2[\text{int}1] + k_6[\text{ketose}]\}(1 - 1/k[\text{aa}])$$

The rate constants k_1, k_2, k_5 and k_6, and 'int1' are defined in Fig. 11 and M represents the Maillard products. An additional rate constant k'_5 is introduced to account for a bimolecular reaction between acrylamide with amino acids. The parameter k is the probability for the conversion of asparagine to acrylamide rather than to 'normal' Maillard products.

4.2 Loss of acrylamide upon extended heating

We now turn our attention to the kinetics of the loss of acrylamide as depicted in Fig. 2. It is evident that the rate of formation of acrylamide is much greater than its subsequent loss, and Elmore et al. (2004) confirm that the concentration of acrylamide reaches a maximum when the amount of asparagine has fallen below 5-10% of its original value. Thus, beyond a heating time of approximately 25 min, there is minimal further formation of acrylamide, and we observe the loss of acrylamide for whatever reason it occurs. Since only a small extent of the reaction is being followed, it is striking that the acrylamide concentration data at 30-60 min fit very well a linear relationship to time. Fig. 12 shows that the measured rate of loss of acrylamide is proportional to the starting acrylamide concentration (*i.e.*, that at 25-30 min of heating), suggesting that,

$$-\frac{d[\text{acryl}]}{dt} = k[\text{acryl}]$$

where k is a first order (most probably pseudo-first order) rate constant.

Whilst this type of behavior is not surprising, it is remarkable that it appears to be independent of whether the loss of acrylamide is in potato, rye or wheat cakes. Hence, we conclude that the measured loss is due either to a reaction of acrylamide with components of the food matrix which are in large excess in the three food products studied, or that acrylamide is lost by some 'physical' process.

The aim of plotting the data in Fig. 12 was to determine a kinetic model for the loss of acrylamide which could then be incorporated into the kinetic model shown in Fig. 11. The data discussed above are limited in their applicability because they do not represent the reactions of acrylamide in the early stages (t<30 min) of the heating process, when amino acids are still present; these can react with acrylamide *via* the Michael addition reaction. Thus, reliance on Fig. 12 potentially can lead to under-estimation of the amount of acrylamide which is lost; current data do not allow us to explore this possibility further.

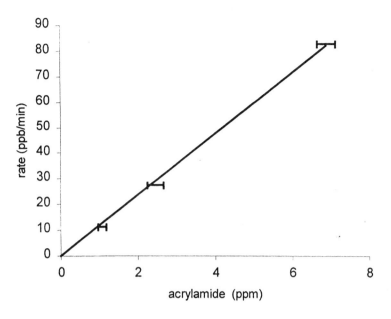

Figure 12. Relationship between the rate of loss of acrylamide (30-60 min from Fig. 1) and the maximum acrylamide concentration reached during the heating of potato (highest acrylamide concentration), rye and wheat (lowest acrylamide concentration) cakes at 180°C.

4.3 Integration of the rate equations

The rate equations were coded for the 'Athena' multi-response modeling software (available from Stewart & Associates Engineering Software Inc., Minneapolis, USA). For potato cakes, the responses used for the derivation of parameters were acrylamide, glucose, fructose, asparagine and total amino acid contents. However, the starting values of a number of parameters could be determined separately from the kinetic data of Elmore et al. (2004) as follows:

1. k_1=1.63 mol^{-1} kg min^{-1} from the initial rate of loss of glucose (3.02 mmol kg^{-1} min^{-1}), the initial concentration of amino acids (64.9 mmol kg^{-1}) and the initial concentration of glucose (28.6 mmol kg^{-1}).
2. k_5=0.012 min^{-1} from the slope of the line in Fig. 12.
3. k_6=0.083 min^{-1} from the initial rate of loss of fructose (2.03 mmol kg^{-1} min^{-1}) and the initial concentration of fructose (24.6 mmol kg^{-1}).
4. k=10200 from a graph of the amount of total amino acid lost *vs.* acrylamide formed.

Elmore et al. (2004) report the loss of moisture from potato, rye and wheat cakes as they are heated at 180°C. It is clear from these data that one reason for the apparent induction time before acrylamide formation begins is the need for the moisture content to fall below a certain critical value (5% of the original value) which corresponds to heating times in excess of 10 min. Based on the assumption that only small amounts of chemical change occur in the first 10 min, the time axis of the kinetic modeling results is displaced by 10 min relative to the experimental time axis.

In order to fit the data, the parameters which had been determined independently were first fixed and the remaining parameters (k_2 and k'_5) were calculated for best fit.: Subsequently, all parameters were allowed to vary and the optimal values returned were as follows (with initial values in brackets): k_1=2.25 (1.63) mol^{-1} kg min^{-1}; k_2=28.4 min^{-1}; k_5=0.06 (0.012) min^{-1}; k'_5=2.94 x 10^{-6} mol^{-1} kg min^{-1}; k_6=0.088 (0.083) min^{-1}; k=9900 (10200).

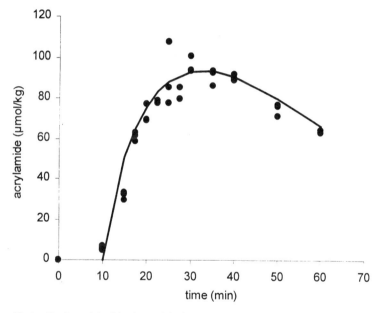

Figure 13. Application of the kinetic model given in Fig. 11 to the formation of acrylamide during the heating of potato cakes at 180°C. The line represents the trend predicted by the model. Experimental data for these plots are taken from Elmore et al. (2004).

Fig. 13 illustrates the quality of fit of the model to experiment for the formation of acrylamide (actual data also shown in Fig. 2), and Fig. 14 similarly illustrates the ability of the model to predict the total amino acid content. Overall, the fit is excellent. The rate constants for the loss of acrylamide k_5 and k'_5 are unreliable because we have no way at present of measuring their values in the early stages of acrylamide formation. Their values are, in any case, highly correlated to the other kinetic parameters. The value of k_2 is new and we have no otherwise derived value for comparison. Whereas the value for k_1 is of similar magnitude to that calculated from the initial rate of loss of glucose, the values of k and k_6 are highly consistent.

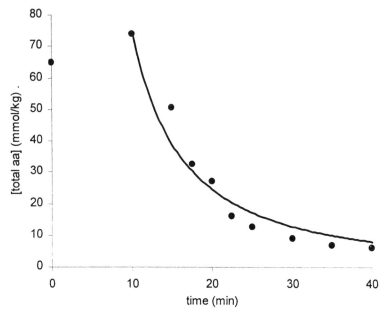

Figure 14. Application of the kinetic model given in Fig. 11 to the loss of total amino acids during the heating of potato cakes at 180°C. The line represents the trend predicted by the model. Experimental data for these plots are taken from Elmore et al. (2004).

5. THE MODEL AS A PREDICTIVE TOOL

An important use of a kinetic model such the one derived here, is to conduct 'what if' experiments, whereby the response to an imposed change in reaction conditions (*e.g.,* concentration, temperature) can be determined without actually doing the experiment on the food product. Thus, Fig. 15 illustrates the effect of added amino acids to the potato cakes and the way that the reactions which compete for intermediate 'int' can effectively reduce the yield of acrylamide. We have now carried out preliminary experiments in which potato cakes were spiked with glycine to confirm this behavior in practice (Barrenechea & Wedzicha, unpublished). Fig. 16 shows the effect of amino acid in more detail by illustrating how the concentration of Maillard products (defined here as all products other than acrylamide which are derived from intermediate 'int') varies relative to the concentration of acrylamide in potato cakes containing their natural concentration of amino acids, and in a sample where the concentration of non-asparagine amino acids had been quadrupled. As expected, the addition of an amino acid to potato speeds up the development of Maillard products; the graph is particularly useful because it now allows us to consider not only the formation of acrylamide but the consequent development of color and flavor when the recipe is changed.

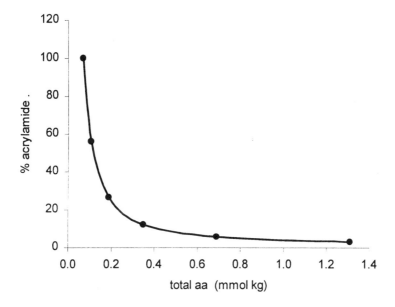

Figure 15. Use of the kinetic model to predict the effect increasing total amino acid (non asparagine) in potato on the acrylamide content of potato cakes heated at 180°C for 25 min. The first point (100 % acrylamide) represents the natural composition of the potato. Successive data points represent x2, x4, x8, x16 and x32 the amount of amino acids (non-asparagine).

6. CONCLUSION

Detailed analysis of the kinetic data obtained by Elmore et al. (2004) has allowed us to propose an overall structure for a kinetic model for acrylamide formation in potato, rye and wheat cakes heated at 180°C. The model is highly self-consistent. We have identified that the stoichiometric ratio between the loss of amino acids and formation of acrylamide is the same for every amino acid. Thus, the reactivity of all amino acids relative to asparagine and to acrylamide formation is the same; only the concentration of each amino acid determines the amount reacted.

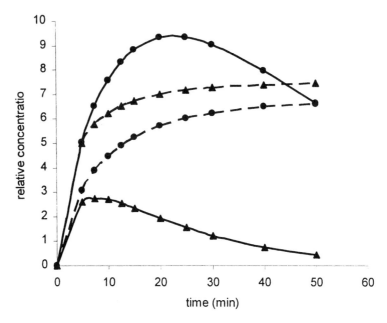

Figure 16. Use of the kinetic model to predict the concentration of acrylamide (solid line) and 'Maillard products' (broken line) formed in potato cakes heated at 180°C containing their natural concentration of amino acids (●) and with the concentration of amino acids (non-asparagine) quadrupled (▲).

This suggests that the competing reactions are all fast; the amino acids simply mop up the precursor intermediate, and the amount of individual amino acids lost depends on the relative numbers of molecules of each present. There is a certain (fixed) probability that asparagine, on reaction, produces acrylamide alongside the 'Maillard' product. The model works very well for potato cakes and can be used to predict the effects of time on acrylamide formation.

Analysis of the data also reveals that model sensitivity to amino acid type seems to be low, for which we can offer no explanation at present. This represents a shortcoming in our understanding of the rate-controlling factors because the data cannot be used to determine how individual amino acids control the formation of reactive intermediates.

The kinetics of acrylamide loss have only been determined for the reactions which take place at long heating times, *i.e.*, after all the amino acids have been exhausted. Since it is envisaged that acrylamide reacts with amino acids, it is possible that the actual yield of acrylamide could be much higher than measured. This possibility needs to be explored in detail and we still require data on the reactivity of acrylamide with food components.

ACKNOWLEDGEMENT

The financial support of a consortium of United Kingdom food producers is greatly appreciated by the authors.

REFERENCES

Elmore, E.J., Koutsidis, G., Dodson, A.T., and Mottram, D.S., 2005, The effect of cooking on acrylamide and its precursors in potato, wheat and rye, in: *Chemistry and Safety of Acrylamide in Food*, M. Friedman and D.S. Mottram, eds, Springer, New York, pp. 255-269.

Gögüs, F., Wedzicha, B.L., and Lamb J., 1998, Modeling of Maillard reaction during the drying of a model matrix, *J. Food Eng.* **35**:445-458.

Leong, L.P., and Wedzicha, B.L., 2000, A critical appraisal of the kinetic model for the Maillard browning of glucose with glycine, *Food Chem.* **68**:21-28.

Mottram, D. S., Wedzicha, B. L., and Dodson, A., 2002, Acrylamide is formed in the Maillard reaction, *Nature* **419**: 448–449.

Mundt, S., and Wedzicha, B.L., 2003, A kinetic model for the glucose-fructose-glycine browning reaction, *J. Agric. Food Chem.* **51**:51-3655.

Swales, S., and Wedzicha, B.L., 1992, Kinetics of the sulphite-inhibited browning of fructose, *Food Add. Contam.* **9**:479-483.

Wedzicha, B.L., and Kedward, C., 1995, Kinetics of the oligosaccharide-glycine-sulphite reaction: relationship to the browning of oligosaccharide mixtures, *Food Chem.* **54**:397-402.

THE EFFECT OF COOKING ON ACRYLAMIDE AND ITS PRECURSORS IN POTATO, WHEAT AND RYE

J. Stephen Elmore[1], Georgios Koutsidis[1], Andrew T. Dodson[1], Donald S. Mottram[1] and Bronislaw L. Wedzicha[2]
[1]*School of Food Biosciences, The University of Reading, Whiteknights, Reading, RG6 6AP;*
[2]*Procter Department of Food Science, University of Leeds, Leeds, LS2 9JT, United Kingdom;*
e-mail: j.s.elmore@reading.ac.uk

Abstract: The relationship between acrylamide and its precursors, namely free asparagine and reducing sugars, was studied in simple cakes made from potato flake, wholemeal wheat and wholemeal rye, cooked at 180°C, from 5 to 60 min. Between 5 and 20 min, large losses of asparagine, water and total reducing sugars were accompanied by large increases in acrylamide, which maximized in all three products between 25 and 30 min, followed by a slow linear reduction. Acrylamide formation did not occur to any extent until the moisture contents of the cakes fell below 5%. A comparison of each type of cake with a commercial product, made from the same food material, showed that acrylamide levels in all three commercial products were well below the maximum levels in the cooked cakes.

Key words: acrylamide, free asparagine, reducing sugars, amino acids, rye, wheat, potato

1. INTRODUCTION

Significant levels of acrylamide (above 200 µg per kg) have been reported in many cooked foods of plant origin, including potato, wheat and rye products (Friedman, 2003), where it is formed from the reaction between free asparagine and intermediates of the Maillard reaction (Mottram et al., 2002; Stadler et al., 2002). The extent of formation is determined by the concentration and types of sugars and amino acids present. In this chapter, data are presented to show how acrylamide forms in simple cakes, made from potato, rye and wheat heated at 180°C, and how free amino acids, reducing sugars and moisture contents change in these products with cooking time. We also compare the levels of precursors and acrylamide in the cooked cakes with those in commercial products.

Using the data from the cakes, a kinetic model can be developed, which can predict the formation of acrylamide during cooking (Wedzicha and Mottram, 2004). Thus, by understanding the relationship between acrylamide, its precursors and the physico-chemical properties of the food, potential means of reducing acrylamide in cooked foods can be derived.

2. EXPERIMENTAL PROCEDURES

2.1 Materials

Amino acids, sugars and ethyl acetate were purchased from the Sigma-Aldrich Company Ltd. (Poole, UK); sodium thiosulfate pentahydrate, hydrobromic acid, hydrochloric acid, potassium bromide and bromine from Fisher Scientific Ltd. (Loughborough, UK); 1,2,3-$^{13}C_3$-acrylamide (1 mg mL^{-1} in methanol (99% ^{13}C)) from Cambridge Isotope Laboratories, Inc. (Andover, MA); methanol from Merck Ltd. (Poole, UK).

Drum-dried potato flake, wholemeal wheat flour and wholemeal rye flour were provided by United Kingdom food producers.

2.2 Cakes made from potato, wheat and rye

Cakes made from potato and water (1:1.3), rye and water (2.2:1), and wheat and water (2.5:1) were cooked in an electric moving band impingement oven at 180°C over a range of times, from 5 to 60 min (Table 1). The uncooked cakes weighed approximately 18 g. They were 3 mm thick and 73 mm in diameter and were pricked, to minimize rising during cooking.

All procedures were carried out in triplicate.

2.3 Measurement of water content

Wheat and rye flours and potato flake samples (2.000±0.0050 g) were heated in an oven for 1 h at 120°C in uncovered metal dishes. The samples were then covered and allowed to cool for 45 min in a desiccator, before being weighed. They were then heated for a further hour at 120°C, allowed to cool and then reweighed. Moisture loss as a result of cooking was measured by weighing the cooked cakes directly before and immediately after cooking. Values were also measured for the moisture content of the cakes at room temperature, so that the concentrations of acrylamide and its precursors in the cakes could be calculated on a dry weight basis.

2.4 Measurement of acrylamide

The method of Castle et al. (1991), with some modifications, was used to extract and derivatize acrylamide from 5.0 g of ground cake, for analysis by gas chromatography-mass spectrometry (GC-MS). The extracting medium was methanol (30 mL), rather than water. After refluxing, [1,2,3-$^{13}C_3$]-acrylamide internal standard (500 ng), rather than methacrylamide, was added to the extract, along with the brominating reagent. Bromination took place overnight at room temperature, rather than in a refrigerator.

The brominated extract (2 µL) was injected onto a Clarus 500 GC-MS system (PerkinElmer, Inc., Boston, MA) in splitless mode at 250°C, the splitter opening after 0.5 min. Pulsed injection was used; helium carrier gas flow rate was 5 mL/min for 0.5 min, followed by a decrease to 1 mL/min over 0.5 min. Flow rate was maintained at 1 mL/min for 10 min and then increased over 0.5 min to 5 mL/min, until the end of the run. A DB-17 MS capillary column was used (30 m × 0.25 mm i.d., 0.15 µm film thickness; Agilent, Palo Alto, CA). The oven temperature was 85°C for 1 min, rising at 8°C/min to 200°C, then 30°C/min to 280°C for 10 min. The transfer line was held at 280°C and the ion source at 180°C. The mass spectrometer was operated in selected ion monitoring mode. Four ions characterized brominated [1,2,3-$^{13}C_3$]-acrylamide (*m/z* 108, 110, 153 and 155) and another four characterized brominated acrylamide (*m/z* 106, 108, 150 and 152). Ions *m/z* 155 and *m/z* 152 were used to quantify brominated [1,2,3-$^{13}C_3$]-acrylamide and brominated acrylamide, respectively.

2.5 Determination of free amino acids

The sample was ground and weighed into a 7 mL vial. Sample size was 1.00 g for wheat, 0.50 g for rye and 0.20 g for potato. Hydrochloric acid (0.01 M) was added (5 mL) to the vial and the sample was stirred for 15 min at room temperature. After stirring, the sample settled for 45 min. An aliquot of supernatant (2 mL) was centrifuged for 30 min and 100 µL of clear liquid were removed for analysis.

The amino acids in the extracts and standards were derivatized using the EZ-Faast amino acid analysis kit (Phenomenex, Torrance, CA; Hušek, 2000). The derivatized amino acids were extracted into isooctane/chloroform (100 µL) and analyzed using the same GC-MS system as before. The derivatized amino acids (2 µL) were injected at 250°C in split mode (5:1) onto a 10 m × 0.25 mm Zebron ZB-AAA capillary column (Phenomenex). The oven temperature was 110°C for 1 min, then increased at 30°C/min to 320°C. Carrier gas flow rate throughout the run was 1.1 mL/min, the ion source was at 220°C and the transfer line at 320°C. A specific mass spectral fragment ion was chosen for quantification of 22 amino acids in the samples. The area of this ion in the peak of each amino acid was measured relative to the area of the m/z 158 ion of norvaline internal standard.

2.6 Analysis of sugars by ion chromatography

Ground cakes (0.50 g) were extracted for 30 min with 10 mL water, containing trehalose internal standard (300 µg). Activated carbon (0.1 g) was added to the extract for 30 min to remove color and was then removed by centrifugation. The extract was further purified using Sep-Pak® Plus C_{18} cartridges (Waters Corporation, Milford, MA), then filtered through a 0.2µm filter disk (Whatman Inc., Clifton, NJ). Flours were extracted using 10 mL of 50% methanol. The methanol was evaporated and the extract was treated as described previously.

Analysis was performed using a Dionex 8220i ion chromatograph with pulsed amperometric detector (Dionex Corporation, Sunnyvale, CA). Twenty microlitres of sample were injected by autosampler onto a Carbopac PA10 column (Dionex) at room temperature. A gradient program was set up using 200 mM sodium hydroxide (solvent **A**) and water at a flow rate of 1 mL/min; 50 % solvent **A**, held for 10 min, then increased to 100% at 35 min. The column was then washed for 10 min with 550 mM sodium acetate in 100 mM NaOH and re-equilibrated with 50 % solvent **A** for 10 min. Glucose, maltose, fructose and sucrose were quantified.

2.7 Analysis of commercial products

Acrylamide, amino acid and sugar levels were measured in three commercial products:
- An extruded potato snack
- A rye-based biscuit
- A wheat-based biscuit

No modifications in analytical procedure for these products were necessary.

3. RESULTS AND DISCUSSION

All quantities were calculated on a dry weight basis and the values at 0 min were for unheated rye flour, potato flake and wheat flour. Analysis on a dry weight basis compensated for any changes in acrylamide and precursor values, due to concentration effects caused by loss of water from the cakes during cooking. Emphasis is placed on the cooking time of 20 min because this approximates to the maximum cooking time in a commercial process. Above this time the cakes were observed to be overcooked.

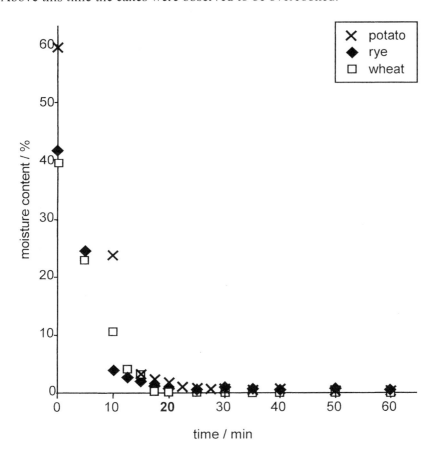

Figure 1. Variation of moisture content with cooking time for rye, potato and wheat cakes cooked at 180°C.

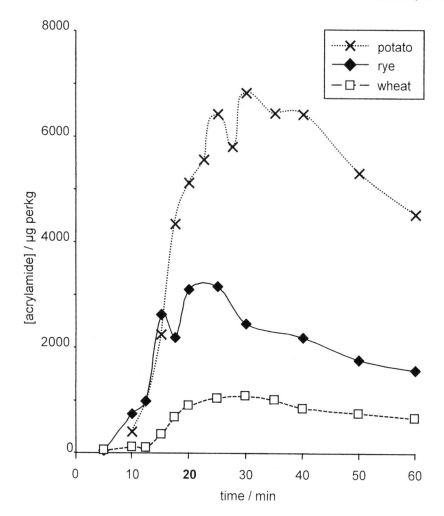

Figure 2. Acrylamide in rye, potato and wheat cakes heated at 180°C.

3.1 Measurement of water content

The moisture contents of the cooked cakes, directly after cooking are shown in Fig. 1. Leung et al. (2003) measured water loss from fritters made from wheat flour, which were deep-fried at 170°C, 190°C and 210°C. They found that moisture content was inversely proportional to acrylamide content at all three temperatures.

3.2 Measurement of acrylamide

The three curves, showing acrylamide formation with time, for each of the products at 180°C, are compared in Fig. 2. All three curves are similar, with a rapid increase in acrylamide formation between 10 and 20 min, reaching a maximum value between 20 and 35 min, followed by a slow linear decrease. Maximum concentrations for acrylamide in potato, rye and wheat cakes were 6800, 3200 and 1100 µg/kg, respectively. Because the cakes were cooked for a relatively long time, compared to commercial products, these values were relatively high. In potato strips cooked at 200°C (Rydberg et al., 2003) maximum acrylamide concentration was similar to the value for the potato cakes, but at a slightly earlier time.

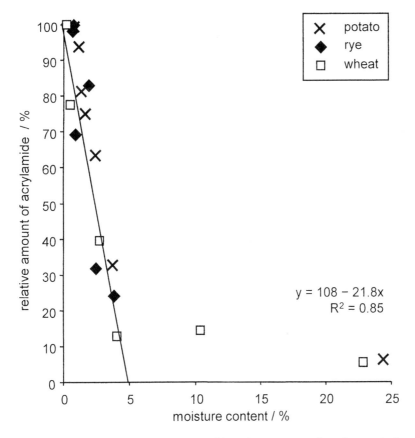

Figure 3. Variation of acrylamide content with moisture content for cakes made from rye, potato and wheat.

Relative acrylamide formation, up to and including the maximum amount formed (maximum acrylamide formation = 100), was plotted against moisture content for all three cakes (Fig. 3). Acrylamide formation only occurred to any extent when moisture levels in the cakes had fallen below 5%. For those data points where moisture was below 5%, acrylamide formation was inversely proportional to moisture content.

In all three foods there was a relatively slow linear decrease of acrylamide, once maximum production had been reached, which could be due to a secondary reaction between acrylamide and other food components or possibly evaporation of acrylamide from the surface of the cakes.

3.3 Determination of free amino acids

The free asparagine contents of the rye, potato and wheat cakes are shown in Table 1. Data for the other free amino acids in rye, potato and wheat are presented elsewhere (Elmore et al., 2004). The concentrations of all of the amino acids in all three foods were much lower after 40 min heating than at 0 min, although some amino acids increased in concentration in some of the products early in the cooking process. Uncooked potato flake contained both the highest concentration of asparagine and total free amino acids of the three foods, followed by rye flour, then wheat flour (Table 2). The asparagine concentration in the potato flake was 3500 mg per kg dry weight, approximating to 700 mg per kg in fresh potato (Souci et al., 2000).

Table 1. Effect of cooking time on free asparagine levels in potato, rye and wheat, and cakes derived from them (μmol per kg dry weight)

time / min	rye	potato	wheat
0	4790	27000	1330
5	7720		1320
10	3520	31300	1080
12.5	4200		1540
15	589	20400	725
17.5	995	11500	272
20	348	8320	102
22.5		3310	
25	188	2380	44
30	153	1280	33
35		874	
40	149	760	26

When comparing rye, potato and wheat together, the decomposition curves of the amino acids alanine, glycine, α-aminobutyric acid, valine, leucine, isoleucine, serine, threonine, proline, aspartic acid, methionine, phenylalanine, tyrosine and tryptophan upon heating were similar. At 20 min, when amino acid losses and formation of acrylamide were at their highest, amino acid levels in rye and wheat for these amino acids were less than 20% of their original value, whereas levels for potato were still over 45% of their original value. The decomposition curves for a typical member of this group of amino acids, valine, are shown in Fig. 4. The concentration of valine at 0 min in all three foods is given a value of 100% and other concentrations are measured relative to this value.

Table 2. Some chemical properties of rye flour, wheat flour and potato flake (dry weight)

	rye	potato	wheat
asparagine (µmol per kg)	4790	27000	1330
total free amino acids (µmol per kg)	17300	64900	7840
asparagine (% total free amino acids)	28%	42%	17%
other major amino acids	aspartic acid 18% alanine 16%	γ-aminobutyric acid 10% aspartic acid 7%	aspartic acid 18% γ-aminobutyric acid 15% tryptophan 12%
total reducing sugars / mmol per kg	57.7	53.2	49.3

Fig. 4 also shows the decomposition curves for asparagine, which differed from those for valine in several ways. Asparagine increased in rye early in the cooking process. Even so, by 20 min in wheat and rye, asparagine had decomposed to a similar extent as valine. However, in potato asparagine was relatively unstable, compared to valine, and, by 20 min, only 30% of its original amount remained. By 40 min, the relative amount of asparagine in all three foods was similar (2-3% of original).

3.4 Analysis of sugars by ion chromatography

The reducing sugars contents of potato flake, rye flour and wheat flour are shown in Table 3. Data for sucrose is presented elsewhere (Elmore et al., 2004). The major difference between the cereals and potato was that maltose was not detected in the potato cakes or flake, whereas in rye flour, maltose comprised 56% and, in wheat flour, 89% of the total reducing sugars. Total reducing sugars, i.e., glucose, fructose and maltose, were similar in all three foods (Table 2) and, in all three foods; glucose levels were similar to fructose levels. Sucrose, a non-reducing sugar, decomposed linearly with heating time for all three foods, with approximately 80% still remaining in each food after 20 min. Leszkowiat et al. (1990) showed that sucrose in a fried potato model system could participate in the Maillard reaction. They suggested that hydrolysis of sucrose to fructose and glucose could occur.

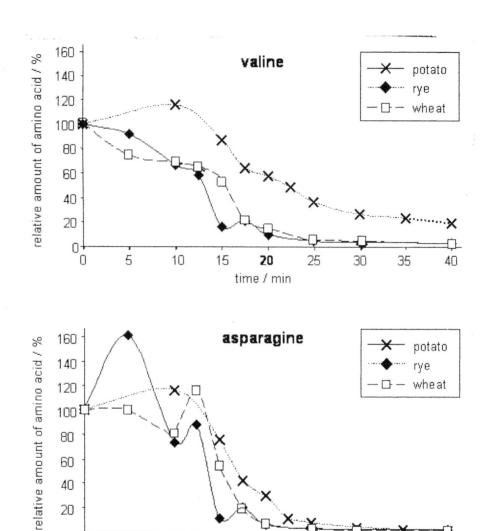

Figure 4. Effect of cooking time at 180°C on valine and asparagine concentrations in cooked rye, potato and wheat cakes.

Table 3. Effect of cooking time on reducing sugar levels in potato, rye and wheat, and cakes derived from them (mmol per kg dry weight)

time / min	rye			potato		wheat		
	glucose	fructose	maltose	glucose	fructose	glucose	fructose	maltose
0	11.9	13.4	32.4	28.6	24.6	3.02	2.36	43.9
5	7.53	7.38	29.0			4.09	2.67	40.9
10	9.62	10.8	19.2	24.0	22.5	3.29	2.64	27.9
12.5	4.79	6.01	12.2			2.83	2.40	28.8
15	5.95	7.73	6.58	7.38	14.1	3.93	2.80	24.1
17.5	3.76	3.38	4.14	1.72	5.56	3.92	2.13	14.4
20	3.56	3.51	3.41	tr	3.18	3.97	2.50	10.5
22.5				tr	1.58			
25	3.48	3.47	1.30	tr	1.24	4.37	2.87	6.49
27.5				tr	1.40			
30	3.36	3.02	tr	–	1.10	3.87	2.82	5.42
35	2.92	2.69	–	–	tr	3.64	2.70	3.50
40	2.39	2.18	–	–	1.17	3.33	2.48	3.01

tr: less than 1 mmol per kg; –: not found above detection limit (0.5 mmol per kg)

The decomposition of maltose in the two cereals was rapid, relative to glucose and fructose, both of which decreased slowly upon heating, particularly in wheat. Hollnagel and Kroh (2000) reported that maltose decomposed to glucose in a Maillard reaction model system, which could explain why glucose levels were maintained in the cereals. Theander and Westerlund (1988) reported an increase in fructose in drum-dried wheat flour and suggested it may arise from the breakdown of fructose-containing carbohydrates, such as raffinose and fructans, which were not measured in this work. In potato, glucose and fructose decomposed readily. Hence, plots of total reducing sugars against cooking time were similar for all three foods (Fig. 5). After 20 min, 34% of total reducing sugars remained in wheat, 18% in rye and 7.5% in potato. Within each food the decomposition curves for fructose and glucose were similar.

3.5 Analysis of commercial products

The asparagine, sugar and acrylamide compositions of the commercial products are shown in Table 4, and Fig. 6 shows the acrylamide and asparagine compositions of each product, relative to the curves obtained for the cooked cakes. The acrylamide content of the potato snack was equivalent to that in the model potato cake, which had been cooked for just over 10 min, whereas its asparagine content corresponded to a potato cake with a cooking time of about 15 min. The reducing sugars content of the potato snack (not plotted) was similar to that of the potato cake cooked for 20 min. Amrein et al. (2003) have shown large variation in both the reducing sugar and free asparagine compositions of fresh potato cultivars. It is likely that different cultivars were used in the potato snack and the cakes. The acrylamide and asparagine contents of the rye biscuit corresponded to cooking times of 9 min and 12 min, respectively, in the model rye cake. The level of acrylamide in the wheat biscuit corresponded to a cooking time of 18 min in the model wheat cake and the level of asparagine a cooking time of 15 min. The level of reducing sugars was equivalent to that found in the model wheat cake cooked for 7 min. The wheat biscuit was made from white flour, which is lower in sugars (Theander and Westerlund, 1988) and asparagine (Elmore, unpublished data) than wholemeal flour. Fat, salt and/or sugar in the commercial products may also have an effect on their levels of acrylamide.

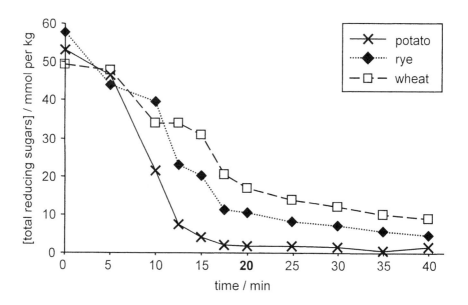

Figure 5. Effect of cooking time at 180°C on the total reducing sugar levels in rye, potato and wheat cakes.

Table 4. Composition of commercial products

	wheat biscuit	rye biscuit	potato snack
acrylamide / µg per kg	320	600	530
asparagine / µmol per kg	260	3200	21000
reducing sugars / mmol per kg	42	71	3.5

3.6 Relationships between acrylamide, asparagine and reducing sugars

In the potato flake, the molar concentration of total free amino acids was slightly greater than the molar concentration of total reducing sugars (Table 2), and the ratio of amino acids to reducing sugars was 1.2. In both cereals, the molar concentrations of total reducing sugars were greater than that of the total free amino acids (amino acids: reducing sugars = 0.30 for rye, 0.16 for wheat). Hence, in rye and wheat, there was a large excess of reducing sugar, compared to free amino acids, and so relative losses of reducing sugars in the two cereals were low. In potato, the slight excess of free amino acids resulted in a greater loss of reducing sugars compared with the cereals, with a correspondingly smaller loss of amino acids. These observations can be related to the formation of acrylamide in all three foods. A plot of asparagine loss against acrylamide formation showed a linear relationship, with a gradient, which equated to the conversion rate of asparagine to acrylamide (data not shown). In potatoes only 0.29% of asparagine was converted to acrylamide, in rye the value was 0.80% and in wheat it was 0.98%.

4. CONCLUSIONS

The relationship between losses of amino acids and reducing sugars and formation of acrylamide can be readily measured using simple cakes made from the food of interest and water. These data can then be used to generate kinetic relationships for acrylamide formation during the cooking process. Future work should examine more closely the first 20 min of heating at 180°C, as well as examining the effects of cooking temperature and the addition of other food ingredients, such as lipid, or increased levels of amino acids other than asparagine.

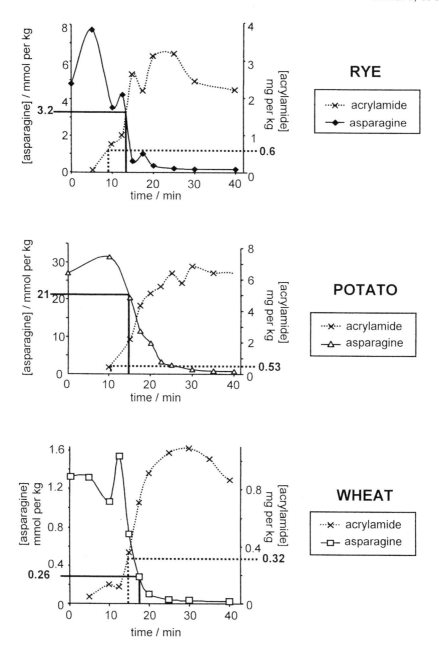

Figure 6. Asparagine and acrylamide levels in commercial products made from rye, potato and wheat, and the variation of asparagine and acrylamide with cooking time for rye, potato and wheat cakes.

REFERENCES

Amrein, T. M., Bachmann, S., Noti, A., Biedermann, M., Barbosa, M. F., Biedermann-Brem, S., Grob, K., Keiser, A., Realini, P., Escher, F., and Amado, R., 2003, Potential of acrylamide formation, sugars, and free asparagine in potatoes: A comparison of cultivars and farming systems, *J. Agric. Food Chem.* **51**: 5556–5560.

Castle, L., Campos, M.-J., and Gilbert, J., 1991, Determination of acrylamide monomer in hydroponically grown tomato fruits by capillary gas chromatography-mass spectrometry, *J. Sci. Food Agric.* **54**: 549–555.

Elmore, J. S., Koutsidis, G., Dodson, A. T., Wedzicha, B. L. and Mottram, D. S., 2005, Measurement of acrylamide and its precursors in potato, wheat and rye model systems, *J. Agric. Food Chem.* **53**: in press.

Friedman, M., 2003, Chemistry, biochemistry, and safety of acrylamide. A review, *J. Agric. Food Chem.* **51**: 4504–4526.

Hollnagel, A., and Kroh, L. W., 2000, Degradation of oligosaccharides in nonenzymatic browning by formation of α-dicarbonyl compounds via a "peeling off" mechanism, *J. Agric. Food Chem.* **48**: 6219–6226.

Hušek, P., 2000, Method of preparing sample for amino acid analysis and kit for analyzing the same. *European Patent Application* EP 1033576.

Leszkowiat, M. J., Barichelo, V., Yada, R. Y., Coffin, R. H., Lougheed, E. C., and Stanley, D. W., 1990, Contribution of sucrose to non-enzymatic browning in potato chips, *J. Food Sci.* **55**: 281–282.

Leung, K. S., Lin, A., Tsang, C. K., and Yeung, S. T. K., 2003, Acrylamide in Asian foods in Hong Kong, *Food Additives and Contaminants* **20**: 1105–1113.

Mottram, D. S., Wedzicha, B. L., and Dodson, A., 2002, Acrylamide is formed in the Maillard reaction, *Nature* **419**: 448–449.

Rydberg, P., Eriksson, S., Tareke, E., Karlsson, P., Ehrenberg, L., and Törnqvist, M., 2003, Investigations of factors that influence the acrylamide content of heated foodstuffs, *J. Agric. Food Chem.* **51**: 7012–7018.

Souci, S.W., Fachmann, W., and Kraut, H., 2000, *Food Composition and Nutrition Tables, 6th ed.*, CRC Press, Boca Raton, pp. 639–641.

Stadler, R. H., Blank, I., Varga, N., Robert, F., Hau, J., Guy, P. A., Robert, M.-C., and Riediker, S., 2002, Acrylamide from Maillard reaction products, *Nature* **419**: 449–450.

Theander, O., and Westerlund, E., 1988, The effects of aqueous ethanol-soluble carbohydrates and protein in heat-processed whole grain wheat and white flour, *J. Cereal Sci.* **7**: 145–152.

Wedzicha, B. L., Mottram, D. S., Elmore, J. S., Koutsidis. G., Dodson, A.T., 2005, Kinetic models as a route to control acrylamide formation in food, in: *Chemistry and Safety of Acrylamide in Food*, M. Friedman and D.S. Mottram, eds, Springer, New York, pp. 235-253..

DETERMINATION OF ACRYLAMIDE IN VARIOUS FOOD MATRICES

Evaluation of LC and GC mass spectrometric methods

Adam Becalski, Benjamin P.-Y. Lau, David Lewis, Stephen W. Seaman, Wing F. Sun
Health Canada, Health Products and Food Branch, Food Research Division, Address Locator 2203D, Ottawa, Ontario, Canada, K1A 0L2; e-mail: Adam_Becalski@hc-sc.gc.ca

Abstract: Recent concerns surrounding the presence of acrylamide in many types of thermally processed food have brought about the need for the development of analytical methods suitable for determination of acrylamide in diverse matrices with the goals of improving overall confidence in analytical results and better understanding of method capabilities. Consequently, the results are presented of acrylamide testing in commercially available food products - potato fries, potato chips, crispbread, instant coffee, coffee beans, cocoa, chocolate and peanut butter, obtained by using the same sample extract. The results obtained by using LC-MS/MS, GC/MS (EI), GC/HRMS (EI) – with or without derivatization – and the use of different analytical columns, are discussed and compared with respect to matrix borne interferences, detection limits and method complexities.

Key words: acrylamide, GC/MS, LC-MS/MS

1. INTRODUCTION

The discovery of acrylamide in the human foods (Tareke et al., 2000, 2002; Rosen and Hellenas, 2002; Ahn et al., 2002; Becalski et al., 2003) led to surveys exploring the levels of that potentially hazardous chemical (IARC, 1994), and spurred a search into suitable analytical procedures for its determination in foodstuffs.

The potential presence of acrylamide in foods was initially investigated, in a tomato and mushroom matrix, by employing derivatization of acrylamide (bromination of a double bond) and subsequent GC-MS detection (Andrawes et al., 1987; Castle et al., 1991; Castle, 1993; Gertz and Klostermann, 2002; Nemoto et al. 2002). Later, Rosen and Hellenas (2002), developed a LC-MS/MS based method, and soon a few more variants of that procedure appeared in the literature (US FDA, 2003; Swiss Feral Office Public Health, 2002; Ono et al. 2003; Takatsuki et al., 2003), some as a simpler LC-MS methodology (Sanders et al., 2002; Cavalli et al., 2003). Also, several groups reported using GC-MS techniques for determination of acrylamide without derivatization (Haase et al., 2003; Biedermann et al., 2002; Tateo and Bononi, 2003; Rothweiler and Prest, 2003). Results from round robin studies indicated that for some matrices such as bread, cereals and potato products, most methods produced comparable data, while for matrices such as cocoa there was a significant divergence of results (Clarke et al., 2002; Fauhl et al., 2002). Moreover, the cleanup procedures for different analytical methodologies varied, thus making direct comparison between methods difficult. One group compared data obtained by LC-MS/MS method to GC-MS (no derivative) (Swiss Federal Office PH, 2002), while Ono et al. (2003) and Takatsuki et al. (2003) compared it to GC-MS (bromo derivative). Only the latter study compared results obtained for a coffee bean matrix.

Our aim of the work reported here was to test the feasibility of using a common extraction process with different food matrices, for the determination of acrylamide via different mass spectrometric techniques (LC-MS/MS, GC/MS (EI), GC/HRMS (EI) – with or without derivatization of acrylamide and through the use of different analytical columns. The resulting data obtained by different techniques was then assessed to indicate which instrumental technique (or techniques) are applicable for determination of acrylamide in a particular matrix.

For sample extraction, we adopted a water extraction procedure known to produce a fairly clean extract. After several trials, the solid phase extraction step was optimized. It consisted of Oasis HLB (Waters) and Accucat (Varian) or Multimode (IST) cartridges.

Using an isotope dilution ($^{13}C_3$-acrylamide) method, extracts of several different matrices: French fries, crisp bread, potato chips, peanut butter, coffee beans, chocolate and cocoa, were analyzed.

A review summarizing analytical procedures for acrylamide has been published (La Calle et al., 2004).

2. MATERIALS AND METHODS

2.1 Chemicals

Dichloromethane (pesticide grade) and methanol (HPLC grade) were obtained from EM Science (Gibbstown, NJ). Water was obtained from a MilliQ Gradient A10 purification system (Millipore). All other reagents were of analytical grade. Acrylamide, 99+%, was from Aldrich and labeled standard of acrylamide, $^{13}C_3$-acrylamide, 99% was from Cambridge Isotope Laboratories (Andover, MA). All stock acrylamide solutions (400 µg/mL and 250 µg/mL) and calibration solutions were prepared in water. Working quantities of standards were stored at 4°C while stock was kept at -18°C.

2.2 Foods

French fries were obtained from a commercial establishment and were stored at -18°C before analysis. Crisp bread sample (T-3001) was obtained through FAPAS and had an assigned consensus acrylamide value of 1213 ng/g; Coffee beans and instant coffee were a gift of R. Stadler (Nestle, Switzerland); baking chocolate was a standard reference material (NIST 2384); all other foods were purchased in the Ottawa area.

2.3 Equipment/Materials

High Pressure Liquid Chromatograph, model 1100 consisting of autosampler, binary pump, degasser and column oven (Agilent, Palo Alto, CA); triple quadrupole tandem mass spectrometer, Quattro-Ultima (Micromass Inc., Manchester, UK); data system, MassLynx version 3.5 (Micromass, UK); analytical column, 2.1 mm ID x 250 mm, 5 µ Aquasil C18, 77505-252130 (Thermo Hypersil-Keystone, Bellefonte, PA), with C18 2 x 4 mm guard column, AJO-4286, (Phenomenex, Torrance, CA); Gas Chromatograph Mass Spectrometer, model 6890/5973 with 7673 injector (Agilent); capillary column 30 m x 0.25 mm x 0.25 µm, Supelcowax, (Supelco, Bellefonte, PA); single taper injection sleeve (Supelco) with carbofrits (Restek, Bellefonte, PA), single taper injection sleeve with glass wool (Agilent); Fisons Autospec Ultima high resolution mass spectrometer coupled with a Fisons GC8000 series gas chromatograph (Micromass); data system, OPUS 3.7/MassLynx 4.0 (Micromass); centrifuge tubes, FEP, 50 ml, 3114-0050 (Nalgene, Rochester, NY); shaker, horizontal (Eberbach, Ann Arbor, MI); centrifuge, fixed angle rotor, RC-2B (Sorvall, Asheville, NC); centrifuge, swinging bucket rotor, RC-3B (Sorvall); centrifuge filter, 15 mL, 5 kDa cut-off, Centricon Plus-20, UFC2BCC08 (Millipore, Bedford, MA);

Oasis HLB polymeric cartridge, 6 mL, 200 mg, 106202 (Waters, Milford, MA); Accucat anion and cation exchange cartridge, 3 mL, 200 mg, 1228-2003 (Varian, Walnut Creek, CA); multimode cartridge (C18, cation, anion exchange) 3 mL, 300 mg, 9040030B (IST, Hengoed Mid Glamorgan, UK).

2.3.1 Liquid Chromatograph Mass Spectrometer Operating Conditions (MS/MS mode)

Mobile phase: 12% methanol in 1 mM aqueous ammonium formate (isocratic); flow rate: 0.175 mL/min; injection volume: 10-20 µL; column temp.: 28°C; autosampler temp.: 10°C; ionization mode: positive ion electrospray; desolvation gas temp.: 250°C; source temp.: 120°C; desolvation gas flow: 525 L/hr; cone gas flow: 150 L/hr; collision gas pressure: 2.9×10^{-3} mbar (argon); resolution settings: around 80% valley separation for both quadrupoles; ion energies: 0.5 V and 1.0 V for quadrupole 1 and 3; precursor ion→product ion transitions in multiple reaction monitoring (MRM): m/z 75 →58 (collision energy 11 eV); m/z 72→55 (11 eV); m/z 72 →54 (11 eV); m/z 72→44 (14 eV); m/z 72→27 (16 eV); cone voltage was 34 V for all MRM transitions, dwell time for each MRM transition: 0.3 sec; mass span: 0.1 Da; interchannel delay: 0.05 - 0.1 sec. APCI parameters: Corona: 3 µA; desolvation gas temp.: 400°C; desolvation gas flow: 143 L/hr; cone gas flow: 101 L/hr.

2.3.2 Gas Chromatograph Mass Spectrometer Operating Conditions (LR MS mode)

Column: Supelcowax-10, (Supelco), (30 m x 0.25 mm x 0.25 µm), carrier gas: helium, constant flow, 1 mL/min, oven temperature profile: initial, 55°C (1 min), rate, 10°C/min. to 210°C, rate, 30°C/min. to 240°C, hold 10 min.; injector temp. 200°C, injection mode: splitless; purge time: 0.75 min; ionization mode: 70eV EI+; scan mode: selected ion monitoring (SIM); condition for underivatized (water) samples: injection volume 1 µL, pressure pulse 40 psi, 0.5 min, ions, (m/z): 55, 71, 74; condition for derivatized (ethyl acetate) samples: injection vol. 2 µL, pressure pulse 30 psi, 0.5 min, ions, (m/z): 70, 106, 110, 149, 151, 154.

2.3.3 Gas Chromatograph Mass Spectrometer Operating Conditions (HR MS mode)

Column: Supelcowax-10, carrier gas: helium, constant pressure, 21 psi; oven temp. profile: initial, 55°C (1 min), rate, 10°C/min. to 210°C, rate, 30°C/min. to 240°C, hold 10 min.; injection mode: cool on-column;

injection vol. 1 µL; mass spectrometer resolution: 8000 (10% valley); ionization mode: 70eV EI+; scan mode: voltage selected SIM; ions (m/z): 148.9476, 150.9456, 151.9577, 153.9556.

2.4 Typical Food Sample Preparation

A sample (100 g) was homogenized in a blender at maximum speed with water (500 mL) for 1 minute and the homogenate (24 g) was transferred to a 50 mL centrifuge tube. (Alternatively, 4 g of sample and 20 mL of water were homogenized directly in a centrifuge tube). Isotopically labeled acrylamide spiking solution (32 µL of 25 µg/mL) and 10 mL of dichloromethane were added to the tube. The mixture was shaken at high speed on a horizontal shaker for 15 min and centrifuged at 15,000 rpm (~24,000 g) in a RC-2B centrifuge for 2 h at 4°C. The top (water) centrifugate layer (about 10 mL) was promptly transferred to a 5 kDa centrifuge filter and centrifuged at 3,500 rpm (~4,000 g) in a RC-3B centrifuge for 4 h at 4°C or longer if necessary.

2.5 Typical Food Sample Cleanup

Oasis HLB cartridges were conditioned with 1 x 5 mL of methanol followed by a 2 x 5 mL of water and Accucat and Isolute Multimode cartridges with 1 x 3 mL of methanol and 2 x 3 mL of water, respectively. Filtrate from RC-3B centrifuge (2 mL) was passed through an Oasis HLB cartridge; the cartridge was rinsed with water (1 mL) and eluted with water (1 mL). That eluate was loaded onto an Accucat cartridge, the first 0.5 mL of eluate was discarded and the remaining portion was collected in a vial. The cartridge was further eluted with 1 mL of water into the same vial and the combined eluate analyzed by LC-MS/MS.

(Alternatively, when the eluate from Oasis HLB was loaded onto the Isolute Multimode cartridge, the first 1 mL was discarded and the Multimode cartridge was eluted with 2 mL of water.)

2.6 Bromination Procedure

The scheme described by Clarke et al. (2002) was generally followed. Standards contained acrylamide at levels of 25, 50, 100, 500, 1000 µg/mL and $^{13}C_3$ acrylamide at 1000 µg/mL. Samples (extracts after SPE purification) or standards (1.5 - 2 mL) were treated with bromination reagent (0.5 mL) overnight at 4°C. After decomposition of excess of bromine with approximately 100 µL of 1 M thiosulphate, sodium sulphate (2 g) was added and the mixture extracted twice with 1 mL of ethyl acetate. Combined extracts were dried over 100 mg of sodium sulphate at -18°C for 1 h. Standards were spiked with 200 µL of triethylamine (TEA), while sample extracts were reduced in volume under a stream of N_2 to 100 µL and then spiked with 10 µL of TEA. Extracts containing TEA were stored at -80°C for two months without apparent degradation.

2.7 Statistical Analysis of Data

Data were processed using SigmaStat software version 2.0, (SPSS, Chicago, Ill.)

3. RESULTS AND DISCUSSION

Before product analysis, we conducted an investigation of the optimization of a SPE cleanup procedure step with the aim of choosing a cartridge that would retain acrylamide and allow for the rinsing off of polar interferences (sugars, and salts). From the cartridges tested, both polymeric (Oasis HLB) and graphiticised carbon (Envirocarb) cartridges showed an appreciable retention of acrylamide from pure aqueous solution. However, the residual dichloromethane content of our aqueous extract caused significant loss of retention of acrylamide on the carbon cartridge. We thus selected the Oasis HLB column as a first purification step. We then established that washing of the column with 1 mL of water before elution was effective in removing most salts and other polar contaminants. The appreciable (~30 %) losses of acrylamide during this step were offset by a much cleaner eluate. Our initial observation that a very large interfering peak in LC eluting close to the solvent peak can be effectively removed by a multimode type cartridge (containing *both* anionic and cationic ion exchange resins), prompted us to evaluate two cartridges – Isolute Multimode (IST) and Accucat (Varian) as a second clean-up step.

The first cartridge contains additional C-18 material while the latter one is based solely on C-3 type resin. The retention of acrylamide (in a water solution) was minimal on the Accucat cartridge (and thus all acrylamide was recovered in 1.5 mL) while Isolute required > 4 mL, resulting in a diluted sample. However, the cleanup using Multimode IST was superior to Accucat treatment in removing interfering compounds from matrices such as cocoa, chocolate and coffee products (Fig. 1).

The advantages of a much cleaner extract obtained by a wash of the HLB cartridge can be utilized by a reduction of the water volume in the final extract from 1.5 mL to 50 µL and a corresponding increase in analyte signal (through evaporation of water in a vacuum desiccator at 20 mm Hg over anhydrous calcium sulphate). Several starchy food extracts and calibration standards were evaporated in that fashion without measurable losses of acrylamide.

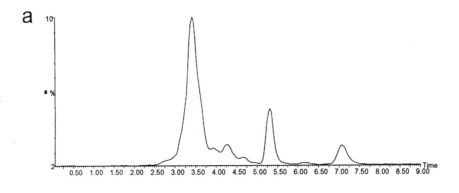

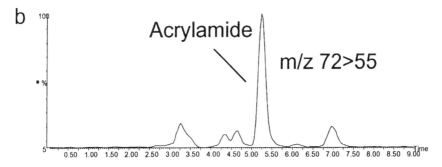

Figure 1. LC-MS/MS chromatogram of baking chocolate sample (acrylamide at ~190 ng/g) obtained after (a) Accucat, (b) IST Multimode cleanup.

The cleanup procedure was assessed on a set of 12 French fries and one of each sample of the following foods: crisp bread, potato chips, peanut butter, coffee beans, chocolate and cocoa powder.

Our modified LC-MS/MS procedure, used for this study as a reference method, was validated for repeatability using in-house reference material made from Pringles potato chips and with a French fries sample containing acrylamide at 802 and 60 ng/g levels, respectively. For recovery experiments, 4 g sub-samples of a French fries sample containing 60 ng/g acrylamide were spiked in triplicate with acrylamide at 500, 1000 and 10,000 ng/g.

The relative standard deviation (RSD) for repeated injection (six days over a 14 day period) of the extract of the Pringles potato chips sample (in-house reference material) containing acrylamide at a level of 774 ng/g was

2.3% (*n=9*). A RSD of 3.4% was obtained when replicates of the same potato chips sample were analyzed on different days over a 2 month period (*n=6*). The RSD for the French fries sample containing acrylamide at the 60 ng/g level was 2.6% (*n=6*). For samples (*n=3*) of a mixture of the above French fries sample spiked at 500, 1000 and 10,000 ng/g levels, the respective blank corrected concentrations and % RSD were: 555 ng/g, RSD 3.3%; 1090 ng/g, RSD 2.6%; 11,000 ng/g, RSD 1.5%. For samples (*n=3*) of a mixture of boiled and mashed potatoes spiked at 0 and 5 ng/g levels, the respective blank corrected concentrations and % RSD were: 3.8 ng/g, RSD 6.9%; 4.9 ng/g, RSD 7.5% (results obtained by vacuum concentration of extracts). The five point calibration curve (concentration of native acrylamide 10, 25, 50, 100 and 500 ng/mL; concentration of $^{13}C_3$-acrylamide 50 ng/mL) was linear with $r^2 = 0.9999$.

An ion transition *m/z* 72→55 was used for the quantification of acrylamide, while ion transition *m/z* 75→58 was used for the $^{13}C_3$ acrylamide. Using criterion signal to noise (S/N) 3:1 (peak-to-peak noise definition) at the *m/z* 72→55 transition, the limit of detection (LOD) was calculated as approximately 6 pg of standard injected on-column.

Later, the same set of samples was tested using three other techniques: GC-MS of aqueous extract, GC-MS and GC-HRMS of brominated aqueous extract. A summary of the data obtained is shown in Table 1. The cocoa sample (for all analyses) and coffee and chocolate samples for GC-MS of aqueous extract had to be processed via Multimode IST cartridge as interferences precluding quantification were obtained during Accucat treatment. The signed rank test indicates that there was a significant difference (P=0.011) between data generated by GC-MS of aqueous extract and LC-MS/MS. Other data sets were consistent with each other.

3.1 Important Parameters and Limitations for each Method

3.1.1 LC-MS/MS of aqueous extract

No improvement in the separation of coffee extract was observed using a microbore column (Genesis AQ 150 x 1 mm, 4 µm, Jones Chromatography). However, interferences occurring in native and $^{13}C_3$- acrylamide channels in the coffee matrix were separated by using two columns in series. In this setup, a 2 x 150 mm Ultracarb (Phenomenex) was placed before a C18 column. We were able to achieve a very good separation in a simpler way than the 4 column switching system described by Takatsuki et al. (2003).

Table 1. Comparison of concentration of acrylamide (ng/g) obtained by use of different analytical methods

Sample Matrix	LC-MS/MS	GC-LRMS	GC-LRMS (Br)[a]	GC-HRMS (Br)
French fries	188	188	169	175
" " "	105	135	98.9	107
" " "	114	154	108	104
" " "	252	339	270	231
" " "	131	168	123	120
" " "	60.1	78.7	52.4	58.0
" " "	292	283	284	259
" " "	84.3	116	87.2	79.9
" " "	211	212	203	194
" " "	115	141	103	103
" " "	221	317	197	211
" " "	112	136	104	99.5
Potato chips	803	844	821	752
Crisp bread	1243	1091	1311	1345
Cocoa powder	*45.0*	[b]	35.6	37.4
Fondue chocolate	70.0	*100*	68.3	80.2
Baking chocolate	191	*197*	164	165
Peanut butter	53.0	53.5	43.5	44.8
Coffee beans	247[c]	*251*	255	227
Instant coffee	753	[b]	706	611

Italics: Multimode IST clean-up
[a] Sample extract derivatized with Br_2
[b] Interferences
[c] Quantitated using tandem LC setup

3.1.2 GC-MS of aqueous extract

An ion *m/z* 71 was used for the quantification of native acrylamide while an ion *m/z* 74 was used for the $^{13}C_3$-acrylamide. A significant amount of chromatographic noise precluded the use of more qualifying ions (this constraint for internal standard might be surmounted by the use of a larger amounts). However, for cocoa and coffee samples, the intensity of the qualifier ion *m/z* 51 was outside the usually acceptable (± 20%) range. The final extract still contained co-extractives and thus on-column injection has proved unreliable. Using splitless injection with single taper or a single taper glass wool packed sleeve also produced spurious data. The quality of injection was improved by using two carbon based Carbofrits (Restek) inserts placed in the single taper sleeve. We have, however, encountered a few injections where the response of the analyte was very low, presumably due to bad vaporization of the sample during these injections. The three point calibration curve (concentration of native acrylamide 100, 500 and

1000 ng/mL; concentration of $^{13}C_3$-acrylamide 50 ng/mL) was linear with $r^2 = 0.997$.

The LOD for French fries sample was calculated as 40 ng/g (S/N =3) and was higher for other matrices. The interferences in the *m/z* 71 channel made quantification impossible for the cocoa sample and for instant coffee with acrylamide level > 500 ppb. No improvement in chromatography was observed by using a medium polarity DB-17 column. The RSD for repeated injection of the extract of potato chips sample was 3.9% (*n=4*). It is of interest to note that while coffee bean samples produced significant interferences during LC-MS/MS analysis, GC-MS trace was relatively free from interference, although with a higher LOD of 100 ng/g.

3.1.3 GC-LRMS of brominated extract

The ion *m/z* 149 was used for the quantification of native acrylamide while the ion *m/z* 154 was used for the $^{13}C_3$ acrylamide. The noise level in the qualifier channel for native acrylamide, *m/z* 106, was acceptable, while the two other channels (*m/z* 70, 110) suffered from significant noise in samples containing coffee and cocoa products. The splitless injection of TEA dehydrobrominated acrylamide extract using a single taper glass wool packed sleeve proved most reliable. While the on-column injection of an extract without TEA treatment did not show any occurrence of decomposition (dehydrobromination) caused by chromatography, it produced only a weak signal for the dibromo derivative molecular ion. However, the splitless injection using glass wool resulted in partial dehydrobromination. We therefore opted for triethylamine-assisted dehydrobromination before analysis to avoid possible problems associated with debromination during injection or chromatography. The five point calibration curve (concentration of native acrylamide 25, 50, 100, 500 and 1000 ng/mL; concentration of $^{13}C_3$-acrylamide 1000 ng/mL) was linear with $r^2 = 0.995$. The detection limit for the French fries sample was calculated as 10 ng/g (S/N =3). The interferences in the *m/z* 154 channel made quantification difficult in samples of instant coffee, coffee beans and chocolate. It appears that an additional cleanup step after bromination of the extract is required for these matrices.

3.1.4 GC-HRMS of brominated extract

The on-column injection of TEA dehydrobrominated acrylamide extract has proven acceptable. The five point calibration curve (concentration of native acrylamide 25, 50, 100, 500 and 1000 ng/mL; concentration of $^{13}C_3$-acrylamide 1000 ng/mL) was linear with $r^2 = 0.994$. The detection limit, which was essentially uniform for all sample extracts, was calculated as 15-20 ppb (injected), which is equivalent to approx. 5 ng/g in the samples (S/N =3). The interferences (minor) were only present in the instant coffee sample in the *m/z* 149 channel.

Chromatograms of the potato chip sample (acrylamide approximately 800 ng/g) obtained by different LC-MS and GC-MS methods are shown in Fig. 2. It is clear that all methods produced acceptable chromatographic data.

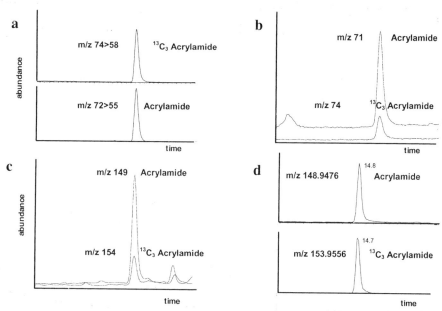

Figure 2. Chromatograms of potato chips sample (at ~800 ng/g) obtained by (a) LC-MS/MS, (b) GC/MS (EI), and after bromination (c) GC/MS (EI), (d) GC/HRMS (EI).

Chromatograms of the coffee bean sample (acrylamide at ~250 ng/g) obtained by different LC-MS and GC-MS methods are shown in Fig. 3. The interferences precluding quantification are seen in the top trace of LC-MS/MS when a single column was used.

Use of an atmospheric pressure chemical ionization (APCI) reduced interferences, when compared to electrospray, for peanut butter and ground coffee samples but this mode ionization mode was not effective for soluble coffee samples (Fig. 4).

It appears that the most robust analytical data were obtained by the use of GC/HRMS after derivatization by bromine.

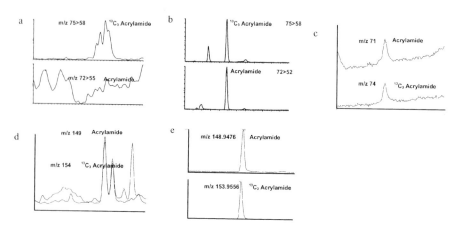

Figure 3. Chromatograms of coffee bean sample (at ~250 ng/g) obtained by (a) LC-MS/MS - single column, (b) LC-MS/MS - tandem columns, (c) GC/MS (EI), and after bromination (d) GC/MS (EI), (e) GC/HRMS (EI).

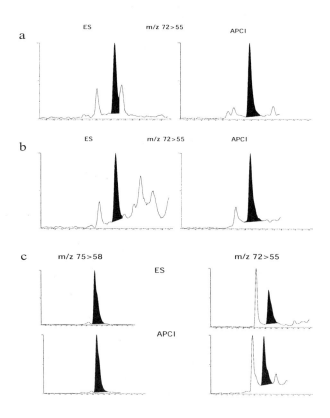

Figure 4. Chromatograms of (a) peanut butter, (b) ground coffee, and (c) instant coffee obtained by using electrospray (ES) or atmospheric pressure chemical ionization (APCI).

4. CONCLUSION

We have shown that a single extract obtained after the cleanup procedure described above is suitable for the determination for acrylamide in many different matrices. It appears that the contaminants present in materials, which underwent high temperature roasting such as coffee, soluble coffee and soluble coffee substitute (containing roasted barley and chicory) produce interferences that particularly affect LC-MS/MS analysis. Cleanup by SPE mixed bed cation and anion exchange resins additionally containing C18 material tends to diminish these interferences. Moreover, the GC analyses of underivatized and derivatized acrylamide can be accomplished without column switching.

REFERENCES

Ahn J.S., Castle L., Clarke D.B., Lloyd A.S., Philo M.R., and Speck D.R., 2002, Verification of the findings of acrylamide in heated foods, *Food Addit. Contam.* **19**:1116-1124.

Andrawes, F., Greenhouse, S., and Draney, D., 1987, Chemistry of acrylamide bromination for trace analysis by gas chromatography and gas chromatography-mass spectrometry, *J. Chrom.* **399**:269-275.

Becalski, A., Lau, B.P-Y., Lewis, D., and Seaman, S., 2003, Acrylamide in foods: occurrence, sources and modeling, *J. Agric. Food Chem.* **51**:802-808.

Biedermann, M., Biedermann-Brem, S., Noti, A., Grob, K., Egli, P., and Maendli, H., 2002, Two GC-MS methods for the analysis of acrylamide in foods, *Mitt. Lebensm. Hyg.* **93**:638-652.

Cavalli, S., Maurer R., and Hoefler F., 2003, Fast determination of acrylamide in food samples using accelerated solvent extraction followed by ion chromatography with UV or MS detection, *LC-GC, application notebook* **35**.

Castle, L., Campos, M.-J, and Gilbert, J., 1991, Determination of acrylamide monomer in hydroponically grown tomato fruits by capillary gas chromatography-mass spectrometry, *J. Sci. Food Agric.* **54**:549-555.

Castle, L., 1993, Determination of acrylamide monomer in mushrooms grown on the polyacrylamide gel, *J. Agric. Food Chem.* **41**:1261-1263.

Clarke, D. B., Kelly, J., and Wilson, L. A., 2002, Assessment of Performance of Laboratories in Determining Acrylamide in Crispbread, *J. AOAC Int.* **85**:1370-1373.

Fauhl, C., Klaffke, H., Mathar, W., Palavinskas, R., and Wittkowski, R., 2002, Acrylamide Interlaboratory Study (February 10, 2002); http://www.bgvv.de/cm/245/proficiency_testing_studie.pdf

Gertz, C., and Klostermann, S., 2002, Analysis of acrylamide and mechanisms of its formation in deep-fried products, *Eur. J. Lipid Sci. Technol.* **104**:762-771.

Haase N.U., Matthaus B., and Vosmann K., 2003, Acrylamide formation in foodstuffs - Minimising strategies for potato crisps, *Deutsche Lebensmittel-Rundschau* **99**:87-90.

International Agency for Research on Cancer, Lyon, 1994, *IARC Monographs on the Evaluation of Carcinogenic Risks to Humans*, **60**:389–433.

La Calle, Beatriz de; Ostermann, Ole; Wenzl, Thomas; Anklam, Elke, 2004, *Report European Workshop on Analytical Methods for the Determination of Acrylamide in Food Products* (June, 2004); http://www.irmm.jrc.be/ffu/acrylamide.html

Nemoto, S., Takatsuki, S., Sasaki, K., and Maitani, T., 2002, Determination of acrylamide in foods by GC/MS using ^{13}C-labeled acrylamide as an internal standard, *J. Food Hyg. Soc. Jap.* **43**:371-376.

Ono, H., Chuda, Y., Ohnishi-Kameyama, M., Yada, H., Ishizaka, M., Kobayashi, H., and Yoshida, M., 2003, Analysis of acrylamide by LC-MS/MS and GC-MS in processed Japanese foods, *Food Addit. Contam* **20**:215-220.

Rosen, J., and Hellenas, K-E., 2002, Analysis of acrylamide in cooked foods by liquid chromatography tandem mass spectrometry, *Analyst* **127**:880-882.

Rothweiler B.,and Prest, H., 2003, Rapid screening for acrylamide in foods using GC-MS with positive chemical ionization, *LC-GC, application notebook* **34**.

Sanders, R. A., Zyzak, D. V., Stojanovic, M., Tallmadge, D. H., Eberhart, B. L., and Ewald, D. K., 2002, An LC/MS acrylamide method and its use in investigating the role of asparagine. *Acrylamide Symposium, 116th Annual AOAC International Meeting*, Los Angeles, September 26, 2002.

Swiss Federal Office of Public Health, 2002, Determination of acrylamide in food, (August 10, 2002); http://www.bag.admin.ch/verbrau/aktuell/d/AA_methode.pdf

Takatsuki, S., Nemoto, S., Sasaki, K., and Maitani, T., 2003, Determination of acrylamide in processed foods by LC/MS using column switching, *J. Food Hyg. Soc. Jap.* **44**:89-95.

Tareke, E., Rydberg, P., Karlsson, P., Eriksson, S., and Tornqvist, M., 2000, Acrylamide: a cooking carcinogen? *Chem. Res. Toxicol.* **13**:517-522.

Tareke, E., Rydberg, P., Karlsson, P., Eriksson, S., and Tornqvist, M., 2002, Analysis of acrylamide, a carcinogen formed in heated foodstuffs, *J. Agric. Food Chem.* **50**:4998-5006.

Tateo, F., and Bononi, M., 2003, *Ital. J. Food Sci.* **15**:149-151.

US FDA, 2003, Detection and quantitation of acrylamide in foods (Feburary 24, 2003); http://www.cfsan.fda.gov/~dms/acrylami.html.

SOME ANALYTICAL FACTORS AFFECTING MEASURED LEVELS OF ACRYLAMIDE IN FOOD PRODUCTS

Sune Eriksson[1,2] and Patrik Karlsson[1]
[1] AnalyCen Nordic AB, Box 905, S-53119 Lidköping, Sweden; [2] Department of Environmental Chemistry, Stockholm University, S-10691 Stockholm, Sweden; e-mail sune.eriksson@analycen.se

Abstract: Acrylamide in food is normally measured as "free water-soluble acrylamide". However, it is shown that certain extraction techniques, like extraction as for dietary fibre or at high pH can affect the result. This has to be accounted for, particularly in exposure assessment and in studies of bioavailability and, in the long run, the health risk assessment.

Key words: Acrylamide, extraction technique, foods

1. INTRODUCTION

1.1 Background

Acrylamide became an issue to the Swedish people as well as for our laboratory in 1997. Due to problems with large water leakage during the building of a railway tunnel through Hallandsås, a mountain ridge in the south west of Sweden, a grouting material had to be used to seal the tunnel walls. The grouting agent, used extensively from August 1997, contained monomeric acrylamide and N-methylacrylamide. In September an acute situation arose, with observation of dead fish and paralysed cattle near the construction area. A large leak of unpolymerized acrylamide and N-methylacrylamide into the environment appeared to be the cause, and the acrylamides spread into rivulets, ground water and wells. At the same time, concern about exposure of the tunnel workers was raised. Through measurement of reaction products (adducts) with hemoglobin in blood, it was shown that many had received high exposures and several showed nerve symptoms similar to those reported from acrylamide poisoning (Hagmar et al., 2001). Also there arose consumer resistance to food products from this area due to concern about contaminated foods, even though, to our knowledge, no acrylamide ever where found in food products produced in that area. As a consequence food products were destroyed and cattle taken away from the area (Boija 1998). The construction of the tunnel was discontinued, and was not started again until 2004. Our laboratory, AnalyCen, was involved in performing analysis of water samples, and had been previously accredited for analysis of acrylamide and N-methylacrylamide in water down to 0.5µg/l, which at that time was sufficient (Swedac 1998).

Analysis of food products, soil and sludge in the Hallandsås area (performed by AnalyCen), was made using methodology modified from the method for analysis of acrylamide in water, and was similar to the method used in a study of fried feed given to rats (Tareke et al 2000). The history of the remarkably high level of acrylamide in certain food products started for AnalyCen and Department of Environmental Chemistry, Stockholm University, on 9 January 2001. When performing follow up analysis on the findings of acrylamide in hamburgers (Tareke et al 2000), we found unexpectedly high levels of acrylamide (around 700 µg/kg) in mashed fried potato.

1.2   Analysis of acrylamide in food

Since the discovery of acrylamide formation during heating of foods (Tareke et al 2000; Tareke et al 2002; Rosen et al 2002) was first published, different analytical methods, using both LC-MS/MS and GC-MS, have been shown by different laboratories and through different proficiency programs to give similar results (Tareke et al 2002, Rosén et al 2002; Ahn et al 2002; Clarke et al 2002; Ono et al. 2003; Wenzl et al 2003; Wenzl et al 2004). Most methods applied to the analysis of acrylamide in food products comprise extraction with water, which means that what is analyzed is *"free water-soluble acrylamide"*.

1.3   Net formation of acrylamide

Different studies have demonstrated that acrylamide added to foods is bound or adsorbed to food components in unknown ways (Biedermann et al 2002a; 2002b; Rydberg et al. 2003). It has also been shown that absorption of acrylamide differs between different food products (Sörgel et al. 2002; Biedermann et al. 2002b). Bologna et al (1999) showed recoveries deviating from that for a control for beans, sugar beet, corn and potato, and found that acrylamide was more stable in potatoes than in the rest of the tested crops. In our laboratory we have seen that reference food samples are not stable at room temperature, the measured content decreasing week by week. The product must be kept in a freezer, which also has been confirmed by others (de la Calle et al., 2003). Similar observations have been with food samples kept at room temperature, in contrast to products like cosmetics (containing polyacrylamide), where from time to time an increase can be seen. An extreme example is soft drinks with a high concentration of sucrose, in which acrylamide added at 100 µg/kg disappeared over night in a fridge.

1.4 Acrylamide chemistry

Acrylamide is a difunctional monomer containing a reactive electrophilic double bond and an amide group, which can also react. It exhibits both weak acidic and basic properties. The electron withdrawing carboxamide group activates the double bond, which as a consequence reacts readily with nucleophilic regents, e.g. by addition. Many of these reactions are reversible, and are usually faster the stronger the nucleophiles. Examples are the addition of ammonia, amines, phosphines and bisulfite. Alkaline conditions permit the addition of mercaptans, sulfides, ketones, nitroalkanes, and alcohols to acrylamide. Examples of alcohol reactions are those involving polymeric alcohols such as cellulose and starch (Habermann 1991).

1.5 Discrepancy between methods

In Sweden, analysis of acrylamide in food products has led to an estimation of a daily intake of acrylamide of about 35 µg/day (Svensson et al 2003), and similar figures have been found in other countries. *In vivo* measurement of acrylamide, through its reaction product (adduct) with haemoglobin in blood, has suggested the daily intake to be about 100 µg (Bergmark 1997, Törnqvist et al 1998). This might mean that the bio-available amount of acrylamide in food products is larger than has been estimated from analysis of foods by normal water extraction.

2. METHODS

Contents of acrylamide in foods were studied with different extraction methods. Commercial food product samples were used, bought in normal food stores in Sweden. Extraction was done with water at different pH and, in the same way as for analysis of dietary fiber (Nordic Committee on Food Analysis 1989), both without and with added enzymes (heat stable α-amylase, protease and amyloglucosidase). The system for measuring dietary fiber is an *in vitro* method imitating the digestion system for carbohydrates.

For extraction at different pH, homogenized samples, normally 10 g, were extracted with 100 mL water, and the internal standard, $^{13}C_3$-acrylamide, added. Adjustment and measuring of pH (pH 2-14) were done, where appropriate, with HCl or NaOH solutions. For reference, analysis using water extraction without adjustment of pH was also done.

The samples were centrifuged, the supernatant filtrated through 0.2 μm syringe filter, to get about 2 mL filtrate. Samples of 2 mL of filtrate from the different work-up procedures were added to an SPE column, Oasis HLB 0.2 g column, Waters (Milford, MA, US), as described by others (Rauch et al 2003, Andrzejewski et al 2004). The recovery was *ca.* 100 % (SD = 7.5%) compared to internal standard for most food matrices, detection level decreased from 10 to 3 μg/kg and the reproducibility is high (CV ≈ 5%) than reported earlier (Tareke et al 2002). The column was washed with one column volume of methanol followed by five volumes water before use. The extract was further cleaned with Isolute SPE column, as described before (Tareke et al 2002). The extract was then used directly for HPLC-MS/MS analysis. For every set of analyses a blank sample of water, which had passed through the analytical work-up procedure, was included. The use of isotope-labeled internal standard is essential for the reproducibility considering that the absolute recovery varies between food matrices. The suppression of the signal in the LC-MS analysis, which has been earlier observed (Rydberg et al 2003), was decreased by adding the OASIS column. It should be remembered that the internal standard does not compensate for a different behavior of the analyte, for example in extraction efficiency, formation from other compounds or reaction with matrix before work-up.

In all experiments there were samples representing potato products, bread products, meat products and some other food stuffs.

3. SUMMARY OF RESULTS

Extraction at pH in the range 2 to about 7.5, gave the same result as normal water extraction. At higher extraction pH the recovery of acrylamide increased, giving a maximum at about 12 and above (n=25). The extent of the increase depended on the type of product. Table 1 shows the factor increase in the measured acrylamide at extraction at pH 12 compared to normal water extraction at pH 6. The reference pH of 6 was chosen to provide comparison with the results for water extraction of normal food products.

Table 1. Increase of measured acrylamide by extraction at pH 12 compared to normal water extraction at pH 6

Food	Number of samples	Factor increase (range of values)	Mean increase
potato products	5	1.1 – 1.6	1.3
bread products	10	1.1 – 4.1	2.0
hamburger	3	3.1 – 3.6	3.3
coffee, cocoa, dried fruits	6	1.7 – 3.6	2.5

Using the extraction procedure according to method for dietary fiber, gave same or higher acrylamide content than extraction with water (range = 1.0-2.8, n = 18, mean = 1.5). This procedure gave about same acrylamide content, whether enzymes are used or not (n=15).

4. FUTURE WORK

- Evaluation of bioavailability: Comparison of adduct measurements and measured acrylamide level in food. Which extraction method corresponds to bio available acrylamide?
- Identification of components which are involved in reactions with acrylamide in food products, and find tools to analyze them.

ACKNOWLEDGMENT

This project was financed through AnalyCen R&D, and is executed in cooperation with Stockholm University (M. Törnqvist with support from FORMAS). Thanks also to Martin Zachrisson and Annika Larsson AnalyCen Nordic, Lidköping.

REFERENCES

Ahn, J. S., Castle, L., Clarke, D. B., Lloyd, A. S., Philo, M. R., and Speck, D. R., 2002, Verification of the findings of acrylamide in heated foods, *Food Addit. Contam.* **19**(12): 1116-1124.

Andrzejewski, D., Roach, J. A. G., Gay, M. L., and Musser, S. M. 2004, Analysis of coffee for the presence of acrylamide by LC-MS/MS, *J. Agric. Food Chem.* **52**: 1996-2002.

Bergmark, E., 1997, Hemoglobin adducts of acrylamide and acrylonitrile in laboratory workers, smokers and nonsmokers, *Chem. Res. Toxicol.* **10**: 78-84.

Biedermann, M., Biedermann-Brem, S., Noti, A., and Grob, K., 2002a, Methods for determining the potetntial of acrylamide formation and its elimination in raw materials for food preparation, such as potatoes, *Mitt. Lebensm. Hyg.* **93**: 653-667.

Biedermann, M., Noti, A., Biedermann-Brem, S.Mozzetti, V., and Grob, K., 2002b, Experimants on acrylamide formation and possibilites to decrease the potential of acrylamide formation in potatoes, *Mitt. Lebensm. Hyg.* **93**: 668-687.

Boija, L., 1998, Friskförklarad! Nu kan vattnet från brunnarna på Hallandsås användas igen, *Var Foda,* Nr 3: 18-19. (in Swedish)

Bologna, L. S., Andrawes, F. F., Barvenik, F. W., Lentz, R. D., and Sojka, R. E., 1999, Analysis of residual acrylamide in field crops, *J. Chrom. Sci.* **37**(July): 240-244.

Clarke, D. B., Kelly, J., and Wilson, L. A., 2002, Assessment of performance of laboratories in determining acrylamide in crispbread, *J. AOAC Int.* **85**(6): 1370-1373.

de la Calle, B., Ostermann, O., Wenzl, T., and Anklam, E., 2003, *Report: European Workshop on Analytical Methods für the Determination of Acrylamide in Food Products,* 28-29 April 2003, Oud-Turnhout, Belgium. European Commission, Joint Research Centre, Institute for Reference Materials and Measurements, Geel, Belgium, pp 1-31.

Habermannn, C. E., 1991, Acrylamide. In: *E. Kirk-Othmer Enzyclopedia of Chemical Technology,* J. J. Kroschwitz, and M. Howe-Grant, ed., 4th ed, vol 1, J. Wiley & Sons, New York, pp 251-266.

Hagmar, L., Törnqvist, M., Nordander, C., Rosén, I., Bruze, M., Kautianen, A., Magnusson, A.-L., Malmberg, B., Aprea, P., Granath, F., and Axmon, A., 2001, Health effects of occupational exposure to acrylamide using hemoglobin adducts as biomarkers of internal use, *Scand. J. Work. Environ. Health,* **27**(4): 219-226.

Nordic Committee on Food Analysis. 1989; AOAC NMKL Method: Total Dietary Fibre. Gravimetric determination after enzymatic degradation in foods, Method 129. 1st ed, http://www.nmkl.org

Ono, H., Chuda, Y., Ohnishi-Kameyama, M., Yada, H., Ishizaka, M., Kobayashi, H., and Yoshida, M., 2003, Analysis of acrylamide by LC-MS/MS and GC-MS in processed Japanese foods, *Food Addit. Contam.* **20**(3): 215-220.

Roach, J. A. G., Andrzejewski, D., Gay, M. L., Nortrup, D., and Musser S. M.,2003, Rugged LC-MS/MS survey analysis for acrylamide in foods, *J. Agric. Food Chem.* **51**: 7547-7554.

Rosén, J., and Hellenäs, K.-E. 2002,. Analysis of acrylamide in cooked foods by liquid chromatography tandem mass spectrometry, *Analyst.* **127**: 880-882.

Rydberg, P., Eriksson, S., Tareke, E., Karlsson, K., Ehrenberg, L., and Törnqvist, M., 2003, Investigations of factors that influence the acrylamide content of heated foodstuffs, *J. Agric. Food Chem.* **51**: 7012-7018.

SWEDAC. Accreditation Swedac Dnr. 97-4101-51.1125; Swedish Board for Accreditation and Conformity Assessment 1998-03-02; Stockholm, Sweden (in Swedish).

Svensson, K., Abramsson, L., Becker, W., Glynn, A., Hellenäs, K.-E., Lind, Y., and Rosén, J., 2003, Dietary intake of acrylamide in Sweden, *Food Chem. Toxicol.* **41**: 1581-1586.

Sörgel, F., Weissenbacher, R., Kinzig-Scheppers, M., Hoffmann, A., Illauer, M., Skott, A., and Landerstorfer, C., 2002;. Acrylamide: increased concentrations in homemade food and first evidence of its variable absorption from food, variable metabolism and placental and breast milk transfer in humans, *Chemotherapy.* **48**: 267-274.

Tareke, E., Rydberg, P., Karlsson, P., Eriksson, S., and Törnqvist, M., 2000, Acrylamide: A cooking carcinogen? *Chem. Res. Toxicol.* **13**: 517-522.

Tareke, E., Rydberg, P., Karlsson, P., Eriksson, S., and Törnqvist, M., 2002, Analysis of acrylamide, a carcinogen formed in heated foodstuffs, *J. Agric. Food Chem.* **50**: 4998-5006.

Törnqvist, M., Bergmark, E., Ehrenberg, L., and Granath, F., 1998, Riskbedömning av akrylamid, *Kemikalieinspektionen PM 7/98*, Kemikalieinspektionen, Stockholm, Sweden, pp 1-28, (in Swedish).

Wenzl, T., de la Calle, M. B., and Anklam, E., 2003,. Analytical methods for the determination of acrylamide in food products: a review, *Food Addit. Contam.* **20**(10): 885-902.

Wenzl, T., de la Calle, B., Gatermann, R., Hoenicke, K., Ulberth, K., and Anklam, E., 2004, Evaluation of the result from an inter-laboratory comparison study of the determination of acrylamide in crispbread and butter cookies, *Anal. Bioanal. Chem.* **379**: 449-457.

ANALYSIS OF ACRYLAMIDE IN FOOD
Dedicated to Professor Dr. Werner Baltes

Reinhard Matissek and Marion Raters
Food Chemistry Institute (LCI) of the Association of the German Confectionery Industry (BDSI), Cologne, Germany; e-mail: reinhard.matissek@lci-koeln.de

Abstract: Since the first discovery of the presence of acrylamide in a variety of food products in April 2002, numerous methods have been developed to determine the acrylamide monomer in heat-treated carbohydrate-rich food. These detection methods are mainly MS-based, coupled with a chromatographic step using LC or GC. The Food Chemistry Institute (LCI) of the Association of the German Confectionery Industry (BDSI) therefore established a detection method by means of aqueous extraction plus a cleaning step and LC-MS/MS detection, making great efforts to ensure internal and external validation. Citing potato crisps as an example, we will in the following show how the German manufacturing companies have gone to great pains to reduce acrylamide levels in their products.

Key words: acrylamide, LC-MS/MS, potato crisps, minimization

1. INTRODUCTION

Publication of the report on acrylamide intake from food, authored by the M. Törnquist and co-workers (2002), confronted food scientists and the food industry with a completely new dimension of toxicologically relevant food components, a so-called foodborne toxicant. Since then, there have been many publications outlining analytical methods for determining acrylamide in diverse matrices (e.g. Rosen and Hellenäs, 2002; Tareke *et al.*, 2002; Biedermann *et al.*, 2002a). The LCI has established the use of an LC-MS/MS method, extensively validating by testing quality assurance sample on a daily basis.

Since the occurrence of acrylamide in food first became known, the German potato chips industry has acted in the interests of preventive consumer protection and has correspondingly quickly introduced and successfully implemented extensive concentration-reducing measures (Matissek, 2002). To this end, the Food Chemistry Institute (LCI) of the Association of the German Confectionery Industry (BDSI) has conducted numerous systematic analyses on behalf of the member companies of the BDSI.

2. EXPERIMENTAL DETAILS

2.1 Materials

Acrylamide (99+%) was supplied by Sigma (Taufkirchen, Germany) and deuterium-labeled acrylamide-d_3 (> 99%) by LCG Promochem (Wesel, Germany). HPLC-grade acetonitrile came from Aldrich (Seeze, Germany) and the syringe filter used was a Rotilabo® Nylon 0.45 µm from Roth (Karlsruhe, Germany). All other reagents used for the analysis of acrylamide were of analytical grade.

CAUTION: acrylamide and acrylamide-d_3 are hazardous and must be handled carefully.

2.2 Sample Extraction

The samples were homogenized, fat-rich samples were defatted with n-hexane by way of extraction and 20 ml of water and 400 µl of internal standard acrylamide-d_3 (5 µg/mL) were added to 2 g of the homogenized sample. The samples were extracted by ultrasonic treatment (15 min, 60°C) and 20 ml of acetonitrile was added. Clean-up of the extracts was performed using 500 µl of Carrez I and II respectively and the samples were then centrifugated (4500 rpm for 10 min, 4°C). Before injection into the LC-MS/MS system, the supernatant was passed through a syringe filter (0.45 µm, Roth Rotilabo® Nylon).

2.3 LC-MS/MS Analysis

Mass spectrometry measurements were performed using a HPLC-system Series 200 (Perkin Elmer, Rodgau, Germany) coupled with a API 2000 mass spectrometer (Applied Biosystems, Darmstadt, Germany). Analytical separation was achieved using a Lichrospher 100 CN 5 µm (250 x 4 mm) with a guard column 5 µm (Merck, Darmstadt, Germany). The elution mode was isocratic, using a mixture of acetonitrile and water (0.5:99.5, v/v) containing 0.1% (v/v) of concentrated formic acid as LC eluent. The flow rate was 0.25 ml/min and the injection volume was 20 µl.

Acrylamide was identified by multi-reaction monitoring (MRM) in positive electrospray ionization mode (ESI+). Three different fragment ion transitions were monitored for both acrylamide (m/z 72→72, m/z 72→52 and 72→44) and the internal standard (m/z 75→75, m/z 75→58 and 75→44), t_R= 11.6 min. The electrospray source had the following settings (with nitrogen): capillary voltage 3 kV; cone voltage 40 V; source temperature 450°C. For all transitions the dwell time was set to 0.15 s. The ion energy was set to 0.5 and 0.7 V for the first and the second quadrupole respectively. The instrument was operated at unit resolution.

2.4 Quantification of Acrylamide

Stock solutions of standard and internal standard, 1 mg/ml, were prepared in water/acetonitrile (50/50, v/v) and stored at -20°C. Standard solutions, for spiking samples as well as for the standard curve, were obtained by dilutions in water/acetonitrile (50/50, v/v).

3. RESULTS AND DISCUSSION

3.1 Performance of Method

A relevant criterion for the quality of an analytical method is the linearity (of the chemical analysis) over a wide concentration range. Quantification of acrylamide was made using the area ratio m/z 55/58 and concentrations were calculated against a standard curve (1 to 500 ng/ml), typically producing correlations coefficients of 0.9997 (Figure 1).

Table 1. Recovery rates of various tested materials

Tested material	Recovery rate [%]
Potato crisps	94
Crisp bread	99
Cookies	105
Cocoa products	100
Average	100

Preliminary estimations of limit of quantification (LOQ) and limit of detection (LOD) for the validated matrices would be no higher than 30 µg/kg and 10 µg/kg respectively. The recovery rates determined for potato crisps, crispbread, cookies as well as cocoa and cocoa products are all within acceptable limits. With a standard additive of 50-1000 µg/kg they fluctuate between 94% in the case of potato crisps and 105% in the case of cookies. Table 1 shows the recovery rates of various tested materials.

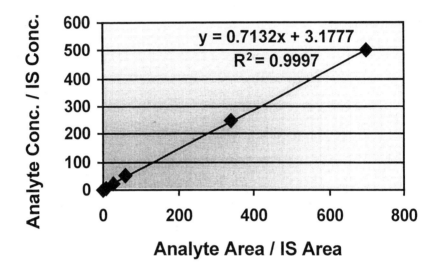

Figure 1. Linearity of standard curve.

As part of the internal quality assurance for the performed analysis, everyone involved in sample preparation analyzed a so-called quality assurance sample each day, since analyzing acrylamide first began at LCI in October 2002. This approach serves to quickly and simply reveal any errors that may exist in sample preparation and/or in the measuring system. Two samples were used for this so-called in-house comparison, namely potato crisps (Sample I; so far analyzed 244 times by seven different people) and crispbread (Sample II; so far analyzed 207 times by four different people).

Table 2. Statistical data of in-house trial: Sample I (Potato Crisps)

		Analyst						
	All	1	2	3	4	5	6	7
N	244	101	45	45	19	7	16	11
Mean [µg/kg]	1091	1068	1107	1093	1108	1125	1141	1085
Median [µg/kg]	1099	1071	1103	1124	1124	1113	1144	1112
Min [µg/kg]	900	900	903	914	917	1041	1019	955
Max [µg/kg]	1326	1221	1326	1247	1208	1194	1267	1173
SD [µg/kg]	87.6	88.2	92.5	86.6	72.7	55.1	72.4	72.1
RSD [%]	8.04	8.26	8.36	7.92	6.56	4.90	6.34	6.65

SD = Standard deviation; RSD = Relative standard deviation

Table 2 shows the statistical data for the potato crisps sample (Sample I) which was tested for acrylamide 244 times in total by the LCI between November 2002 and July 2003, firstly as a mean of all seven analysts (Column 1) and secondly as a respective mean of each individual (Columns 2-8, Nos. 1-7). The acrylamide levels determined over all seven analysts ranged from 900 to 1320 µg/kg, with an average value of 1091 µg/kg and a median of 1099 µg/kg. This results in a relative standard deviation of 8.0%. The data in Table 2 show there are comparatively large differences between the relative standard deviations for the individual analysts.

Table 3 shows the same statistical data for the crispbread sample (Sample II), already tested for acrylamide 207 times in total by four different analysts at the LCI. It is a sample with a noticeably lower acrylamide level, close to the minimum detection limit.

Table 3. Statistical data of in-house trial: Sample II (Crispbread)

		Analyst			
	All	1	2	3	4
N	207	101	19	30	57
Mean [µg/kg]	41	43	41	37	38
Median [µg/kg]	41	43	43	37	36
Min [µg/kg]	30	32	32	30	31
Max [µg/kg]	52	52	50	45	50
SD [µg/kg]	5.45	4.62	6.28	3.16	5.07
RSD [%]	13.36	10.64	15.40	8.57	13.28

As expected, the acrylamide level had some influence on the statistical data. The acrylamide concentrations determined over all four analysts ranged from 30 to 52 µg/kg, with an average value and median of 41 µg/kg. This results in a relative standard deviation of 13%.

As a comparison to our in-house-validation data, Figure 2 shows the result of two interlaboratory tests, a FAPAS[1]-Test in 2003 (Kelly, 2003) and a BfR[2]-test in 2002 (Faul *et al.*, 2002). The boxes each show the range of the measured values as still being acceptable on the basis of z-score assessment ($< |2|$ = chemical analysis is O.K.). This means that measured values of between 97 and 237 µg/kg acrylamide are to be seen as acceptable or "correct" for the potato crisps sample used in the FAPAS test. This corresponds to a percentage deviation from the calculated "true value" for

---

[1] Food Analysis Performance Assessment Scheme, CSL Central Science Laboratory, York, UK
[2] Bundesinstitut für Risikobewertung / Federal Institute For Risk Assessment, Berlin, Germany (former BGVV)

the sample of approx. 42% in each direction. For the biscuit sample in the BfR-test, this "true value" lies between 327 and 737 µg/kg acrylamide, representing a percentage deviation of approx. 39% in each direction.

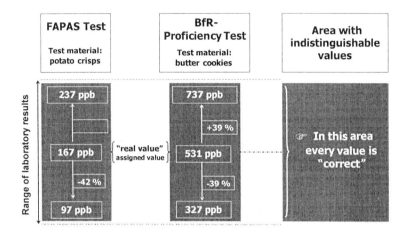

Figure 2. Range of results of two inter-laboratory tests.

3.2 Minimization of Acrylamide in Potato Crisps

Since the occurrence of acrylamide in foodstuffs first became known, the German potato crisps industry has acted in the interests of preventive consumer protection and quickly implemented or initiated appropriate measures to reduce acrylamide levels. The BDSI's own food chemistry institute, the LCI, has carried out around 12,000 systematic analyses for member companies. The LCI is the service centre for all BDSI member companies and makes solution proposals for preventing the formation of acrylamide in food while also assisting the members in implementing such internal measures. To this end, so-called coordination groups were set up for the individual specialized fields within the BDSI, so as to monitor and support empirical scientific research at the LCI for the individual specialized fields.

There are two basic ways to reduce acrylamide in potato crisps. Firstly, through the raw material itself, the potato. Potatoes have a considerable potential to form acrylamide. Fried potato is the food category which has probably chalked up the highest concentrations of acrylamide recorded so far (Friedmann, 2003). The level varies depending on the type of potato (Biedermann *et al.*, 2002b; Haase *et al.*, 2003), several other factors (e.g. fertilization, climate, location), or mainly (it is supposed) due to the specific asparagine levels. With regard to the formation of acrylamide, the industry is currently in the process of selecting the most suitable types of potatoes. Storage of the potatoes also has a very big impact on the formation of acrylamide. By optimizing storage conditions (in particular, maintaining a storage temperature of below 10°C), the levels of reducing sugars released due to starch degradation during storage can be minimized (Noti et al., 2003; Grob et al., 2003).

Another fundamental way to reduce acrylamide concentrations in the end product is by influencing the technology used in the production of potato crisps. The industry has therefore adopted the following key technological measures:
- optimizing the temperature/time profile during the deep-frying process,
- increasing end-product moistness, and
- using an opto-electronic sorting process to remove dark potato crisps.

By maintaining and regularly optimizing these measures, the industry has so far succeeded in reducing acrylamide concentrations in its end products by 10-15%.

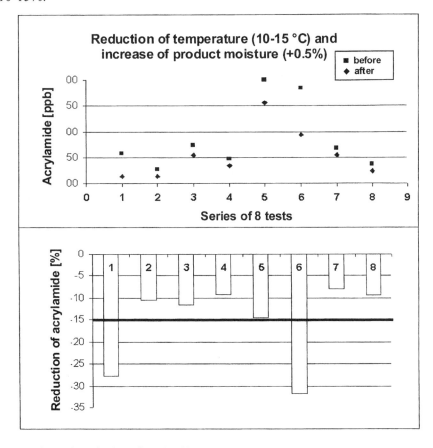

Figure 3. Reduction of acrylamide in potato crisps using technological means.

Figure 3 gives an example of what the industry can achieve by adopting such technological measures. The deep-frying temperature was reduced by 10-15°C while product moistness was simultaneously increased by 0.5%. The result of the series of eight tests carried out on different production lines was a significant reduction in the acrylamide concentrations by an average of 15%.

Figure 4 provides a summary of the acrylamide-reducing effects of the production optimization measures implemented by the German industry in the case of potato crisps. The average weekly values are shown in each case. The above-described technological measures, which have been implemented from May/June 2002 onwards, have had a very significant impact, made evident by the steeply dipping curve in the first four months of the period under consideration. This effect is eclipsed by seasonal, harvest-related conditions, meaning that towards the end of the storage period we see an increase in acrylamide concentrations, which drop again significantly after the new potatoes come onto the market. This diagram is in fact constantly updated (cf. www.lci-koeln.de).

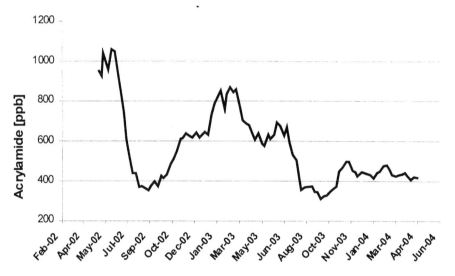

Figure 4. Weekly average – Trend line according to production date.

Figure 5 provides a schematic overview of (i) the assumed concentrations in potato crisps before the occurrence of acrylamide first became known (broken line), (ii) the impact of steps then taken to minimize acrylamide levels (continuous line), and (iii) those steps which are expected to be taken in the future (transparent line), allowing for seasonal fluctuations caused by

Analysis of Acrylamide in Food

the raw material itself. The results of tests carried out during the period under review (cf. Figure 4) confirm this sinus-like curve.

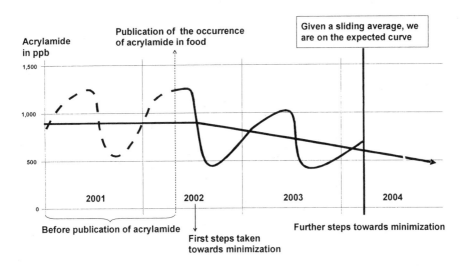

Figure 5. Minimization of acrylamide in potato crisps - hypothetical trend line.

3.3 Conclusions

In summary it can be said that since the first surprising discovery of acrylamide in a number of different foodstuffs, the LCI has undertaken great efforts to establish a method for determining acrylamide in several foods while taking great pains to ensure in-house and external validation. In addition to this, the German industrial potato crisp producers have taken a number of steps to minimize acrylamide concentration levels. These steps towards minimization have proven successful and are indeed having a big impact. Joint efforts still need to be undertaken by scientists, the authorities, and the industry if further minimization targets are to be achieved. In addition to this, potato growers are called upon to select and/or develop suitable types of potatoes as a first preventive measure. Clearly beneficial results are expected from the ongoing basic research work.

ACKNOWLEDGEMENTS

We would like to thank Brigitte Brombach and Sabine Gotzner (both LCI, Cologne) for their technical support and Dr. Rolf Nilges (Intersnack GmbH & Co KG, Cologne) for his statistical evaluations.

REFERENCES

Biedermann M., Biedermann-Brem S, Noti A, Grob K, Egli P, Mändli H 2002a. Two GC-MS methods for the analysis of acrylamide in food. *Mitt Lebensm Hyg* **93**: 638-652.

Biedermann M., Noti A., Biedermann-Brem S., Mozzetti V., Grob K., 2002b. Experiments on acrylamide formation and possibilities of decreasing the potential of acrylamide formation in potatoes. *Mitt Lebensm Hyg* **93**: 668-687.

Faul C., Klaffke H., Mathar W., Palvinskas R., Wittkowski R., 2002. ‚Acrylamide Interlaboratory study 2002. http://www.bfr.bund.de/cm/208/acrylamide___interlaboratory_study_2002.pdf

Friedmann M., 2003. Chemistry, biochemistry, and safety of acrylamide. *J Agric Food Chem* **51**: 4504-4526.

Grob K.,, Biedermann M.,, Biedermann-Brem S.,, Noti A.,, Imhof D., Amrein T., Pfefferle A., Bazzocco D., 2003: French fries with less than 100 µg/kg acrylamide. A collaboration between cooks and analysts. *Eur Food Res Technol*. **217**: 185-194.

Haase N. U., Matthäus B., Vosmann K., 2003: Acrylamide formation in foodstuffs – minimizing strategies for potato crisps. *Deutsche Lebensm Rund* **99**: 87-90.

Kelly J., 2003. FAPAS Food analysis performance assessment scheme, Acrylamide Series 30 Round 2, 2003, Report No. 3002.

Matissek R., 2002. Verbraucherinformation zur Thematik Acrylamid bei Kartoffelchips – Fortschritte der industriellen Kartoffelchipshersteller im Rahmen des Minimierungskonzeptes. http://www.lci-koeln.de/media/Acrylamid_Matissek.pdf

Noti A., Biedermann-Brem S., Biedermann M., Grob K., Albisser P., Realini P., 2003. Storage of potatoes at low temperature should be avoided to prevent increased acrylamide during frying or roasting. *Mitt Lebensm Hyg*. **94**: 167-180.

Rosen J., Hellenäs K.-E., 2002. Analysis of acrylamide in cooked food by liquid chromatography tandem mass spectrometry. *The Analyst* **127**: 880-882.

Tareke E. P., Rydberg P., Karlsson S., Erikson M., Törnqvist M., 2002. Analysis of acrylamide, a carcinogen formed in heated foodstuffs. *J Agric Food Chem* **50**: 4998-5006.

ON LINE MONITORING OF ACRYLAMIDE FORMATION

David J. Cook, Guy A. Channell, and Andrew J. Taylor
Samworth Flavour Laboratory, Division of Food Sciences, The University of Nottingham, Sutton Bonington Campus, Loughborough, Leicestershire LE12 5RD, U.K.; e-mail: David.Cook@nottingham.ac.uk

Abstract: A system to monitor the formation of acrylamide in model systems and from real food products under controlled conditions of temperature, time and moisture content has been developed. By humidifying the gas that flows through the sample, some control over moisture content can be affected. Results are presented to show the validity and reproducibility of the technique and its ability to deliver quantitative data. The effects of different processing conditions on acrylamide formation and on the development of color, due to the Maillard reaction, are evaluated.

Key words: Acrylamide, Food, APCI-MS, Maillard reaction

1. INTRODUCTION

Atmospheric Pressure Chemical Ionization Mass Spectrometry (APCI-MS) is a technique which is ideally suited to the analysis of volatile molecules in the gas phase (Taylor et al., 2000). Our research group has a long-standing interest in applying the technique to monitor the generation, physical partitioning and release of flavor volatiles from food and model food systems. We have developed two approaches to studying the volatile products of thermal reactions in foods using APCI-MS. Channell (2003) described fundamental studies on the Maillard reaction, using a film reactor, wherein primary reactants are generated in-situ and conditioned to various moisture levels prior to heating. Secondary reagents may then be injected through the reactor to investigate their effect on end-product profiles. Turner (2002a; 2000) reported the on-line monitoring of thermal flavor generation from skimmed milk powder at a range of initial moisture contents. This version of the thermal flavor generation apparatus has been refined over a number of years, from a capped bottle within a GC oven, to the optimized, more controlled system described by Turner (2002b). Sample is heated in a packed bed reactor, through which nitrogen gas is purged at 5 mL/min, to carry volatiles into the APCI-MS source at a steady rate.

The latter method has been developed here to study the evolution of acrylamide when a hydrated potato flake product was processed under controlled conditions of temperature and humidity.

The history and toxicology behind the current activity in acrylamide research are described in detail elsewhere in this volume. Suffice to say, that since the Swedish National Food Authority and the University of Stockholm reported ppm levels of acrylamide (a known neurotoxin, rodent carcinogen and potential human carcinogen) in heat-treated foods (Tareke et al., 2002), there has been an intense interest in developing scientific understanding of this complex area. The extraction and analysis of acrylamide from food matrices was reviewed recently (Wenzl et al., 2003). Techniques are broadly categorized as either GC or LC-MS-based and in either case can be time-intensive. On-line monitoring of acrylamide production has the potential to generate data rapidly and to provide an insight into the kinetic and physical processes involved. To date there has been just one report of the application of on-line monitoring to acrylamide analysis (Pollien et al., 2003). This paper demonstrated proof of concept and reported detection of acrylamide during the thermal treatment of potato as well as in Maillard model systems (asparagine plus reducing sugars). However, a systematic investigation of the influence of temperature and humidity on acrylamide formation, using on-line methodology, has not previously been reported.

One important distinction to make is that the on-line monitoring technique measures acrylamide release into the gas phase, which is governed by the acrylamide concentration at the food surface and its physicochemical partition between the two phases. One objective of this initial work was to establish the relationship between the observed gas phase concentration of acrylamide and the actual concentration in a sample. A further objective was to investigate whether the generation of color due to Maillard reaction products was related to the amount of acrylamide formed.

2. MATERIALS AND METHODS

2.1 Apparatus for on-line monitoring of thermally generated volatiles.

In brief outline, the apparatus (Figure 1) consisted of a sample cell (stainless steel HPLC column, 55mm x 10 mm ID; Phenomenex, UK) within a GC oven (Sigma 3B, Perkin-Elmer, UK).

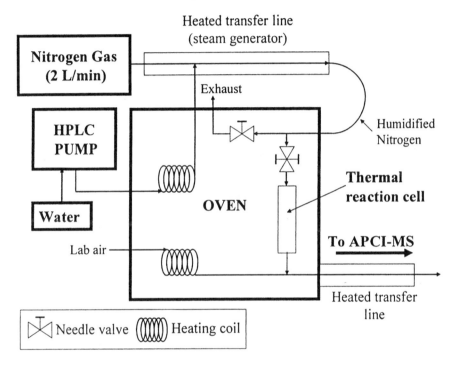

Figure 1. Apparatus for on-line monitoring of volatile generation under conditions of controlled temperature and humidity.

Nitrogen purge gas (2 L/min) was heated to 5°C above the oven temperature and a series of valves used to regulate the flow through the reaction cell to 5 mL/min. The APCI-MS sampling rate was between 10-11 mL/min, so that the entire output of the cell was analyzed (to enhance sensitivity). 'Make-up' flow was laboratory air, which had first passed through a heating coil inside the GC oven, so that it did not condense volatiles in the sample cell effluent. Nitrogen purge gas could be humidified in a controlled fashion by injecting pre-heated water (0 – 0.8 mL/min) into the flow at a point 2 m along the heated transfer line towards the nitrogen supply.

2.2 APCI-MS analysis

2.2.1 Instrumental detail

A Thermo Finnigan LCQ Deca Xp Ion Trap Mass Spectrometer (ITMS) was used with a custom built APCI gas phase interface (Jublot et al., 2004). The source was designed to form predominantly protonated molecular ions (positive ion mode) with only minor fragmentation.

Out flow from the thermal reaction cell was sampled directly into the APCI source via 0.53mm I.D. deactivated fused silica capillary tube within a heated transfer line (180°C). For the analysis of acrylamide, the ITMS was set up to work in MS/MS mode in the low mass range, from m/z 40 to 75. The parent ion m/z 72 (acrylamide) was fragmented to produce predominantly m/z 55 (*Isolation width*: 1.2; *Collision energy*: 25%; *Activation Q*: 0.35; *Activation time*: 30ms; *Max inject time*: 200ms). Raw data were averaged over 10 micro scans.

Acrylamide concentrations in the headspace were calculated relative to the signal (m/z 55) for a calibrant of known concentration (Taylor et al., 2000). Calibration was found to be linear within the range of 3 - 3000 ppbv acrylamide in the gas phase, whilst the range typically used for calibrating the release from potato cakes was 3 - 300 ppbv. The detection limit (signal to noise 3:1) was in the region of 1 - 2 ppbv acrylamide in the gas phase.

2.2.2 Avoiding suppression effects in APCI-MS

Ion suppression can result from competition for charge between compounds during ionization. In positive ion mode, a compound with a high proton affinity, if present in sufficient quantity, can saturate the ionizing capacity of a source, resulting in the incomplete ionization of compounds with a lower proton affinity. When a complex mixture of volatiles are analyzed, such as in the Maillard reaction, ion suppression is a distinct possibility. The aim here was to maximize sample size (to enhance sensitivity), without causing suppression of the acrylamide response.

Trials were conducted to ensure that the amount of sample which was placed in the reaction cell did not cause ion suppression. The acrylamide response was found to be robust with respect to suppression, due to its' high gas phase basicity (affinity for protons during ionization). Hence, when excessive quantities of sample were introduced, suppression effects were first observed for ion species with a lower proton affinity. On a dry weight basis, approximately 0.35 g of material could be loaded without causing suppression of acrylamide ionization in the APCI source.

2.3   Potato flake experiment

Drum-dried potato flake (50% Pentland Dell / 50% Maris Piper) was hydrated (1 part by weight of flake to 6.6 parts water) and filled (1.5 g) into a semi-circular tray, which had been engineered to fit snugly inside the thermal reaction cell. Samples were treated under controlled conditions of temperature and humidification using the apparatus of Figure 1. The MS/MS technique (Section 1.2.1) was used to measure acrylamide release to the gas phase from 1.5 g of potato cake during isothermal treatments at 150, 160, 170, 180, 190 & 200°C. Each treatment was performed in duplicate using humidification at three steam ratios (0, 170 & 340 g steam / kg N_2).

2.4   Correlation of gas phase acrylamide concentration with sample content

2.4.1   Calibration

Samples of potato, rye and wheat cakes, which had been baked for in excess of 50 minutes at 180°C in a convection oven, were supplied by collaborators at the University of Reading, U.K. The samples had been analyzed for acrylamide content, using a minor adaptation of the bromination and GC-MS protocol described by Castle (1993). Six samples, with as broad a range of acrylamide contents as possible, were used to 'calibrate' the on-line monitoring technique, by correlating the observed gas phase concentration with their known acrylamide content. Finely ground sample (200 mg) was placed in the on-line reaction cell and heated isothermally at either 160, 180 or 200°C. In most cases, the acrylamide response rose to a maximum within 8 minutes (thermal equilibrium time) and then declined, since the samples had already been cooked to beyond the point of generating acrylamide. This maximum value was recorded as the gas phase concentration.

2.4.2   Acrylamide content and color of potato cakes after 40 minutes in the on-line system

Twenty of the 36 samples from the potato flake experiment (spread across the design space) were snap frozen in liquid Nitrogen on removal from the reaction cell, after 40 minutes in the on-line apparatus. Finely ground sample was subsequently analyzed for acrylamide content by placing it back in the on-line cell and (as for the calibrants in Section 2.2.1) observing the maximum gas phase acrylamide response. The calibration developed from samples of known acrylamide content was then used to estimate the residual acrylamide content of the potato cake samples.

The CIE L*a*b* color co-ordinates of each of the twenty samples were recorded using a Hunter Lab Color Quest v 3.5, in reflectance mode with an observer angle of 2°.

3. RESULTS AND DISCUSSION

3.1 Effects of temperature and humidity on acrylamide generation in potato cakes

An example of on-line data from the potato cake experiment is shown in Figure 2. This single run illustrates the power of the technique in rapidly generating data with which to follow the time-course of acrylamide production. In this example, the potato cake was effectively dried out after around eight minutes at 190°C, at which point the interior temperature began to increase above 100°C. Thereafter, the acrylamide concentration in the gas phase rose sharply with increasing sample temperature, reaching a maximum shortly before the sample equilibrated to oven temperature. Such clear maxima were only exhibited for samples treated at 190°C and above. At lower oven temperatures the acrylamide release curves were broader and less intense.

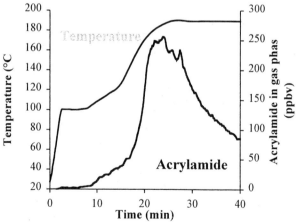

Figure 2. Example of an acrylamide release curve from potato cake. *(Isothermal treatment at 190°C and 340 g/kg steam ratio. Interior sample temperature was measured during the process using a thermocouple).*

Parameters of the release curves were extracted and modeled using Design Expert software v6.02 (Statease, Minneapolis, U.S.). These were the maximum intensity of acrylamide release (I_{max}), the time to reach maximum (T_{max}) and the total area under the release curve for the first 40 minutes of treatment (Table 1).

Table 1: Summary data from the modeling of acrylamide release curves in the potato cake experiment

Parameter	CV (%)	Significant Model factors	Model R^2
I_{max} (ppbv)	14.6	Temperature ($P < 0.0001$)	0.87
T_{max} (min)	3.7	Temperature ($P < 0.0001$)	0.76
Total Area (5 – 40 min)	14.9	Temperature ($P < 0.0001$)	0.87

I_{max} = maximum gas phase concentration, occurring at time T_{max}. CV = coefficient of variation. Model R^2 = the correlation coefficient of the model developed for the design space.

Temperature was a significant factor ($P < 0.0001$) determining each of the acrylamide release parameters, but the level of humidification was not a significant factor in any of the models ($P < 0.05$). In this instance, the effect of humidification was probably small in relation to the high moisture content of the samples. Further experiments are being conducted using drier food systems and greater degrees of humidification, to see if humidity can be used to manipulate acrylamide production under certain circumstances.

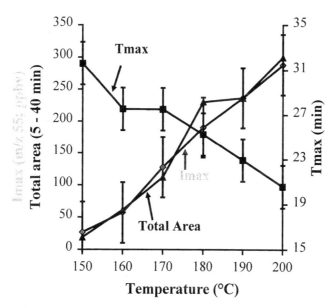

Figure 3. Effect of temperature on acrylamide release parameters. *(Data have been averaged across all humidity treatments and are the mean ± SD of six measurements).*

As the process temperature was increased from 150-200°C, there were steady increases in the mean values of both I_{max} (27-288 ppbv) and of the total area beneath the acrylamide release curve (Figure 3). Over the same range, the mean time taken to reach maximum acrylamide release (T_{max}) fell from 31 to just 20 minutes.

The maximum gas phase concentration of acrylamide increased in an approximately linear way with temperature. However, since the partition of acrylamide between the potato cake surface and the gas phase is temperature dependent, it was necessary to investigate the relationship between concentrations in the two phases at a range of temperatures (Figure 4). The calibration was indeed found to be temperature dependent over the range investigated. The gas phase acrylamide concentration resulting from a specified sample increased 6-fold when the temperature was raised from 160 to 200°C. This means that, whilst the temporal profile of acrylamide release is recorded accurately by the on-line technique, the intensity axis must be interpreted with caution when comparing samples that were treated at different temperatures.

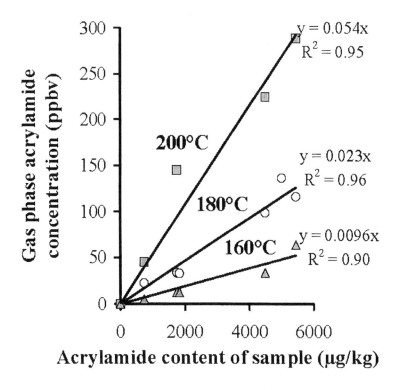

Figure 4. Temperature dependence of the gas phase concentration observed above samples of known acrylamide content.

When the calibration curves were used to estimate the residual acrylamide in samples from the potato cake on-line experiment (Figure 5), it was clear that the maximum acrylamide formation had occurred in samples treated at 160-180°C. At 150°C much less acrylamide was produced, whilst at 190-200°C, similarly low levels were the result of a more rapid production, followed by a decay in concentration (e.g. see Figure 2). The steady decline in acrylamide concentration observed after the maximum is thought to be a chemical effect, as opposed to a physical loss, since experiments in sealed systems have shown similar trends (e.g. see paper by Sadd, this publication). Acrylamide is hypothesized to be consumed by a competing reaction, which becomes evident when most of the primary reactants (asparagine, reducing sugars and their Maillard products) are depleted. At higher temperatures, this depletion occurs more rapidly (reference Elmore et al., this publication) and the consumption of acrylamide may, in addition, be accelerated.

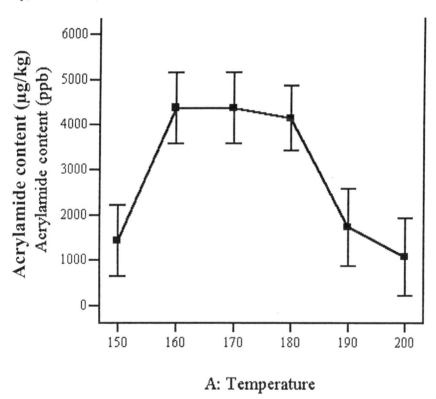

A: Temperature

Figure 5. Design Expert model graph showing the effect of process temperature on the residual acrylamide content of potato cakes.

Similarly to the results presented here, Mottram et al. (2002) observed a maximum temperature of acrylamide formation (170°C), when reacting asparagine and glucose in phosphate buffer at pH 5.5. However, the observed temperature dependence of acrylamide production will depend on both the food system concerned (e.g. due to the concentration of precursors, initial moisture content etc.) and on the severity of the thermal process (position along the path of acrylamide production and depletion). Most commercial food processes utilize shorter and much less severe heat treatments than the full 40-minute process employed here, which ultimately pyrolyzed samples.

3.2 Relationship between acrylamide formation and other Maillard reaction products (color)

It is widely acknowledged that the amino acid asparagine is the primary precursor of acrylamide in heated foods (Becalski et al., 2003; Mottram et al., 2002; Zyzak et al., 2003) and that a carbonyl or dicarbonyl source (principally a reducing sugar in foodstuffs) is required. It appears that several inter-related pathways of acrylamide formation exist in complex food systems. The unifying feature of the principal mechanisms involved is that they share primary reactants, pathways and common intermediates with the Maillard reaction. Since the reactions are postulated to occur simultaneously on heating and compete for common reactants, it was of interest to see if the extent of the Maillard reaction (characterized by the degree of melanoidin pigment formation) was related to the amount of acrylamide formed in the potato cake samples.

The CIE L*a*b* color co-ordinates of samples removed from the on-line reactor after 40 minutes became darker (lower L*) and less yellow (lower b*) as the isothermal process temperature increased (Figure 6). There was a negative correlation ($R^2 = 0.78$) between L* and I_{max} for acrylamide release. However, the two phenomena are unlikely to be related causally (samples cooked at higher temperatures were darker in color and the physical partition of acrylamide to the gas phase increased with temperature).

The green-redness of samples (a*, Figure 6) peaked at around 180°C and declined at higher or lower temperatures; in profile at least, this showed some similarity to the trend in residual acrylamide content (Figure 5); however, correlation between the two was poor ($R^2 = 0.16$). Hence, in this instance, the degree of color formation was not adequately correlated with acrylamide content. A better correlation may have been observed had color been measured at shorter process times (on the up slope of acrylamide generation). However, as described elsewhere in this publication (e.g. see paper by Wedzicha) kinetic modeling of the reactions involved in acrylamide formation suggests that the ratio of Maillard products to acrylamide can be manipulated by changing the ratios of key reactants. Hence, color may be a good marker of acrylamide content in some systems, but (fortunately for those wishing to reduce acrylamide whilst retaining desirable color and flavor characteristics) the two are not implicitly linked.

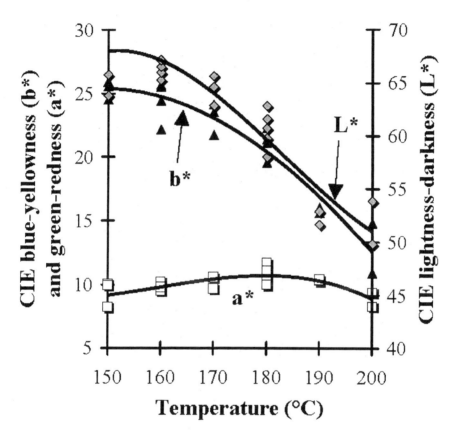

Figure 6. CIE color co-ordinates of the treated potato cakes after 40 minutes in the on-line reactor. *(Polynomial curves were fitted to the data using Microsoft Excel software).*

4. VALIDITY AND APPLICATION OF THE ON-LINE MONITORING TECHNIQUE

It was important to assess the validity of the on-line technique by comparing the observed results with established (and more time-intensive) analytical techniques. Collaborators at the University of Reading analyzed the acrylamide content (as for Section 2.2.1) of potato cakes prepared from the same batch of potato flake and baked in a convection oven for a range of temperature / time processes. Each data point was analyzed in triplicate, resulting in an accurate depiction of the evolution of acrylamide in the samples over time at temperatures ranging from 150-200°C (Figure 7).

When comparing the two systems, it is important to bear in mind the principal differences between them:

- The on-line system measured the proportion of acrylamide released into the gas phase, as opposed to the actual acrylamide content
- Sample size and shape differed, with implications for heat transfer (1.5 g in the on-line reactor v 22.5 g of oven-baked potato cake)
- Faster rates of heat transfer in the on-line system.

In spite of these differences, there was a striking similarity between the profiles of acrylamide production measured by the two techniques at a specified temperature (Figure 7). For example, at 190 and 200°C each system demonstrated a rapid rise to a maximum, followed by a relatively steep decline in acrylamide levels.

Judging from the residual acrylamide contents of the on-line potato cakes (Figure 5), a 40 minute on-line process was roughly equivalent to 70 minutes treatment in a convection oven (dotted line in Figure 7; acrylamide content at 160-180°C >190°C > 150°C > 200°C).

At 180°C the profiles were, likewise, remarkably similar, this time showing a leveling off in acrylamide concentration after a maximum. In each case, treatment at 170°C showed a plateau, without substantial decline thereafter. Whilst the profiles of acrylamide production resulting from each method were similar, they were off-set in both the intensity dimension (due to the effect of temperature on the partition of acrylamide to the gas phase) and in the time dimension (due to differences in sample size and heat transfer rates). Judging from the residual acrylamide contents of the on-line potato cakes (Figure 5), a 40 minute on-line process was roughly equivalent to 70 minutes treatment in a convection oven (dotted line in Figure 7; acrylamide content at 160-180°C >190°C > 150°C > 200°C).

Provided that the appropriate checks and calibrations are in place, the on-line system can be a convenient and powerful tool for following the time-course of acrylamide production in food or model systems. If kinetic data is to be ascertained (one obvious application) then the partition behavior of acrylamide at different temperatures must first be taken into account. However, when working at a specified temperature, the technique could, for instance, provide rapid data on the efficacy of various treatments or formulation changes in reducing the acrylamide content of food systems.

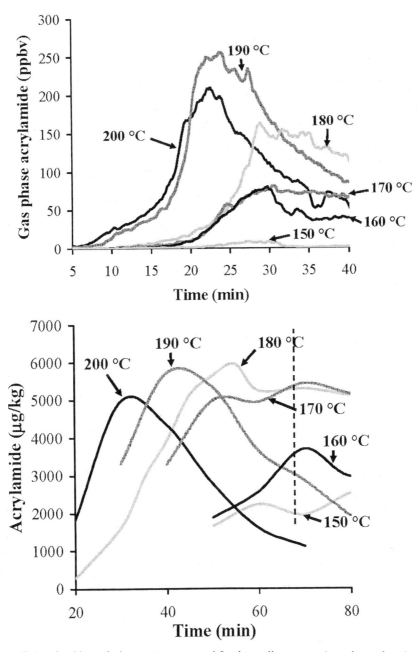

Figure 7: Acrylamide evolution curves measured for the on-line system (gas phase, above) and during convection oven baking with multiple time-point sampling, extraction and GC-MS analysis (below).

ACKNOWLEDGEMENTS

This research was conducted in collaboration with partners at the University of Reading, U.K. (Don Mottram, Steven Elmore, Giorgos Koutsidis and Andy Dodson) and the University of Leeds, U.K. (Bronek Wedzicha).

LITERATURE CITED

Becalski, A., Lau, B. P. Y., Lewis, D., and Seaman, S. W., 2003, Acrylamide in foods: Occurrence, sources, and modeling, *J. Agric. Food Chem.* **51**:802-808.

Castle, L., 1993, Determination of acrylamide monomer in mushrooms grown on polyacrylamide-gel, *J. Agric. Food Chem.* **41**:1261-1263.

Channell, G. A., and Taylor, A. J., 2003, On-line monitoring of the Maillard reaction using a film reactor coupled to ion trap mass spectrometry, *Abstracts of Papers, 226th ACS National Meeting, New York, NY, United States, September 7-11, 2003*:AGFD-070.

Jublot, L., Linforth, R. S. T., and Taylor, A. J., 2004, Direct coupling of supercritical fluid extraction to a gas phase atmospheric pressure chemical ionization source ion trap mass spectrometer (SFE-APCI-ITMS) for fast extraction and analysis off food components, *J. Chromatog. A*: in press.

Mottram, D. S., Wedzicha, B. L., and Dodson, A. T., 2002, Acrylamide is formed in the Maillard reaction, *Nature* **419**:448-449.

Pollien, P., Lindinger, C., Yeretzian, C., and Blank, I., 2003, Proton transfer reaction mass spectrometry, a tool for on-line monitoring of acrylamide formation in the headspace of Maillard reaction systems and processed food, *Anal. Chem.* **75**:5488-5494.

Tareke, E., Rydberg, P., Karlsson, P., Eriksson, S., and Törnqvist, M., 2002, Analysis of acrylamide, a carcinogen formed in heated foodstuffs, *J. Agric. Food Chem.* **50**:4998-5006.

Taylor, A. J., Linforth, R. S. T., Harvey, B. A., and Blake, A., 2000, Atmospheric pressure chemical ionization mass spectrometry for in vivo analysis of volatile flavor release, *Food Chem.* **71**:327-338.

Turner, J. A., Linforth, R. S. T., and Taylor, A. J., 2002a, Real-time monitoring of thermal flavor generation in skim milk powder using atmospheric pressure chemical ionization mass spectrometry, *J. Agric. Food Chem.* **50**:5400-5405.

Turner, J. A., Sivasundaram, L. R., Ottenhof, M.-A., Farhat, I. A., Linforth, R. S. T., and Taylor, A. J., 2002b, Monitoring chemical and physical changes during thermal flavor generation, *J. Agric. Food Chem.* **50**:5406-5411.

Turner, J. A., Taylor, A. J., and Linforth, R. S. T., 2000, Monitoring aroma generation by the Maillard reaction in real time, *Czech J. Food Sci.* **18**:289-290.

Wenzl, T., de la Calle, M. B., and Anklam, E., 2003, Analytical methods for the determination of acrylamide in food products: a review, *Food Addit. Contam.* **20**:885-902.

Zyzak, D. V., Sanders, R. A., Stojanovic, M., Tallmadge, D. H., Eberhart, B. L., Ewald, D. K., Gruber, D. C., Morsch, T. R., Strothers, M. A., Rizzi, G. P., and Villagran, M. D., 2003, Acrylamide formation mechanism in heated foods, *J. Agric. Food Chem.* **51**:4782-4787.

FACTORS THAT INFLUENCE THE ACRYLAMIDE CONTENT OF HEATED FOODS

Per Rydberg[1], Sune Eriksson[2], Eden Tareke[1], Patrik Karlsson[2], Lar Ehrenberg[1], and Margareta Törnqvist[1]
[1]Department of Environmental Chemistry, Stockholm University, SE-106 91 Stockholm, Sweden and [2]AnalyCen Nordic AB, Box 905, SE-531 19 Lidköping, Sweden; e-mail: per.rydberg@mk.su.se

Abstract: Our finding that acrylamide is formed during heating of food initiated a range of studies on the formation of acrylamide. The present paper summarizes our follow-up studies on the characterization of parameters that influence the formation and degradation of acrylamide in heated foods. The system designed and used for studies of the influence of added factors was primarily homogenized potato heated in an oven. The net content of acrylamide after heating was examined with regard to the following parameters: heating temperature, duration of heating, pH and concentrations of various components. Higher temperature (200°C) combined with prolonged heating led to reduced levels of acrylamide, due to elimination/degradation processes. At certain concentrations, the presence of asparagine or monosaccharides (in particular fructose, glucose and glyceraldehyde) was found to increase the net content of acrylamide. Addition of other free amino acids or a protein-rich food component strongly reduced the acrylamide content, probably by promoting competing reactions and/or covalently binding of formed acrylamide. The pH-dependence of acrylamide formation exhibited a maximum around pH 8; lower pH enhanced elimination and decelerated formation of acrylamide. In contrast, the effects of additions of antioxidants or peroxides on acrylamide content were not significant. The acrylamide content of heated foods is the net result of complex reactions leading to both the formation and elimination/degradation of this molecule.

Key words: Acrylamide; amino acids, cooking; food; heating; Maillard reaction; mono saccharides; potato.

1. INTRODUCTION

The finding that acrylamide is formed in various foods cooked at elevated temperatures (Tareke *et al.*, 2000; Tareke *et al.*, 2002), at levels as high as milligrams per kilogram in the case of carbohydrate-rich foods such as potatoes, caused considerable alarm (WHO, 2002). This study examines various factors that may affect the formation of acrylamide. This was done in order to design strategies to prevent or decrease the extent of formation of acrylamide.

Much effort was devoted to develop a general and reproducible test system. Our choice was to use potatoes as the main food. In order to achieve accurate test results, various methods of heating were examined including microwave heating, heating in a temperature controlled frying pan and heating in a programmable GC-oven. Heating in a temperature programmed GC-oven gave the most reproducible results. Microwave heating gave rise to uncontrollable variations, mainly due to inhomogeneous heating resulting in local pyrolysis. The majority of the experiments were carried out at home, in a standardized Swedish kitchen (Svensk byggnorm, 1967).

It had been proposed (Tareke *et al.*, 2002) that formation of acrylamide during cooking involves the Maillard reaction. The content of acrylamide detected in a food is the net result of processes leading to both formation and degradation of this compound. Several research groups have examined acrylamide formation in model systems involving reducing sugars and amino acids (Mottram *et al.*, 2002; Stadler *et al.*, 2002; Becalski *et al.*, 2003; Zyzak *et al.*, 2003). Asparagine, which together with glutamine is the most abundant free amino acid in potatoes (Davies *et al.*, 1977; Eppendorfer *et al.*, 1996), has been suggested to be a likely source of acrylamide, particularly in potato products.

This study was designed to elucidate the effects of certain additives which can alter the pH and concentrations of potato constituents and that might influence the net production of acrylamide. In particular, the effects of natural components (sugars, amino acids, ascorbic acid, lean meat, fish) were evaluated. The possible role of antioxidants, free radicals and reactive oxygen species in the formation of acrylamide were also evaluated.

2. MATERIALS AND METHODS

The materials and instrumentation used have been described previously (Rydberg *et al.*, 2003).

The choice of chemicals were selected as far as possible in their pH neutral forms, in order to cause minimal alterations of pH of the prepared potato samples. Potatoes were purchased in large amounts (normally 15 kg) and carefully selected for similar tuber sizes, stored in the dark until use. The same cultivars, including controls, were used in the each experiment, within as short time period of purchase as possible. In studies of effects of amino acids and sugars, the controls were heated in the beginning, in the middle and at the final stages of the experiment.

In order to minimize artifacts after heating, the samples were immediately frozen at −20°C and stored at this temperature until analysis. The samples were prepared in Stockholm and subsequently transported in dry ice to Lidköping for analysis of acrylamide content.

The instrumentation and equipment employed for the work-up of samples and liquid chromatographic-tandem mass-spectrometric (LC-MS/MS) analysis of acrylamide are described elsewhere (Tareke et al., 2002). Cooking of foodstuffs was performed by heating as described earlier (Tareke et al., 2002) in a thermostated frying pan (Övervik et al., 1984).

In order to minimize individual variations in sample composition, 5 or 6 potatoes weighing approximately 500 g were peeled and thereafter chopped into small pieces. Four hundred grams of these pieces were placed in the mixer, various additions (amino acids, carbohydrates, etc.) were made and the potatoes were homogenized. The upper layer of this homogenate, including the foam, was used for pH measurement, while the remainder was poured gently into ice-cube trays directly frozen at −20°C as described earlier (Rydberg et al., 2003). Each heated sample consisted of three such cubes (weighing in average 15 g). The samples were heated in a temperature-programmed oven as described earlier (Rydberg et al., 2003).

It should be noted that the samples containing mixtures of fish and potato were minced (not homogenized as were the other samples) before heating in a temperature controlled frying pan.

The contents of acrylamide in the heated samples were measured, after work-up, using LC-MS/MS (ESI+) (Tareke et al., 2002; Rydberg et al., 2003). The analytical procedure is characterized by a high sensitivity (level of detection ≈ 10 μg/kg) and good reproducibility (CV ≈ 5 %) (Tareke et al., 2002; SWEDAC, 2002).

3. RESULTS

3.1 Experimental Reproducibility

The analytical procedure utilized here has been described earlier (Tareke *et. al*, 2002 and Rydberg *et. al.*, 2003). The applied procedure is associated with a high degree of reproducibility and stability, as demonstrated by several thousand analyses performed in our laboratory (at AnalyCen). The small variations that arise in the analytical chain were of minor importance.

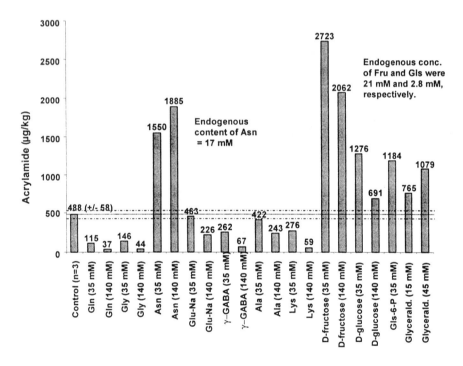

Figure 1. Effects of addition of amino acids and sugars on the net content of acrylamide in samples of homogenized potato slurry (Frieslander, batch 1, single samples) heated in an oven at 180°C for 25 min (data from Rydberg *et al.*, 2003). Data corrected for weight loss, which was 66.6 ± 1.6% (mean ± SD).

The development of a reproducible test system ranging from preparation to heating of samples was a challenge. In the beginning of this work, with a preliminary test system, *e.g.*., microwave heating of homogenized potatoes, it was clear that the observed variations did not permit more sophisticated studies. Much effort was therefore placed on finding a general and reproducible test system. Potatoes were found to be suitable to work with; this food was also important from a public health perspective. By applying

Factors Influencing the Acrylamide Content of Heated Foods

the described procedure for preparation and heating of frozen homogenized potatoes (Rydberg et. al., 2003), it was possible to investigate effects from added components, where values exceeding the limits ±25% (estimated error = 13.7% in *Figure 1* and 11.3% *in Figure 2*) were estimated not to be due to measurement error.

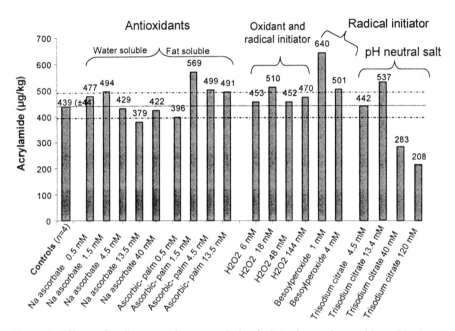

Figure 2. Effects of sodium ascorbate, ascorbyl palmitate, benzoyl peroxide and hydrogen peroxide on the acrylamide content of homogenized potato slurry (Ukama, single samples) heated in an oven at 180°C for 25 min (data from Rydberg et al., 2003). Data corrected for weight loss.

3.2 Effects from Temperature and Duration of Heating

The influence of temperature on the formation of acrylamide has been repeatedly demonstrated (Tareke et al., 2002; Mottram et al., 2002; Stadler et al., 2002; Becalski et al., 2003). It has been shown that the acrylamide content of potato strips first increased exponentially with time at a constant oven temperature (e.g., at 200°C, Rydberg et. al. 2003) then, after prolonged heating a decrease in the acrylamide content was observed, evidently because degradation of acrylamide (which also occurs in simple model systems; Mottram et al., 2002; Becalski et al., 2003) becomes predominant.

The formation of acrylamide in mashed potato samples was studied in a microwave heated oven (Fig. 3). In this experiment, the increase of the acrylamide content was 140 times higher after prolonging the heating time from 100 s to 150 s., This result shows that the formation of acrylamide progresses with the pyrolysis of the sample. Fig. 4 depicts the influence on the acrylamide content, with and without correction for weight loss, during heating French fries in an oven for 15 min at different temperatures of up to 220°C (data from Tareke et al., 2002).

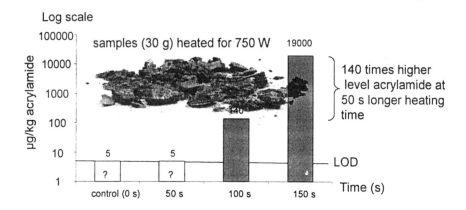

Figure 3. Formation of acrylamide in mashed potato samples (30 g) heated for given times in a microwave oven at 750W (data from Rydberg et al., 2003). (LOD = limits of detection)

3.3 Effects of Addition of Protein-Rich Components

Addition of lean meat of fish to potato samples was carried out in order to examine the possibility that the relatively low levels of acrylamide detected in heated beef or fish products might reflect competing or degrading reactions involving proteins. For this purpose, grated potato was mixed with different relative amounts (0 – 100%) of cod meat and the patties thus formed heated in a frying pan or oven (data presented in Rydberg et al., 2003). These experiments revealed that the decrease in acrylamide content (by 70% at equal amounts of potato and fish) was considerably greater than would be expected in the case of a linear relationship of formed acrylamide content between the measured values in pure potato and pure fish. This may reflect a protective action of protein, e.g., by elimination of acrylamide formed, perhaps via reaction with nucleophilic groups (-SH, $-NH_2$) on amino acid side-chains.

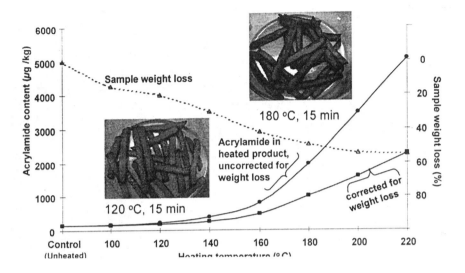

Figure 4. Formation of acrylamide (expressed as µg/kg) in French fries heated in an oven at different temperatures for 15 min. The temperature program is described in and the data taken from Tareke *et al.* (2002). Values both with (squares) and without (circles) are compensated for weight loss (indicated by the dotted line with triangles).

3.4 Effects of Adding Amino Acids

When free amino acids were added to homogenized potato heated in the oven, the net content of acrylamide was reduced by all of the amino acids tested (glycine, alanine, lysine, glutamine and glutamic acid), with the exception of asparagine. Additions of 35 mM of these amino acids resulted in an average decrease in acrylamide levels to approximately 1/2 of that observed in the absence of such addition (Fig. 1), with glutamine and glycine showing the largest reduction (about 70%). A further lowering occurred with 140 mM concentrations [75% ± 19% (mean ± SD, n=5) (Fig. 1)]. In contrast, 35 mM asparagine strongly enhanced the net acrylamide content and caused an additional small increase at a concentration of 140 mM. The endogenous level of free asparagine in this potato sample was found to be 17.0 mM [as measured (according to Davies, 2002) at AnalyCen AB on frozen potato homogenate and calculated on a wet weight basis]. This is in good agreement with reported values of 15-20 mM free asparagine in potatoes (Davies *et al.*, 1977; Eppendorfer *et al.*, 1996). Thus, the amounts of asparagine added here increased the level of this amino acid by factors of 3 and 9.

3.5 Effects of Addition of Carbohydrates

Addition of 35 mM glucose to the frozen homogenized potato system elevated the net acrylamide content (by 160%), in agreement with findings on model systems (Mottram *et al.*, 2002; Stadler *et al.*, 2002; Becalski *et al.*, 2003; Zyzak *et al.*, 2003). Surprisingly, this enhancement was considerably lower with 140 mM glucose (+41%, Fig. 1). Addition of fructose at 35 mM caused an even greater elevation (+460%), which was again attenuated at the higher concentration (+420%). The endogenous contents of free glucose and fructose in this potato sample were found to be 21 and 2.8 mM, respectively (measured according to Sullivan *et al.*, 1993) at AnalyCen AB (calculated on a wet weight basis). Thus, the additions made here corresponded to increases in the glucose level by 3- and 8-fold, and in the fructose level by 14- and 50-fold.

Addition of glyceraldehyde at concentrations of 15 and 45 mM also gave rise to a moderate increase (approximately 60%) in acrylamide content (Fig. 1). Furthermore, 35 mM glucose 6-phosphate enhanced this content to about the same extent as 35 mM glucose.

3.6 pH Effects

The effects of variations of pH on the acrylamide content of homogenized potato after heating were also explored. Lower pH values were attained by addition of HCl, ascorbic acid or citric acid; while the pH was raised by addition of NaOH. Plotting acrylamide content as a function of pH revealed a maximum value at approximately pH 8 (see Rydberg *et. al.*, 2003). The recovery of acrylamide added to these samples prior to heating was reduced at lower pH's (Fig. 5), indicating more rapid degradation of this compound in acidified matrices. This effect is probably accompanied by a reduction in the rate of formation. It should be noted that the pH of these samples was measured at ambient temperature prior to heating.

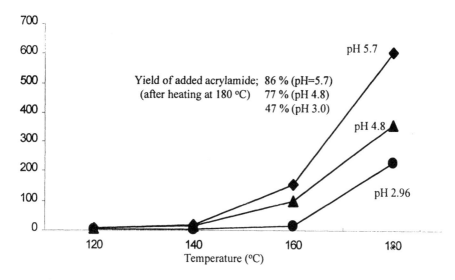

Figure 5. Effect of pH on the content of acrylamide (expressed as µg/kg with compensation for the weight loss) in homogenized potatoes heated in an oven at different temperatures. The pH of the original sample was 5.72 and acidification was achieved by addition of HCl. Heating was carried out for 25 min (data from Rydberg *et al.*, 2003).

In order to study the effects on acrylamide content of additives that modify the pH of foods, citric acid was added to homogenized potato prior to heating in the oven. The effects of such addition resembled those observed upon acidification with HCl (data not shown).

The results of these experiments emphasize the importance of controlling sample pH, which can vary between individual tubers and batches of potatoes. This is especially important when additions are made that may modify the pH. For this reason, the additives studied here (e.g., amino acids) were, insofar as possible, used in their neutral form.

3.7 Influence of Antioxidants, Oxidants and Radical Initiators

Ascorbic acid has been shown to behave as a reducing sugar in the Maillard reaction, where its degradation products react with amino acids and peptides (Löscher *et al.*, 1991; Vernin *et al.*, 1998) and inhibit browning (Matheis, 1987; Ziderman *et al.*, 1989).

In order to examine the effects of antioxidants on the acrylamide content of potato samples heated in the oven, two antioxidants that are accepted food additives were admixed into these samples. Possible influence of the polarity of the molecules was elucidated by comparing effects of the lipophilic ascorbyl palmitate, and the hydrophilic sodium ascorbate. At the lowest concentration tested (1.51 mM), both of these antioxidants caused a slight increase in the acrylamide content, exceeding the estimated error (Fig. 2).

The effects of hydrogen peroxide (which is both an oxidant and a radical initiator) and of benzoyl peroxide (radical initiator) were also examined. A possible real increase (+46%) in acrylamide content was obtained in the presence of 1 mmol/g of benzoyl peroxide ($n=1$), whereas the effect of addition of 4 mmol/kg was in the range of the estimated error. Furthermore, addition of increasing concentrations of hydrogen peroxide did not alter the content of acrylamide observed following heating (Fig. 2). These findings indicate that peroxidation or formation of free radicals (in the absence of frying oils) plays no major role in the acrylamide formation associated with heating.

4. SUMMARY/CONCLUSION

Based on the present investigation, Table 1 summarizes observed influences of various factors on the acrylamide content of heated foods. In general, the present studies involving foods support to the conclusions drawn from simple model experiments, i.e., that acrylamide can be formed by reaction of the amino acid asparagine with glucose or fructose. In the experiments with potato reported here, fructose was seen to be a more efficient precursor than glucose or glucose-6-phosphate. The observed influence of added amino acids on the acrylamide content illustrates the complexity of the underlying processes in heated foodstuffs, in terms of the reactions that are rate limiting with respect to the formation of acrylamide, and the presence of inhibitors and/or competing reactions. In addition to these reactions, other nucleophilic components might also eliminate acrylamide. The reactivity of acrylamide has recently been discussed by Friedman (2003).

Table 1. Parameters examined with respect to their influence on the formation/elimination of acrylamide during cooking.

Parameter	Net effect	Comments/Interpretation
Increased temperature	+	Increase
pH	+/-	Maximum around
Addition of monosaccharides	+	Increase
Addition of Asparagine	+	Increase
Addition of other aa than Asn	-	Up to 93% reduction
Addition of antioxidants/oxidants and radical initiators	+/-?	Relatively small effects
Addition of components that binds water	-	Decrease
Addition of fish meat	-	Decrease

Furthermore, our present findings lead us to propose that additives that have the capacity to bind water reduce the net formation of acrylamide, probably because water inhibits pyrolysis reactions. In this context, studies designed to elucidate the influence of the initial water content of samples on the formation of acrylamide during heating should be further explored.

Acrylamide is classified as a probable human carcinogen (IARC, 1994). An improved understanding of the reactions leading to acrylamide formation and degradation/elimination during heating will facilitate development of methods for decreasing the acrylamide content of cooked food.

ACKNOWLEDGEMENTS

We are grateful to Dr. Ulf Olsson for critical reading of the manuscript. This work was supported financially by the Swedish Research Council for Environment, Agricultural Science and Spatial Planning (FORMAS) and faculty resources at Stockholm University.

REFERENCES

Becalski, A., Lau, B. P.-Y., Lewis, D., and Seaman, W. S., 2003, Acrylamide in foods: Occurrence, sources and modelling, *J. Agric. Food Chem.* **51**:802-808.

Davies, A. M. C., 1977, The free amino acids of tubers of potato varieties grown in England and Ireland, *Potato Res.* **20**:9-21.

Davies, M., 2002, *The Biochrom Handbook of Amino Acids*, Issue 1, Biochrom Ltd, Cambridge. pp. 51-56 and 69-77.

Eppendorfer, W. H., and Bille, S. W., 1996, Free and total amino acid composition of edible parts of beans, kale, spinach, cauliflower and potatoes as influenced by nitrogen fertilisation and phosphorus and potassium deficiency, *J. Sci. Food Agric.* **71**:449-458.

Friedman, M., 2003, Chemistry, biochemistry, and safety of acrylamide. A review, *J. Agric. Food Chem.* **51**:4504-4526.

IARC, 1994, Acrylamide, in: *IARC Monographs on the Evaluation of Carcinogen Risk to Humans: Some Industrial Chemicals*, Vol. 60, International Agency for Research on Cancer: Lyon, pp 389-433.

Löscher, J., Kroh, L., Westphal, G., and Vogel, J., 1991, L-Ascorbinsäure als Carbonylkomponente nichtenzymatischer Bräunungsreaktionen: 2. Mitteilung Amino-Carbonyl-Reaktionen der L-Ascorbinsäure, *Z. Lebensm. Unters. Forsch.* **192**:323-327.

Matheis, G., 1987, Polyphenol oxidase and enzymatic browning of potatoes (Solanum tuberosum): II. Enzymatic browning and potato constituents, *Chem. Mikrobiol. Technol. Lebensm.* **11**:33-41.

Mottram, D. S., Wedzicha, B. L., and Dodson, A. T., 2002, Acrylamide is formed in the Maillard reaction, *Nature* **419**:448-449.

Övervik, E., Nilsson, L., Fredholm, L., Levin, Ö., C.E., N., and Gustafsson, J. Å. 1984, High mutagenic activity formed in pan-broiled pork, *Mutat. Res.* **135**:149-157.

Rydberg, P., Eriksson, S., E., T., Karlsson, P., Ehrenberg., and Törnqvist, M., 2003, Investigations of factors that influence the acrylamide content of heated foodstuffs, *J. Agric. Food Chem.* **51**:7012-7018.

Stadler, R. H., Blank, I., Varga, N., Robert, F., Hau, J., Guy, P. A., Robert, M.-C., and Riediker, S. D., 2002, Acrylamide from Maillard reaction products, *Nature* **419**:449-450.

Svensk Byggnorm, 1967 (SBN 67), http://www.boverket.se.

SWEDAC, 2002, Accreditation SWEDAC Dnr. 01-4262-51.1125; Swedish Board for Accreditation and Conformity Assessment, Stockholm, Sweden.

Sullivan, D.M.; Carpenter, D.E., 1993, *Methods of Analysis for Nutrition Labeling*, AOAC International, Arlington, VA, pp. 455-533.

Tareke, E., Rydberg, P., Karlsson, P., Eriksson, S., and Törnqvist, M., 2000, Acrylamide: a cooking carcinogen?, *Chem. Res. Toxicol.* **13**:517-522.

Tareke, E., Rydberg, P., Karlsson, P., Eriksson, S., and Törnqvist, M., 2002, Analysis of acrylamide, a carcinogen formed in heated foodstuffs, *J. Agric. Food Chem.* **50**:4998-5006.

Vernin, G., Chakib, S., Rogacheva, S. M., Obretenov, T. D., and Párkányi, C., 1998, Thermal decomposition of ascorbic acid, *Carbohydrate Res.* **305**:1-15.

World Health Organization, 2002, Health Implications of Acrylamide in Food. Report of a Joint FAO/WHO Consultation, June 25-27, Department of Protection of the Human Environment, WHO, Geneva, Switzerland.

Ziderman, I. I., Gregorski, K. S., Lopez, S. V., and Friedman, M., 1989, Thermal interaction of ascorbic acid and sodium ascorbate with proteins in relation to nonenzymatic browning and Maillard reactions of foods, *J. Agric. Food Chem.* **37**:1480-1486.

Zyzak, D. V., Sanders, R. A., Stojanovic, M., Tallmadge, D. H., Eberhart, B. L., Ewald, D. K., Gruber, D. C., Morsch, T. R., Strothers, M. A., Rizzi, G. P., and Vilagran, M. D., 2003, Acrylamide formation mechanism in heated foods, *J. Agric. Food Chem.* **51**: 4782-4787.

MODEL SYSTEMS FOR EVALUATING FACTORS AFFECTING ACRYLAMIDE FORMATION IN DEEP FRIED FOODS

R.C. Lindsay and S. Jang
Department of Food Science, 1605 Linden Drive, University of Wisconsin-Madison, WI 53706; e-mail: rlindsay@wisc.edu

Abstract: Simulated food pieces constructed from fiberglass pads (models for French fries and chips) were used as carriers for defined aqueous solutions, dispersions of test substances and ingredients to evaluate acrylamide formation. The pads were loaded with a solution containing asparagine and glucose (10 mM each) plus selected reaction modulators before deep fat frying and analysis for acrylamide. Data from fiberglass models along with companion sliced potato samples were used in developing hypotheses for the mechanisms involved in the suppression of acrylamide formation by polyvalent cations, polyanionic compounds, pH, and altered food polymer states in fried potato products.

Key words: acrylamide, model systems, formation, suppression, mechanisms, food polymer states, asparagine, glucose, fried foods, potatoes.

1. INTRODUCTION

The emergence of the carbonyl-amino Maillard-type reaction between asparagine and reducing sugars (and other carbonyl compounds) as the primary mechanism responsible for most of the acrylamide formed in susceptible fried and other thermally-processed foods (Mottram et al., 2002; Stadler et al., 2002; Yaylayan et al., 2003; Zyzak et al., 2003) has prompted the use of simplified models to study various aspects of acrylamide formation. These have included the use of aqueous buffered solutions of asparagine and reducing sugars for studying acrylamide formation under elevated a_w and controlled thermal conditions (Mottram et al., 2002;

Becalski et al., 2003; Zyzak et al., 2003), and the exposure of several combinations of either dry or moistened mixtures of amino acids, sugars and other reactants to various thermal treatments approaching and including pyrolytic conditions (Stadler et al., 2002; Yaylayan; Becalski et al., 2003). A more complex fabricated fried potato snack (chip) model composed of potato starch, water, maltodextrin, glucose, selected amino acids, and monoglycerides has been described by Zyzak et al. (2003). Additionally, several others have employed preparation and processing of susceptible foods for acrylamide formation studies (Grob et al., 2003; Biedermann et al., 2002; Jung et al., 2003; Zyzak et al., 2003).

Although these earlier reports have advanced the understanding of acrylamide in heated foods, roles of the factors believed to influence the chemical process often are not clear because of complex physical and chemical features associated with the inherent properties of foods and the processing employed. In some instances a lack of exposure of the samples in model systems to frying conditions, along with the inclusion of complex natural ingredients, somewhat limits their utility. In this study simulated food pieces (modeling French fries or chips) constructed with fiberglass pads were used as carriers for defined aqueous solutions and dispersions of test substances and ingredients to evaluate acrylamide formation without the confounding influences of other food constituents during frying.

2. MATERIALS AND METHODS

2.1 Supplies and reagents

Glass fiber sample pads for preparing model units were obtained from CEM Corp (ca 10 cm x 10 cm x 0.1 cm; 0.8 g for microwave oven moisture analysis; prod. #200150, Matthews, NC). Food grade soy lecithin and corn starch were obtained from Archer Daniels Midland Company (Decatur, IL) and A. E. Staley Manufacturing Company, Decatur, IL), respectively. Intact chipping potatoes (unspecified variety) were provided by Frito-Lay (Beloit, WI). Commercial partially hydrogenated vegetable oil was purchased from a local food service supplier (Sysco Food Services, Baraboo, WI). Olive oil was purchased from a local grocery store. Labeled 1,2,3-^{13}C-acrylamide for use as a quantitative analytical standard was purchased from Cambridge Isotope Laboratories (Andover, MA). All other reagents and chemicals were purchased from Aldrich Chemical Co. (Milwaukee, WI) and were used without further purification.

2.2 Fiberglass model systems

2.2.1 Stacked glass fiber pad bundle model

Simulated French fried potato strips were constructed with glass fiber pads (98 x 103 mm, polymer-free, product #200250, CEM Corp., Matthews, NC). Fiberglass pads were cut into rectangular-shaped strips (10 x 50 mm), and then sufficient fiberglass strips were stacked and tied with tight-twist cotton string to yield 10 x 13 x 50 mm bundles (tare *ca* 1.0 g).

An aqueous solution containing 10 mM glucose plus 10 mM asparagine was employed as a standard (control) for generating acrylamide during subsequent frying. Other treatment ingredients were added to the base test solution as needed. Bundles of stacked fiberglass pad strips were loaded by submerging in desired test solutions. Excess solution was removed by gentle squeezing to yield nearly saturated fiberglass bundles. Absorbed solution weights were determined by difference between total and tare weights. After frying at 204 ± 2°C for 45 sec, intact individual stacked glass fiber pad bundles were analyzed for acrylamide. Calculations of acrylamide quantities were based on weights of test solutions initially retained in fiberglass pad bundles.

2.2.2 Single sheet glass fiber model

A single sheet glass fiber pad model was employed for simulation of potato chips. Weighed amounts of aqueous solutions containing 10 mM glucose plus 10 mM asparagine and any additional chemicals were loaded by pipette onto individual glass pads (10 cm x 10 cm x 0.1 cm; 0.8 g; approximately saturated). Each sheet was then fried at 182 ± 2°C for 30 sec, and the entire pad was subsequently analyzed for acrylamide. Quantifications for acrylamide in glass fiber sheets were based on weight of test solution initially contained in the pad before frying (e.g., μg acrylamide/kg test solution).

2.3 Deep frying equipment

Controlled deep frying was carried out with a small-size, commercial batch fryer (Model 1421 Toastmaster, McGraw-Edison, Algonquin, IL) containing *ca.* 10 L of fresh food service grade, partially hydrogenated canola oil (Imperial, containing TBHQ, citric acid, and dimethyl-polysiloxane; Sysco Corp., Houston, TX). The fryer was fitted with a high-capacity stainless steel stirrer controlled with a variable speed motor (Stir-Pak; Cole-Parmer, Vernon Hills, IL) along with a thermocouple probe and electronic thermometer (type k; Fisher Scientific) for continuous monitoring of oil temperature. Samples were introduced into the frying oil uniformly at a point 2 - 3°C before the maximum (selected) temperature of the heating cycle was achieved.

2.4 Potato handling and preparation

Stored, commercial in-use chip potato tubers were transferred (<8 h) to similar storage (8°C, 92% relative humidity) under controlled conditions in a well-ventilated room located in the campus Biotron facility where they were held (<3 mo) until removal to ambient conditions immediately before used in experiments. Potato tubers were processed by first washing in tapwater, and then slicing (1.5 mm thick) with a rotary-blade food slicer (Hobart Corp., Troy, OH). Where possible, 5 adjacent slices from the same potato were used for each treatment within an experiment. Individual slices within a lot were manually trimmed to remove the periderm and some adjacent tissue to yield approximately circular pieces 65-90 mm diameter, and then the lot was rinsed in cool water (15-20°C) before subjecting it to an experimental treatment. Slices less than 60 mm diameter were discarded. Potato slices were rinsed in deionized water for *ca.* 20 sec, and then drip-drained before frying. When potato slices were blanched as part of the process, they were held at 80°C for 1 min before immersing in tap water (15°C) for cooling. Where potato slices were further treated with process chemicals, they were immersed in various aqueous solutions additionally for 3 min at 21°C. Finally, potato slices were fried at 180°C in partially hydrogenated vegetable oil until steam bubbles ceased (*ca.* 2 min).

2.5 Acrylamide analysis

Samples were analyzed using an adaptation of the US EPA procedure for GC-MS detection of brominated acrylamide (US EPA 1996; Tareke et al., 2002). Initial aqueous extraction of samples (2-10 g each for foods or entire model unit) was carried out by placing either into 100 g of distilled water containing 1.0 µg of 1,2,3-^{13}C-acrylamide as internal standard (Cambridge Isotope Laboratories, Inc.). Each sample was then held for *ca.* 10 min at ambient temperature before blending thoroughly with a probe homogenizer (Tekmak Co., Cincinnati, OH) at speed 7 for *ca.* 1 min. The cloudy aqueous layer of each was transferred to a 50 mL of disposable centrifuge tube, and then centrifuged (model CR312, Jouan, Winchester, VA) at 3820 *g* for 20 min at 5°C. After centrifugation, the floating lipid layer was removed with a pipette, and the aqueous extract (ca. 50 mL) was poured into a 200 mL glass-stoppered flask in preparation for bromine derivatization.

Next, 7.5 g of KBr was dissolved into each aqueous extract, and each was acidified with conc. HBr (40 wt % in H_2O) to pH 1-3 which was verified with pH paper (EM Science, Gibbstown, N.J). Each sample flask was wrapped with aluminum foil, and 2.5 mL of saturated bromine water reagent was added, tightly closed with a glass-stopper, and then placed in a dark room at 5°C for overnight. Excess bromine reagent in each sample was inactivated by adding 1.0 M sodium thiosulfate dropwise until the solution became colorless. Then, 15 g of sodium sulfate was added into each sample flask followed by vigorous shaking, and a further addition of 25 mL ethyl acetate. The upper organic layer of ethyl acetate upper layer was transferred to a glass tube (16 x 150 mm), and then placed in a water bath (40 °C) for concentration under a slow stream of nitrogen to ca 0.5 mL.

Instrumental analyses were performed using a Hewlett-Packard 6890 series GC system equipped with an Agilent 7683 series injector (set for 1 µL) and an Agilent 5973 Network Mass Selective Detector. Separation parameters included the use of a DB-17 column (30 m x 0.25 mm i.d.), and the GC was programmed from 65°C to 215°C with a rate of 15°C/min, and then to 235°C with a rate of 2°C/min at which point it was held for 2 min. Injector and detector temperatures were both at 250°C. The selective ion monitoring mode was used for identification of brominated acrylamide with mass 150, 152, and 106 used for acrylamide analyte, and 153, 155, and 110 used for the internal standard (1,2,3 ^{13}C-acrylamide). Ion masses of 150 for acrylamide and 155 for the internal standard were used for quantification calculations. The method overall minimum detection limit under the conditions of these studies was 10 µg/kg.

3. RESULTS AND DISCUSSION

3.1 Overall performance of fried fiberglass pad models

Both fried stacked pad (French fry) and single sheet (chip) fiberglass models provided an overall repeatability well within ± 20% during routine usage, and were relatively easy to use.

However, researchers are cautioned that individual lots of fiberglass pads may yield dissimilar acrylamide levels from a simple glucose plus asparagine (10 mM each) solution as well as from some other experimental treatments. The most likely source of the variability has been traced to varying water-extractable alkalinity from the pads which apparently results from hydration of alkaline earth oxides (Na_2O, K_2O, MgO, CaO) that are incorporated into the glass composition. As a result, care should be exercised when interpreting data relating to acidity treatments, and where solubilization of Ca^{++} and Mg^{++} ions might affect reactions occurring in the glass fiber matrix. Nevertheless, regular inclusion of a valid glucose plus asparagine (10 mM each) background control sample within experimental series appears to provide a reliable means for circumventing problems associated with manufacturing variations in the fiberglass pads.

Caution also should be exercised to select polymer-free fiberglass pad stock because the high temperatures associated with frying can result in high background acrylamide levels presumably because of the liberation of acrylamide from binding polymers.

When the single sheet format was tested as a baking model, acrylamide was not detected after exposure to baking regimes employed for cookies or bread. Although the model readily browned, it is surmised that acrylamide which was formed apparently vaporized from the matrix under these conditions. Thus, the single sheet fiberglass model in the current iteration is not applicable for baking studies.

3.2 Evaluation of contributions of lipid oxidation to acrylamide formation in fried potato products

The oxidation and/or thermal degradation of lipids in fried foods has been discussed as a possible mechanistic route contributing to the formation of acrylamide via an acrylic acid intermediate (Yashhara et al., 2003). In testing the hypothesis that accelerated oxidation during frying would elevate acrylamide formation, ferric chloride was introduced to the surface of sliced potatoes before frying and typical results are shown in Table 1.

Table 1. Comparison of exposure of potato slices to Fe^{+++} and Ca^{++} ions before frying upon acrylamide formation.

Potato Chip Sample Treatment	Acrylamide Concentration (µg/kg)	Acrylamide Reduction (% versus Untreated)
Untreated - -cut, wash & fry	616	–
Treated with 1000 ppm $FeCl_3$ [a]	143	77
Treated with 1000 ppm $CaCl_2$ [a]	259	59

[a] Treatment involved - cut, wash; blanch in deionized water at 80°C for 1 min; soak in 1000 ppm $FeCl_3$ or 1000 ppm $CaCl_2$ at 21°C for 3 min; fry

Although transition state metal cations (especially iron and copper) are well known for accelerating lipid oxidation via catalysis of hydroperoxide decompositions (Fennema, 1996), treatment of potato slices with 1000 ppm Fe^{+++} ions instead reduced acrylamide formation by 77% compared to the untreated control. Somewhat unexpectedly the introduction of Ca^{++} as a non-transition state cation control treatment also substantially suppressed acrylamide formation during frying of potato chips (Table 1). Thus, the results of these experiments clearly indicated that accelerated oxidation of lipids in potatoes during frying did not result in enhanced acrylamide formation.

Data presented in Table 2 was obtained using the stacked glass pad model, and it shows that when the glucose-asparagine browning reaction was isolated from other potato constituents, 1000 ppm of either Fe^{+++} or Ca^{++} ions suppressed over 90% of the acrylamide formed in the untreated control.

Table 2. Influence of Fe^{+++} and Ca^{++} ions upon acrylamide formation in fried stacked pad model systems containing glucose plus asparagine (10 mM each).

Stacked Glass Pad Sample Treatment	Acrylamide Concentration (µg/kg)	Acrylamide Reduction (% versus Untreated)
Untreated	173	–
Treated with 1000 ppm $FeCl_3$	11	93
Treated + 1000 ppm $CaCl_2$	13	93

Similarly, when only 25 ppm of either Fe^{+++} or Ca^{++} ions were incorporated into the single sheet glass pad model system, acrylamide formation was reduced about 50% compared to the untreated control (Table 3). These results indicated that the multivalent cations effected considerable inactivation of the glucose-asparagine Maillard reaction. Furthermore, since both Fe^{+++} and Ca^{++} ions behaved similarly in the model systems, a common mechanistic basis for the activity of the two cations was suggested.

Table 3. Effect of 25 ppm Fe^{+++} and Ca^{++} on acrylamide formation in fried single sheet models containing glucose plus asparagine (10 mM each).

Single Sheet Glass Pad Sample Treatment	Acrylamide Concentration (µg/kg; ppb)	Acrylamide Reduction (% versus Untreated)
Untreated	418	–
Treated with 25 ppm $FeCl_3$	226	46
Treated with 25ppm $CaCl_2$	204	51

3.3 Ionic complexations as contributing mechanisms for the suppression of acrylamide formation in foods

The notable suppressions of acrylamide formation observed for multivalent cations (e.g., Fe^{+++}, Ca^{++}, etc) in glucose-asparagine model systems led to the hypothesis that ionic associations involving the ions and charged groups on asparagine and related intermediates were likely to be involved. Further experiments employing chelating substances along with Ca^{++} ions were expected to suppress or negate the effects of the cations through chelation, but instead they complemented the acrylamide suppressing effects of the Ca^{++} ions. The complementary suppression effect appeared to be especially pronounced for chelating substances exhibiting marginal solubility, and phytic acid (phytate) was found to be especially effective.

Data in Table 4 from glucose-asparagine fried stacked pad model system experiments show that incorporation of 250 ppm of phytate suppressed acrylamide formation by about 80% compared to the untreated control. The low pH (2.8) provided by the phytate was expected to contribute to the suppression because protonation of asparagine amino groups would interfere with the initiation of the Maillard reaction (cf. Jung et al., 2003). When the glucose-asparagine model system was similarly adjusted to pH 2.8 with HCl, somewhat less acrylamide suppression than that provided by 250 ppm phytate was observed (63% reduction compared to the control). When 1000 ppm Ca^{++} ions were introduced with 250 ppm phytate, their combined effects along with an accompanying pH suppression yielded 95% acrylamide reduction compared to the control.

Table 4. Effect of pH, calcium, and phytate on the suppression of acrylamide formation in glucose-asparagine (10 nM each) fried stacked pad model systems.

Stacked Glass Pad Sample Treatment	Acrylamide Concentration (μg/kg; ppb)	Acrylamide Reduction (% versus Untreated)
Untreated (standing pH 4.3)	173	–
Treated - pH 2.8 with HCl	64	63
Treated -250 ppm phytate (pH 2.8)	30	82
Treated - 250 ppm phytate + 1000 ppm $CaCl_2$ (pH 2.8)	10	95

Results from experiments testing the effects of pH and treatments on potato slices with 1000 ppm Ca^{++} ions plus 250 ppm phytate (Table 5) showed that the influence of non-buffered pH (3.0) adjustment was minimal (ca. 20% reduction), but the combined effects of non-buffered pH reduction and Ca^{++} plus phytate gave about 70% reduction in acrylamide compared to the untreated control. However, although potato surfaces exhibit considerable buffering capacity, caution must be exercised in applying acidic treatments to fried potato products because when the final pH of the finished product is below about 4.0, sourness may detract notably from the expected flavor profile of the final product.

A hypothesis of integrated ionic complexations involving salt formation, multivalent cation crosslinking, and chelation has been developed to account for the suppressing effects Ca^{++} and phytate exert upon acrylamide formation (see Fig. 1). Both asparagine (Fig. 1A) and N-glucosylasparagine (Fig. 1B) possess charged groups and available coordination electrons that could participate in ionic multimolecular associations. Chen et al. (1989) have reported that multivalent transition and alkaline metal ions appear to stabilize N-glycosides in aqueous model systems.

Table 5. Effect of pH and 1000 ppm Ca^{++} plus 250 ppm phytate upon acrylamide formation in potato chips.

Potato Chip Sample Treatment	Acrylamide Concentration (μg/kg; ppb)	Acrylamide Reduction (% versus Untreated)
Untreated -cut, wash & fry	558	–
Treated with water at pH 3.0 [a]	433	22
Treated with $CaCl_2$ + phytate [b]	161	71

[a] Cut, wash; blanch in deionized water adjusted to pH 3.0 with HCl at 80°C for 1 min; -rinse at 21°C for 1 min in deionized water adjusted to pH 3.0 with HCl; fry
[b] as above but without pH adjustment and rinse with 1000 ppm $CaCl_2$ + 250 ppm phytate

The schematic for the complexation of asparagine by Ca^{++} plus phytate (Fig. 1A) depicts both an ionic association via a cross-linking by the calcium cation and a stabilizing monovalent salt formation between the amino group of asparagine and a phosphate anion moiety on a phytate molecule. Such phytate complexes would be expected to be marginally soluble in water, and when they actually precipitate, the asparagine would be unavailable for participation in early stage Maillard browning.

Because asparagine molecules possess both a potentially positively charged group (the primary amino group) and a potentially negatively charged group (the carboxyl group), and since reducing sugars do not possess either positive or negative charges, it seems reasonable to conclude that the asparagine molecule or reacted moiety is responsible for interactions with Ca^{++} ions and chelating molecules (e.g., phytate) which lead the inhibition of acrylamide formation. These ionic and electron associations appear to stabilize the molecular complexes during the high heat conditions encountered during frying, and thereby suppress early stage Maillard reactions.

3.4 Application of food state diagram concepts for studying the effects of physical influences on acrylamide formation in foods

During studies of various heating treatments of potato slices upon acrylamide formation, it became apparent that physical conditions within the matrix of slices influenced the rate and extent of acrylamide formation. When treatments favored high molecular mobility because of water-plasticized polymer conditions in the rubbery state (Blansford and Lillford, 1993; Fennema, 1996), acrylamide formation was greatly amplified.

A) A phytate—asparagine complexation schematic

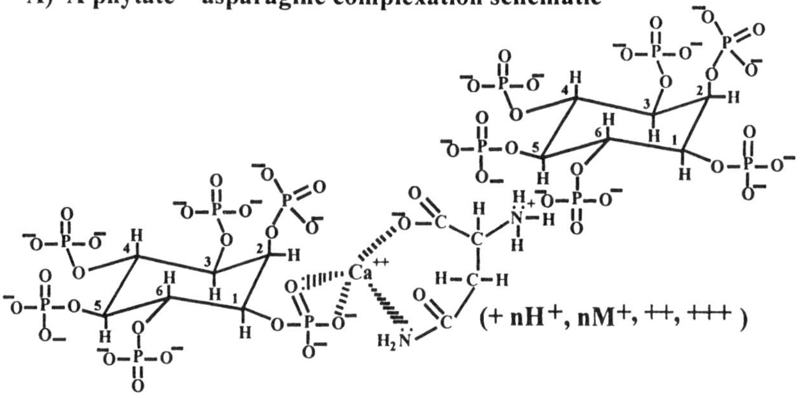

B) An N-Glycosylasparagine—phytate complexation schematic

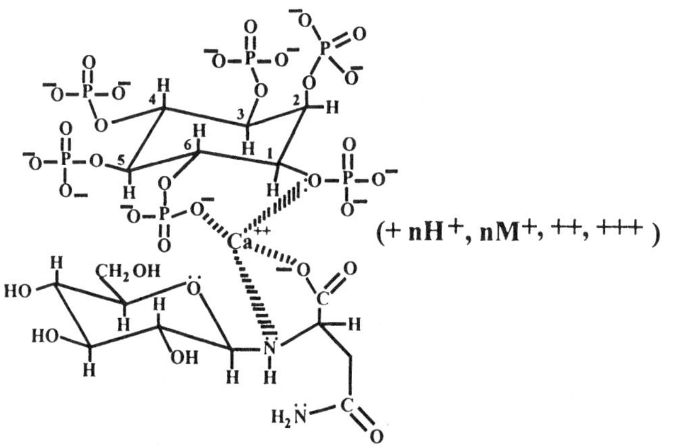

Figure 1. Some proposed ionic complexations via salt formation, multivaltent cation crosslinking, and chelation hypothesized to contribute to the suppression of acrylamide formation in foods.

Results of fried stacked glass pad model systems employing glucose-asparagine (10 mM each) and various amounts of soluble corn starch or insoluble (raw, native) corn starch are summarized in Tables 6 and 7. When soluble starch alone (5 and 10%) was present (Table 6), acrylamide concentrations were greatly amplified over the untreated control. On the other hand, when insoluble native starch (5%) was present (Table 7), acrylamide formation was greatly suppressed (*ca.* 80%). Recently, Shih et al. (2004) reported that use of pre-gelatinized rice flour increased acrylamide formation in fried long grain rice batters.

Model Systems for Factors Affecting Acrylamide in Deep Fried Foods   339

Table 6. Effect of soluble food polymers (soluble corn starch) on acrylamide formation in fried stacked glass pad model systems containing 10 mM each glucose and asparagine.

Stacked Glass Pad Sample Treatment	Acrylamide Concentration (μg/kg)	Acrylamide Reduction (versus Untreated)
Untreated-	258	–
Treated + 5% soluble corn starch	500	INCREASE 193 %
Treated + 10% soluble corn starch	642	INCREASE 249 %

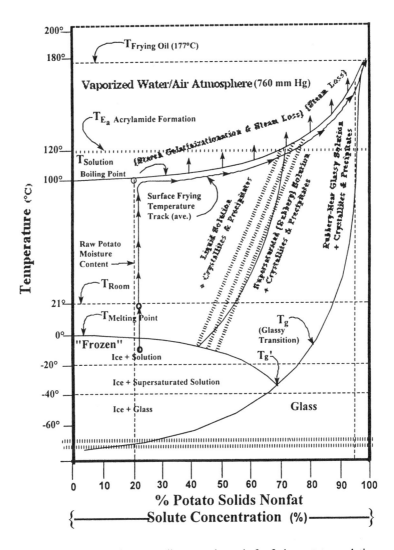

Figure 2. Proposed dynamic water-solute state diagram schematic for frying potatoes relating process conditions to factors affecting acrylamide formation via glucose-asparagine Maillard reactions.

Table 7. Effect of insoluble food polymers (raw native corn starch) on acrylamide formation in fried stacked glass pad model systems containing 10 mM each glucose and asparagine.

Stacked Glass Pad Sample Treatment	Acrylamide Concentration (μg/kg)	Acrylamide Reduction (% versus Untreated)
Untreated	262	–
Treated + 5% raw corn starch	55	79
Treated +10% raw corn starch + 1.5% sodium chloride	36	86

The schematic shown in Fig. 2 was developed in an attempt to extend the food state diagram concept to include conditions encountered in the preparation of potato products, especially in the critical frying zone (low moisture contents, i.e., <5%) on the surface of pieces where acrylamide formation is especially rapid. Experimental acrylamide data points based on final moisture contents for potato chips appear in agreement with expected acrylamide values. Further research on this aspect and other factors influencing acrylamide formation is currently underway.

REFERENCES

Becalski, A., Lau, B. P.-Y., Lewis, D., and Seaman, S., 2003, Acrylamide in foods; occurrence and sources, and modeling, *J. Agric. Food Chem.* **51**:802-808.

Biedermann, M., Biederman, B. S., Noti, A., and Grob, K., 2002, Methods for determining the potential of acrylamide formation and its elimination in raw materials for food preparation, such as potatoes, *Mitt. Lebensm. Hyg.* **93**:638-652.

Blanshard, J.M.V. and Lilliford, P. J., 1993, *The Glassy State in Foods*, Nottingham University Press, Loughborough, UK.

Chen, J., Pill, T., and Beck, W. Z., 1989, *Naturforsch* **44**:459-464.

Fennema, O., 1996, *Food Chemistry*, 3rd ed., Marcel Dekker, NY.

Grob, K., Biedermann, M., Biedermann, B. S., Noti, A., Imhof, D., Amrein, T., Pfefferle, A., and Bazzocco, D., 2003, French fries with less than 100ug/kg acrylamide. A colloboration between cooks and analysts, *Eur. Food Res. Technol.* **217**:185-194.

Jung, M. Y., Choi, D. S., and Ju, J. W., 2003, A novel technique for limitation of acrylamide formation in fried and baked corn chips and in French fries, *J. Food Sci* **68**:1287-1290.

Mottram, D. S., Wedzicha, B. L., and Dodson, A. T., 2002, Acrylamide is formed in the Maillard reaction, *Nature* **419**:448-449.

Shih, F. F., Boue, S. M., and Shih, B. Y., 2004, Effects of flour sources on acrylamide formation and oil uptake in fried batters, *J. Am. Oil Chem. Soc.* **81**:265-268.

Stadler, R. H., Blank, I., Varga, N., Robert, F., Hau, J., Guy P. A., Robert, M. C., and Riediker, S., 2002, Acrylamide from Maillard reaction products, *Nature* **419**:449-450.

Tareke, E., Rydberg, P., Karlsson, P., Eriksson, S., and Tornqvist, M., 2002, Analysis of acrylamide, a carcinogen formed in heated foodstuffs, *J. Agric. Food Chem.* **50**:4998-5006.

US EPA, 1996, *SW 846, Test Methods for Evaluating Solid Waste, Acrylamide by Gas Chromatography, Method 8032A*:U.S. Environmental Protection Agency, Washington, DC.

Yasuhara, A., Tanaka, Y., Hengel, M., and Shibamoto, T., 2003, Gas chromatographic investigation of acrylamide formation in browning model systems, *J. Agric. Food Chem.* **51**:3999-4003.

Yaylayan, V. A., Wnorowksi, A., and Locas, C. P., 2003, Why asparagine needs carbohydrate to generate acrylamide, *J. Agric. Food Chem.* **51**:1753-1757.

Zyzak, D. V., Sanders, R. A., Stojanovic, M., Talmadge, D. H., Eberhart, B. L., Ewald, D. K., Gruber, D. C., Morsch, T. R., Strothers, M. A., Rizzi, G. P., and Villagran, M. D., 2003, Acrylamide formation mechanism in heated foods, *J. Agric. Food Chem.* **51**:4782-4787.

CONTROLLING ACRYLAMIDE IN FRENCH FRY AND POTATO CHIP MODELS AND A MATHEMATICAL MODEL OF ACRYLAMIDE FORMATION

Acrylamide: Acidulants, phytate and calcium

Yeonhwa Park[1], Heewon Yang[1], Jayne M. Storkson[1], Karen J. Albright[1], Wei Liu[1], Robert C. Lindsay[2], and Michael W. Pariza[1]

[1]*Food Research Institute, University of Wisconsin-Madison, 1925 Willow Dr., Madison, WI 53706;* [2]*Department of Food Sciences, University of Wisconsin-Maidson, 1605 Linden Dr., Madison, WI 53706; e-mail:ypark2@wisc.edu*

Abstract: We previously reported that in potato chip and French fry models, the formation of acrylamide can be reduced by controlling pH during processing steps, either by organic (acidulants) or inorganic acids. Use of phytate, a naturally occurring chelator, with or without Ca^{++} (or divalent ions), can reduce acrylamide formation in both models. However, since phytate itself is acidic, the question remains as to whether the effect of phytate is due to pH alone or to additional effects. In the French fry model, the effects on acrylamide formation of pH, phytate, and/or Ca^{++} in various combinations were tested in either blanching or soaking (after blanching) steps. All treatments significantly reduced acrylamide levels compared to control. Among variables tested, pH may be the single most important factor for reducing acrylamide levels, while there were independent effects of phytate and/or Ca^{++} in this French fry model. We also developed a mathematical formula to estimate the final concentration of acrylamide in a potato chip model, using variables that can affect acrylamide formation: glucose and asparagine concentrations, cut potato surface area and shape, cooking temperature and time, and other processing conditions.

Key words: Acrylamide; pH; French fries; phytate; calcium; acidulants; preservatives

1. INTRODUCTION

The discovery of acrylamide in cooked food, particularly starch-based foods (Tareka et al., 2000; Tareka et al., 2002), caused concern over the safety of such foods. The concern was based on three considerations: acrylamide is categorized as a probable human carcinogen; foods that contain acrylamide are widely consumed; and the amounts of acrylamide found in food are higher than many other known foodborne carcinogens (Tareke et al., 2002; Friedman, 2003). While the potential carcinogenicity of acrylamide for humans remains a matter of debate (Pelucchi et al., 2003; Granath and Tornqvist, 2003; Mucci et al., 2003), research has focused on understanding the mechanism(s) of acrylamide formation in food as well as elimination/reduction strategies to minimize possible human health risk.

Mottram et al. (2002) showed that asparagine, one of the major free amino acids in plant sources, can be the substrate for acrylamide formation in reactions involving reducing sugars. This has been supported by others (Becalski et al., 2003; Yasuhara et al., 2003; Yaylayan et al., 2003; Zyzak et al., 2003). However this may not be the only mechanism of acrylamide formation in food. Yasuhara et al. (2003) reported that triolein, glycerol, or acrolein can react with asparagine to form acrylamide. Acrylamide can also form by reaction between ammonia and acrylic acid.

Given this information one may propose three possible approaches to reducing acrylamide: remove/reduce substrates before cooking, inhibit reactions that produce acrylamide during processing, and remove/reduce acrylamide after processing. The first approach is problematic in that it involves product reformulation, including the potential use of genetically modified plants, while the last, which might be achieved by the use of enzymes that degrade acrylamide, may be limited to aqueous systems. The second approach, inhibiting reactions that produce acrylamide during processing, is the focus of this report. There are several reports that controlling pH during food processing, such as use of citric acid (Jung et al., 2003; Rydberg et al., 2003) can be useful in reducing acrylamide.

In this paper, first we report the development of an improved method for sample preparation for acrylamide analysis in food using GC/MS. While based on the EPA protocol (US EPA, 1996), which involves a time-consuming bromination procedure and large quantities of solvents, our simplified procedure substantially reduces sample preparation time and solvent use. Using this improved method we studied the effects of cooking oil on acrylamide formation and the use of food grade acidulants and preservatives in potato chips and fry models. We further expanded our research to study the effect of phytate and/or calcium ion on acrylamide formation. Our approach combined with efforts to reduce asparagine and reducing sugar in potatoes by selecting appropriate potato varieties, particularly those with low reducing sugar (a key component of acrylamide formation in potato based products, Biedermann et al, 2002; Biedermann-Brem et al, 2003), and avoidance of overcooking, especially homemade products (Biedermann et al. 2002; Grob et al. 2003), may help to reduce acrylamide in foods. Finally, we will briefly describe our mathematical model that may have application for predicting acrylamide levels in potato chips.

2. IMPROVED SAMPLE PREPARATION FOR GC/MS

Analytical methods (GC/MS, LC/MS/MS, or other detection methods) for detecting acrylamide have been extensively studied since the discovery of acrylamide in foods.

The GC/MS analytical method for acrylamide detection involves bromination of acrylamide based on EPA's method (1996; Tareke et al., 2002). A more recently developed method of acrylamide analysis includes LC/MS/MS, which has no need for bromination of samples and may give more accurate results from a variety of food matrixes. However, the availability of this equipment can be a limitation, as well as the variety of sample preparation steps needs to be justified for analysis. When those two methods were compared, both LC/MS/MS and GC/MS (with derivatization) gave comparable results (Tareke et al., 2002; Ono et al., 2003; Wenzl et al., 2003; Wenzl et al., 2004).

Our approach was to modify EPA's method to increase sample throughput, since EPA's sample preparation method is labor intensive and uses a relatively large quantity of sample (10 g) and volume (starts with 100 ml water) (Fig. 1A). With this method, one person can handle 6-8 samples per day. To improve sample preparation throughput and efficiency, we modified the method by reducing the scale to 1/5, starting with a 2 g sample instead of 10 g, and simplified the procedure (Fig. 1B). The main modifications were reducing the centrifuge step to 1 high speed from 2 spins, brominating and extracting in the same tube, and one extraction with ethyl acetate. Using our method, one person can handle about 30-40 samples per day.

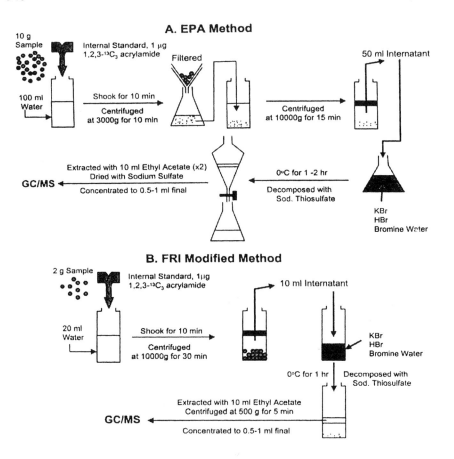

Figure 1. Schematic diagram of two methods, EPA (A) and FRI modified method (B). Samples were ground by using either a coffee grinder (for chips) or a food processor (for fries and baked products). In both methods, the ethyl acetate layer was concentrated to approximately 1 ml using a Brinkmann rotavapor, and filtered through a 0.45 µm syringe filter, before being subjected to GC/MS. Control tubes containing 1, 2 or 5 µg acrylamide plus internal standard were treated identically to the samples and used for quantification. Hewlett-Packard 6890 Series GC System with Agilent 7683 Series Injector and Agilent 5973 Network Mass Selective Detector were used. A DB-17 column (30 m × 0.25 mm i.d.) was used and running conditions were 65°C to 215°C with a rate of 15°C/ min and then 235°C with a rate of 2°C/ min held for 2 min. Injector and Detector temperatures were 250°C. Selective Ion Monitoring was used for mass 150, 152, and 106 for acrylamide and 155, 153, and 110 for internal standard. For quantification, 150 for acrylamide and 155 for internal standard was used.

This modified method was linear in the 0-1,000 µg/kg range (r^2=0.9998). The limits of detection and quantification were 4 µg/kg and 12 µg/kg, respectively, in potato and grain based products. Recovery of acrylamide from spiked samples were 108%, 105%, and 93% (±10%, n=12-14), in potato chip, French fry and baked models, respectively. We repeatedly tested 3 "check samples", designated as JMS 1 to 3, to determine the validity of our method (Table 1). The results in this table indicate that the method is reproducible. The "check samples" were also analyzed by LC/MS/MS and results were similar (Table 1). Thus, we concluded that this modified method is comparable to current LC/MS/MS method for acrylamide detection and also is repeatable and sensitive enough to detect low levels of acrylamide.

Table 1. Repeated analysis of "check samples" using FRI modified method and LC/MS/MS[a]

Assay No.	Test Samples		
	JMS 1	JMS 2	JMS 3
		µg/ kg	
1	27, 27	21, 19	256, 306
2	31, 29	17, 18	272, 272
3	27	17, 17	278, 264
4	26	15, 15	282, 293
5	26		258
6	27		273
7	27		273
8	27		285
9	29		287
10	25		268
Mean±std	27±2	17±2	276±13
Results of LC/MS/MS	27.5	<20	287

[a]JMS1 was a standard solution of 28 µg acrylamide per kg in water. JMS2 was baked potato (baked at 177°C for 1 hr) and JMS3 was commercial potato chips. Samples were analyzed using FRI method or by LC/MS/MS (performed by Covance Laboratories, Madison, WI).

3. EFFECTS OF COOKING OIL AND pH

Our laboratory tested the effects of various cooking oils on acrylamide formation using potato chip and French fry models. Our data indicate that cooking oil type (canola, cotton seed, olive, peanut, safflower, shortening, soybean and sunflower) *per se* is not an important variable for acrylamide formation, with the exception of olive oil, which had about a 300% increase of acrylamide compared to control (corn oil) chips (data not shown).

Olive oil has different characteristics than other commercial oil because it is less refined than other oils, and is not commonly used for frying. Nonetheless, olive oil significantly increased formation of acrylamide and this observation is consistent with that by Becalski et al. (2003). It is of interest that recent FDA data indicated that olives, both green and black, had relatively high levels of acrylamide, the reason for which is not clear at this point (FDA, 2004). Fat oxidation products may be important factors in acrylamide formation in food, possibly due to physical affects that influence the chemical reactions leading to acrylamide formation at high temperatures. It has been suggested by Gertz and Klostermann (2002) that palm oil and the presence of silicone in oil increased acrylamide formation, possibly due to increased heat transfer.

However, when olive oil was added to dry potato before heating at 150°C for 30 min, the acrylamide level was not increased compared to control (Biedermann et al. 2002). This discrepancy may possibly be due to cooking temperature, 150°C *vs.* 191°C in our potato chip model.

As reported in Rydberg et al. (2003), it is known that the Maillard reaction is affected by pH. We also tested various pH conditions, 0.1M HCl (pH 1.3), vinegar (pH 2.35), and 0.1M NaOH solution (pH 12.4) to determine the effect of pH extremes on acrylamide formation. Chips treated with these solutions had no detectable or very low levels of acrylamide (<20 µg/kg), which is consistent with observations by Rydberg et al. (2003). These results prompted us to further investigate the effect of pH on acrylamide formation in chip and fry models.

4. USE OF ACIDULANTS AND PRESERVATIVES

Based on the above results that lower pH can reduce acrylamide in final products, we tested acidulants as well as preservatives that can lower pH. The pHs of these solutions as well as the acrylamide formation are shown in Tables 2 and 3. Since solutions for fries contained 1.5% salt and 0.5% SAPP (sodium acid pyrophosphate), the pHs of these solutions were slightly different from those for chips.

All acidulants tested, except gluconic acid in chips, significantly reduced acrylamide compared to control. Fumaric, phosphoric and tartaric acids were quite effective in both chips and fries and also had pH values lower than or close to 2. The reason gluconic acid did not work may be due to the conversion of gluconic acid to δ-gluconolactone (a neutral substance, Fennema, 1996) at high temperature.

Among preservatives, benzoic, propionic and sorbic acids lowered acrylamide formation in chips and fries. Among derivatives of hydroxybenzoates, only ethyl esters gave consistent results on acrylamide reduction. The other derivatives, especially methyl derivatives, increased acrylamide in chips but reduced it in fries. It is not apparent what causes the variation of hydroxybenzoate derivatives on acrylamide formation, but it may be due to the pH range of these derivatives, which is above pH 4. Further clarification is needed.

Table 2. Effects of selected acidulants on acrylamide formation[a]

	Chips		Fries	
	Acrylamide (% of control)	pH	Acrylamide (% of control)	pH
Adipic acid (1%)	$79^a \pm 11$	2.54	$78^a \pm 16$	2.88
Fumaric acid (sat. about 0.6%)	$24^a \pm 4$	2.16	$28^a \pm 8$	2.25
Gluconic acid (1%)	111 ± 7	2.58	$71^a \pm 9$	2.72
Malic acid (1%)	$39^a \pm 5$	2.22	$54^a \pm 21$	2.33
Phosphoric acid (1%)	$40^a \pm 9$	1.72	$25^a \pm 9$	1.68
Succinic acid (1%)	$55^a \pm 1$	2.43	$50^a \pm 13$	2.68
Tartaric acid (1%)	$37^a \pm 6$	2.09	$29^a \pm 14$	2.12

[a]Potato slices for chips (potatoes were provided by FritoLay, Beloit, WI) were soaked in solutions for 20 min, dried, and then fried at 177°C for 90 sec. Fries (Russet Burbank potatoes provided by McCain Foods USA, Inc., Plover, WI) were cut (10 mm thick and 76 mm long, no skin), blanched at 80°C for 10 min, soaked in these solutions for 10 min, dried, par-fried at 191°C for 1 min, frozen overnight and then finish fried at 171°C for 2 min 45 sec. Corn oil was used for both chips and fries. Numbers are mean ± standard errors of two independent experiments. Means with superscripts are significantly different ([a] indicates lower) from corn oil control. All solutions used for fries also contained 1.5% salt and 0.5% sodium acid pyrophosphate.

Table 3. Effects of selected preservatives on acrylamide formation[a]

	Chips		Fries	
	Acrylamide (% of control)	pH	Acrylamide (% of control)	pH
Benzoic acid (1%)	$57^a \pm 18$	2.85	$55^a \pm 3$	2.87
Propionic acid (1%)	$78^a \pm 10$	2.68	$65^a \pm 22$	2.88
Sorbic acid (1%)	n.d.	2.75	$88^a \pm 4$	3.36
Methyl hydroxybenzoates (0.025%)	$177^b \pm 18$	4.16	$73^a \pm 1$	4.16
Ethyl hydroxybenzoates (0.025%)	$60^a \pm 20$	4.23	$88^a \pm 31$	4.15
Propyl hydroxybenzoates (0.025%)	$65^a \pm 4$	4.51	107 ± 40	4.16

[a]See Table 2 legends for the experimental conditions. Numbers are mean of two independent experiments. Numbers with superscripts are significantly different ([a] indicates lower and [b] indicates higher) from corn oil control. n.d.: not determined.

Others have reported the use of citric and ascorbic acids to reduce pH and as a result reduce acrylamide formation (Biedermann et al., 2002; Rydberg et al., 2003; Jung et al., 2003).

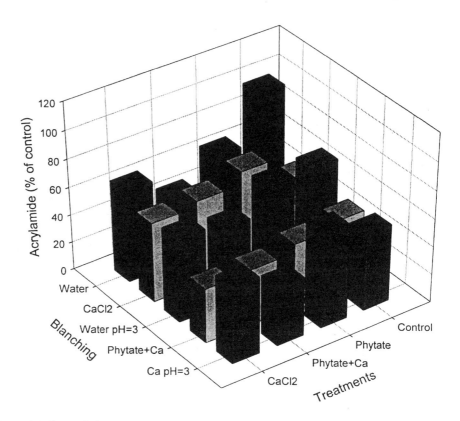

Figure 2. Effects of phytate and calcium ion on acrylamide formation in French-fry model. French fries were prepared from fresh potatoes as described in Table 2. Briefly, potatoes were cut (10 mm thick and 76 mm long) and blanched at 80°C for 10 min in one of the treatment solutions: water, 0.1% $CaCl_2$, water with pH adjusted to about 3, 0.025% phytate and 0.1% $CaCl_2$, or 0.1% $CaCl_2$ with pH adjusted to about 3. Then potatoes were subsequently soaked in one of the treatment solutions for 10 min; water (control), 0.025% phytate, 0.025% phytate and 0.1% $CaCl_2$, or 0.1% $CaCl_2$. After par-frying in corn oil at 191°C for 1 min and frozen overnight, samples were finish fried at 171°C for 2 min 45 sec. Data are mean of 3-5 replicates expressed as % of control (water blanching and water soaking).

5. USE OF PHYTATE AND CALCIUM ION

Food analysis data for acrylamide suggested that corn or whole wheat products contain relatively low amounts of acrylamide compared to products made with potatoes or bleached flour (FDA, 2004). This prompted us to question whether the relatively high levels of phytate in corn and whole-wheat products (especially compared to potatoes, Graf, 1983; Phillippy et al., 2003) may have contributed to the low levels of acrylamide. Phytate is a naturally occurring chelator, which is known to bind multivalent cations and may inhibit the absorption of those ions in the intestine. Since the formation of acrylamide involves interaction between asparagine and carbonyl compounds, the presence of phytate may interfere with these reactions. In fact, when Lindsay and Jang (see the chapter by Lindsay & Jang in these proceedings) tested phytate in potato chip and glassfiber filter models, they found phytate significantly reduced acrylamide formation. These effects of phytate can also be enhanced by the co-presence of appropriate amounts of calcium ions (or any multivalent cation) by forming phytate and calcium ion complexes. However, since the pH of phytate solution used is low (pH ≈ 2.6), the question arose whether the effect of phytate on acrylamide formation is due to reduced pH or phytate's chelating ability. Thus we tested combinations of calcium and phytate treatments using the French fry model. We used a fry model since it allowed us to integrate two treatments, blanching and/or soaking, as shown in Fig. 2.

All treatments significantly reduced acrylamide formation compared to control. The most effective treatment was phytate plus $CaCl_2$ blanching followed by $CaCl_2$ soaking treatment, although there were minimal differences between treatment groups. Among blanching treatments, phytate plus $CaCl_2$ had a 40% reduction of acrylamide formation compared to control blanching solution. Similarly, $CaCl_2$, pH=3, had a 26% reduction compared to control. However, $CaCl_2$ alone or water (pH=3) had no significant effect on acrylamide formation. Meanwhile, among soaking treatments, there were significant effects of $CaCl_2$ solution and phytate plus $CaCl_2$ solution but not phytate itself.

We have estimated the effects of those variables on acrylamide formation. Based on our results, pH ≈ 3 may have the effect of reducing acrylamide about 40% compared to control. Treatment with calcium ions had more variation (10-15% reduction), although additional work is needed to give further support to these findings. When calcium ion that had been pH adjusted to ≈ 3 was used as a treatment, it reduced acrylamide formation about 40%. Phytate alone can reduce acrylamide about 40% while phytate and calcium can inhibit up to 50% of acrylamide formation. We would like to point out that about 30% of the inhibition of phytate, phytate plus calcium ion, or calcium ion at low pH may be due to pH effects. Beyond that, whether inhibitory effects on acrylamide formation are due to chelation or ion complexation between intermediates that may form during acrylamide formation remains unclear. Although we were trying to compare the effects of pH alone from phytate's other functions, such as chelating, our experiments did not compare the same pH between our control and phytate (pH=2.6). However, these results suggest that chelators and/or multivalent ions may be useful for the reduction of acrylamide formation in foodstuffs.

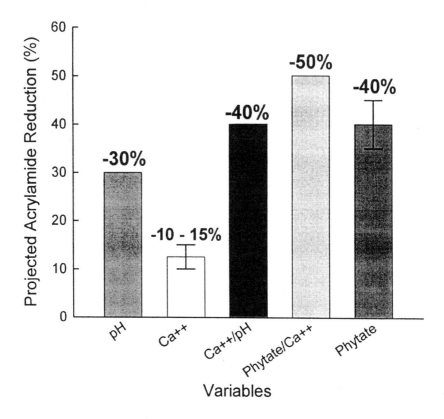

Figure 3. Summary of French fry experiments for phytate and calcium ion. Bar represents percentage of contribution of each variable on inhibition of acrylamide.

6. MATHEMATICAL MODEL

One of our goals is to develop a mathematical formula that utilizes information on the various known contributors for acrylamide formation to predict the level of acrylamide in final products. Variables known to contribute to acrylamide formation are the concentrations of asparagine and reducing sugar in raw materials, cooking temperature, cooking time, product shape (a reflection of surface area exposed to oil), and pH of sample preparation steps. Since the concentrations of asparagine and sugar are critical factors, the production of acrylamide and its rate may be written as (Smith, 1981):

A Mathmatical Model of Acrylamide Formation

$$\text{Asparagine} + \text{Reducing sugar} \rightarrow \text{Acrylamide} \; (+ \text{Byproducts})$$

$$\frac{dC_{Acr}}{dt} = kC_{RS}^{\alpha}C_{Asn}^{\beta} \qquad (1)$$

$$= k(C_{RS,0} - C_{Acr})^{\alpha}(C_{Asn,0} - C_{Acr})^{\beta}$$

where changes of acrylamide concentration (dC_{Acr}) over a designated time period (dt) correlate to the concentration of reducing sugar (C_{RS}) and asparagine (C_{Asn}) where α and β indicate order of reactions with respect to reducing sugar and asparagine with reaction rate constant (k). $C_{RS,0}$ and $C_{Asn,0}$ are the concentrations of reducing sugar and asparagine at time 0. After the designated time period (dt), concentrations of reducing sugar and asparagine will be changed to concentration of acrylamide less the initial concentrations. However, because less than 0.1% of asparagine and reducing sugar will be converted to acrylamide (Becalski et al., 2003; Zyzak et al., 2003), we can assume,

$$C_{RS,0} - C_{Acr} \approx C_{RS,0}, \quad C_{Asn,0} - C_{Acr} \approx C_{Asn,0} \qquad (2)$$

Then, the equation simplifies to

$$\frac{dC_{Acr}}{dt} = kC_{RS,0}^{\alpha}C_{Asn,0}^{\beta} \qquad (3)$$

Integrating the above equation with respect to time will yield

$$C_{Acr} = kC_{RS,0}^{\alpha}C_{Asn,0}^{\beta}t$$

$$= A\exp(-E/RT)C_{RS,0}^{\alpha}C_{Asn,0}^{\beta}t \qquad (4)$$

Reaction rate constant can be replaced with $A \cdot \exp(-E/RT)$, where A is pre-exponential factor, E is activation energy, R is gas constant, and T is absolute temperature (K). With the data from our potato chip model, we have calculated A' (modified pre-exponential factor, $= A \cdot C_{RS,0}^{\alpha} \cdot C_{Asn,0}^{\beta}$). Fig. 4 shows our modified equation using actual data along with real data points (filled circles) and predicted acrylamide value in a sheet. This is based on limited data of cooking temperatures between 160-190°C. We anticipate improving this model as more data become available, especially the possibility of extending it beyond potato based products. More discussion of the kinetics of acrylamide formation as well as degradation can be found elsewhere in these proceedings.

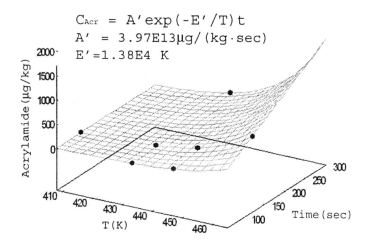

Figure 4. Comparison between actual and predicted values of acrylamide. Filled circles indicate actual acrylamide values from potato chip experiments. The surface contour indicates predicted acrylamide values at indicated temperature and time combinations

7. SUMMARY

We used a modified EPA method which can handle up to 30-40 assays per person per day and can be used for potato- and grain-based products. With this method, we found that there are minimal effects of frying oil with the exception of olive oil, which increased acrylamide formation.

Since lowering pH can significantly reduce acrylamide formation, we used potato chip and French fry models to test a number of food grade acidulants and preservatives that can lower pH. All acidulants tested were able to reduce acrylamide formation with the exception of gluconic acid in the chip model. Preservatives tested had less effect on acrylamide formation and more variation among sample replicates. In addition, the presence of phytate along with calcium ion may be an effective tool to reduce acrylamide formation when applied during food processing steps. Our mathematical model is product specific for potato chips and temperature range of 160 to 188°C.

ACKNOWLEDGEMENTS

The authors thank Drs. Hyuk Yu, Phil Simon and Erin Silva and Mr. Sungjoon Jang for their helpful discussion. This research was supported in part by gift funds administered through the Food Research Institute, University of Wisconsin-Madison.

REFERENCES

Becalski, A., Lau, B. P.-Y., Lewis, D., and Seaman, S. W., 2003, Acrylamide in foods: Occurrence, sources, and modeling, *J. Agric. Food Chem.* **51**:802-808.

Biedermann, M., Noti, A., Biedermann-Brem, S., Mozzetti, V., and Grob, K., 2002, Experiments on acrylamide formation and possibilities to decrease the potential of acrylamide formation in potatoes, *Mitt. Lebensm. Hyg.*, **93**;668-687.

Biedermann-Brem, S., Noti, A., Grob, K., Imhof, D., Bazzocco, D., and Pfefferle, A., 2003, How much reducing sugar may potatoes contain to avoid excessive acrylamide formation during roasting and baking?, *Eur. Food Res. Technol.* **217**:369-373.

FDA, 2004, Exploratory Data on Acrylamide in Food; http://www.cfsan.fda.gov/~dms/acrydata.html.

Fennema, O. R., 1996, *Food Chemistry,* 3rd ed., Marcel Dekker Inc., New York, pp. 767-823.

Friedman, M., 2003, Chemistry, biochemistry, and safety of acrylamide. A review, *J. Agric. Food Chem.* **51**:4504-4526.

Graf, E., 1983, Application of phytic acid, *J. Am. Oil Chem. Soc.*, **60**:1861-1867.

Gertz, C., and Klostermann, S., 2002, Analysis of acrylamide and mechanisms of its formation in deep-fried products, *Eur. J. Lipid Sci. Technol.*, **104**:762-771.

Granath, F., and Tornqvist, M., 2003, Who knows whether acrylamide in food is hazardous to humans?, *J. Nat'l. Cancer Inst.* **95**:842-843.

Grob, K., Biedermann, M., Biedermann-Brem, S., Noti, A., Imhof, D., Amrein, T., Pfefferle, A., and Bazzocco, D., 2003, French fries with less than 100 μg/kg acrylamide. A collaboration between cooks and analysts, *Eur. Food Res. Technol.* **217**:185-194.

Jung, M. Y., Choi, D. S., and Ju, J. W., 2003, A novel technique for limitation of acrylamide formation in fried and baked corn chips and in French fries, *J. Food Sci.* **68**:1287-1290.

Mottram, D. S., Wedzicha, B. L., and Dodson, A. T., 2002, Acrylamide is formed in the Maillard reaction, *Nature* **419**:448-449.

Mucci, L. A., Dickman, P. W., Steineck, G., Adami, H.-O., and Augustsson, K., 2003, Dietary acrylamide and cancer of the large bowel, kidney, and bladder: Absence of an association in a population-based study in Sweden, *Br. J. Cancer* **88**:84-89.

Ono, H., Chuda, Y., Ohnishi-Kameyama, M., Yada, H., Ishizaka, M., Kobayashi, H., and Yoshida, M., 2003, Analysis of acrylamide by LC-MS/MS and GC-MS in processed Japanese foods, *Food Add. Contam.*, **20**:215-220.

Pelucchi, C., Franceschi, S., Levi, F., Trichopoulos, D., Bosetti, C., Negri, E., and Vecchia, C. L., 2003, Fried potatoes and human cancer, *Int. J. Cancer* **105**:558-560.

Phillippy, B. Q., Bland, J. M., and Evens, T. J., 2003, Ion chromatography of phytate in roots and tubers, *J. Agric. Food Chem.*, **51**:350-353.

Rydberg, P., Eriksson, S., Tareke, E., Karlsson, P., Ehrenberg, L., and Tornqvist, M., 2003, Investigations of factors that influence the acrylamide content of heated foodstuffs, *J. Agric. Food Chem.*, **51**:7012-7018.

Smith, J. M., 1981, *Chemical Engineering Kinetics*, 3rd ed., McGraw-Hill Inc., New York, pp 39-51.

Tareke, E., Rydberg, P., Karlsson, P., Eriksson, S., and Tornqvist, M., 2000, Acrylamide, a cooking carcinogen?, *Chem. Res. Toxicol.* **13**:517-522.

Tareke, E., Rydberg, P., Karlsson, P., Eriksson, S., and Tornqvist, M., 2002, Analysis of acrylamide, a carcinogen formed in heated foodstuffs, *J. Agric. Food Chem.* **50**:4998-5006.

US EPA, 1996, *SW 846, Test methods for evaluating solid waste, Acrylamide by gas chromatography, Method 8032A*: U.S. Environmental Protection Agency, Washington, DC.

Wenzl, T., de la Calle, B., Gatermann, R., Hoenicke, K., Ulberth, F., and Anklam, E., 2004, Evaluation of the results from an inter-laboratory comparison study of the determination of acrylamide in crispbread and butter cookies, *Anal. Bioanal. Chem.*, **379**;449-457.

Wenzl, T., de la Calle, M. B., and Anklam, E., 2003, Analytical method for the determination of acrylamide in food products: a review, *Food Add. Comtam.* **20**:885-902.

Yasuhara, A., Tanaka, Y., Hengel, M., and Shibamoto, T., 2003, Gas chromatographic investigation of acrylamide formation in browning model systems, *J. Agric. Food Chem.* **51**:3999-4003.

Yaylayan, V. A., Wnorowski, A., and Locas, C. P., 2003, Why asparagine needs carbohydrates to generate acrylamide, *J. Agric. Food Chem.* **51**:1753-1757.

Zyzak, D. V., Sanders, R. A., Stojanovic, M., Tallmadge, D. H., Eberhart, B. L., Ewald, D. K., Gruber, D. C., Morsch, T. R., Strothers, M. A., Rizzi, G. P., and Villagran, M. D., 2003, Acrylamide formation mechanism in heated foods, *J. Agric. Food Chem.* **51**:4782-4787.

QUALITY RELATED MINIMIZATION OF ACRYLAMIDE FORMATION - AN INTEGRATED APPROACH

Knut Franke, Marco Sell, and Ernst H. Reimerdes
German Institute for Food Technology, P.O. Box 1165, D-49610 Quakenbrueck, Germany; e-mail: e.reimerdes@dil-ev.de

Abstract: An integrated approach is described with respect to acrylamide minimization in heated foodstuffs. All relevant variables have to be considered and the main focus is on maintaining the expected product quality. The role of the processes at the interface between product and heating medium during processing is characterized for the case of frying operations. Examples of parameters influencing these processes with respect to minimizing acrylamide and maintaining product quality (e.g. brown color) are described. First, the local distribution of acrylamide in a French fries type model food was investigated. Lowering water activity at the surface of French fries before frying contributes to a reduction of acrylamide without lowering product quality. Both pre-drying of the potato sticks before frying and an increasing of salt concentration at the product surface by coating with a salt solution showed positive effects. Additionally, it was demonstrated by simulation that combined effects of these measurements may enable a reduction of up to 80% in the acrylamide content.

Key words: acrylamide, Maillard reaction, French fries, quality, frying process, model system

1. INTRODUCTION

Since the first warnings about the formation of acrylamide in many heated foodstuffs, numerous efforts have been made by research and industry to minimize acrylamide contents through technological measures. In some food categories e.g. crisp bread, potato chips and also French fries, a significant reduction has been achieved by modifications of traditional processing of these foods (Grob et al., 2003). However, to continue the success in minimizing acrylamide contents and maintaining product quality expected by the consumer, an integrated approach is necessary taking into account relevant variables from raw material to process/equipment and product quality. An integrated approach means that instead of single parameters (e.g. temperature) the whole complex is considered and relationships between the variables have to be described focusing on food quality (Franke et al., 2003).

Deep fat frying, for example of French fries, has been identified as one of the food heating processes leading to considerable acrylamide formation in the product. The substance exchange and related reactions at the interface product/heating medium (oil) and their influence on quality and acrylamide formation during processing is of particular interest. During frying, water is evaporated from the product surface and the steam bubbles rise through the fat to the environment. Parts of the frying fat penetrate the product and contribute to its unique taste. The outer layer undergoes a further temperature raise in the dry state and reaches 120°C and higher (Vitrac et al., 2000). Maillard reaction takes place resulting in desired color (browning) and taste (Krokida et al., 2001). The integrated approach for the frying process of French fries is outlined schematically in Fig. 1.

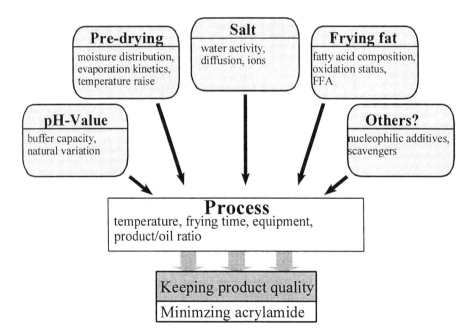

Figure 1. Integrated approach for minimizing acrylamide contents.

Knowledge of these processes is the basis for influencing the reaction mechanisms during heating, especially with respect to Maillard reaction, which is related to acrylamide formation (Stadler et al., 2002; Friedman, 2003). This includes the role of precursors, e.g. free asparagine and reducing sugars, on the product, as well as the influence of additives to product and/or frying fat.

2. LOCAL DISTRIBUTION OF ACRYLAMIDE IN FRENCH FRIES

Ready-to-eat French fries consist of a brown crispy crust at the outer layers and a soft yellow interior of cooked potato tissue. These two different taste-imparting properties are the unique attribute of this product. It can be assumed that most of the acrylamide is formed in the dry crust, because only in this region are the temperatures higher than 100°C during frying, and the local moisture content becomes low after evaporation. These conditions promote acrylamide formation.

Therefore, a distinct spatial distribution of acrylamide can be assumed due to these internal differences in structure and temperature load in French fries. To test these effects, a model potato food having a composition similar to French fries, but a larger size, was prepared and fried. After frying thin layers were sliced off and analyzed for acrylamide, water and fat.

2.1 Materials and methods

Potato blocks sizing 20 mm in width, 20 mm in height and about 100 mm in length were formed from a firm potato dough. The dough consisted of 240 g dried potato flakes (commercial product) and 750 g water. After mixing and shaping, the blocks were air-dried at 50°C for 30 min in a drying chamber to form a firm skin at the surface of the blocks, preventing decomposition during frying. The blocks were fried in a catering fryer at 168°C for 12 min to obtain a desired brown color similar to French fries and a dry crust. The fried blocks were frozen before slicing. A household bread-slicer fixed to slice thickness of 1 mm was used to cut off successive crust layers. Every block was sliced consecutively from all four long sides. The sliced material from all blocks was collected for every sliced layer. The residual soft core was immediately analyzed. Due to the uneven surface of the fried blocks, the thickness of the sliced layers varied. The weight of every layer as the sum of slices from all blocks was determined and the thickness was estimated from the ratio to the initial weight of all blocks.

The acrylamide content was determined by means of a mass spectroscopic method after liquid chromatographic separation from residual matrix and sample preparation. D_3-acrylamide was used as internal standard for quantification. Sample preparation was done by extraction with acetonitrile/water mixtures and ultra-centrifugation. Proteins were precipitated after addition of Carrez I and II reagents (Rosen and Helenas, 2002). The water and fat content was determined according to AOAC methods (AOAC, 2002).

2.2   Results and discussion

Fig. 2 shows the contents of acrylamide, water and fat in the model potato block from the product core (left) to the surface (right). The vertical lines show the thickness of the truncated crust layers. The measured values are drawn in the middle of every layer.

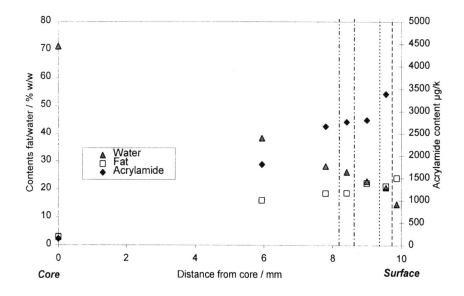

Figure 2. Spatial distribution of acrylamide, water and fat in model French fries.

The total water content of the fried model system was about 49% and slightly above the water content of the smaller French fries. This is due to the larger size of the model system having a higher part of moist core. The fat content was 11%, which is in the range of French fries.

As can be expected, the water content decreased from the core part from about 70% to about 15% at the outer surface. The sharp decrease is due to the evaporation of water beginning in the outer layers. On the other hand, fat content increased from the inner core to the outer surface. If the water evaporated from the surface, pores were formed and the fat could penetrate the fried product (Moreira et al., 1999). However, the largest changes were found in the acrylamide content. The acrylamide content in the core was low (about 150 µg/kg) but detectable. The reason for this may be the long frying time and the fact that acrylamide formed in the outer layers may diffuse into the core. The content at the outer layers was high, about 3500 µg/kg. This is more than 20 times higher than in the core and confirms the importance of surface processes for acrylamide formation (Franke and Reimerdes, 2004). Fig. 2 shows that the layers high in acrylamide also contained high amounts of frying fat. This fact demonstrates the influence of the frying process on acrylamide formation.

3. INFLUENCE OF PRE-DRYING ON THE QUALITY AND ACRYLAMIDE CONTENTS OF FRENCH FRIES

The water activity at the product surface plays an important role in the Maillard reaction pathways as well as for acrylamide formation. One of the important processes during frying is water evaporation from the outer layers at about 100°C. A lower water activity in the outer layers is necessary for the formation of a dry crisp crust. However, the evaporation of water requires energy, extends the frying process and leads to a temperature load on the product surface in the dry state. Therefore, we also investigated the influence of an additional pre-drying process of frozen, par-fried French fries on the frying process and on acrylamide formation.

3.1 Materials and methods

Commercial frozen, par-fried French fries were supplied by a local potato factory and pre-dried according to the procedure shown in Fig. 3 in a tray dryer at 70°C. The drying times were 60 and 180 min, respectively; the corresponding decreases in moisture were about 7% and 20%. The standard frying process was carried out in a catering fryer with 13.5 liters hardened frying fat at 168°C for 3 min with a sample weighing 680 g. The pre-dried French fries were fried at 3 different frying times adapted to the state of pre-drying. This means that the total frying time was reduced to obtain products with a color range which is similar to French fries without pre-drying and fried under standard conditions.

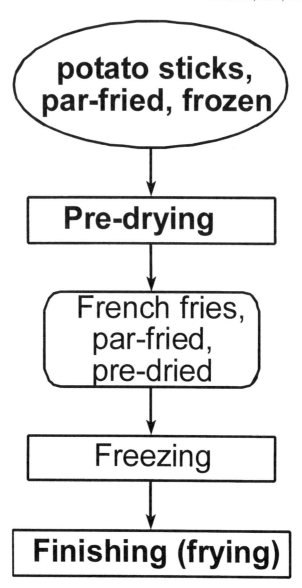

Figure 3. Sequence of operations for pre-drying of par-fried French fries.

The color of the French fries was measured using a Chromameter CR-100 D 65 (Minolta) in the L^*, b^*, a^* color system. The value of b^* characterizes the degree of browning of French fries during frying. The higher the b^*-value the darker and more brown the fried products.

3.2 Results and discussion

Fig. 4 shows the acrylamide content and the color of the French fries fried for different time periods.

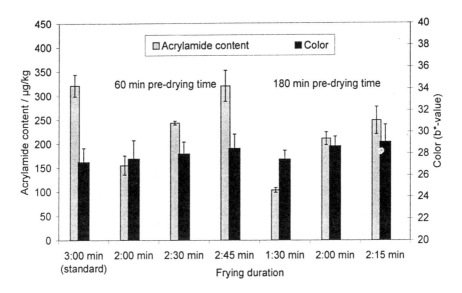

Figure 4. Acrylamide content and color (browning) of French fries pre-dried before finish frying.

Fig. 4 shows that pre-drying results in French fries with a slight darker color in spite of reduced frying time. Color levels comparable to the standard frying process were reached for the shortest frying duration at each pre-drying time. In these cases, the acrylamide content was much lower for the pre-dried French fries compared to the standard fries. Therefore, pre-drying of par-fried French fries enables shorter frying times resulting in the same quality with respect to the desired color and crispness (results not shown). The acrylamide content could be lowered from 300 µg/kg (standard frying) down to 100 µg/kg after 180 min pre-drying or about 160 µg/kg after 60 min pre-drying.

4. INFLUENCE OF SALT COATING ON QUALITY AND ACRYLAMIDE FORMATION

As demonstrated above, water evaporation and, therefore, water activity at the surface of the French fries influences the mechanisms of acrylamide formation during frying. Another possibility to decrease water activity was to increase the local salt concentration. This seems to be a very useful way to lower acrylamide content because the French fries are often salted before consumption.

4.1 Materials and methods

The frozen fries were kept at room temperature for 1 hour to allow surface thawing. Each portion of French fries was dipped for 5 min in 1 L aqueous salt solution (NaCl) (from 3% up to 15% w/w) and then left to drip off for 1 min before being frozen at -20°C. In addition to the salt solutions, a control with demineralized water was included to study the effect of enrichment of water in the outer layers and leaching of components during coating. After coating and freezing, the French fries were fried the next day using standard conditions (see 3.1 above). After frying, the complete portion was used for sample preparation for acrylamide determination (see 2.1).

4.2 Results and discussion

The acrylamide content and color of the fried products *vs.* salt concentration in the coating is presented in Fig. 5. A comparable quality of French fries was obtained.

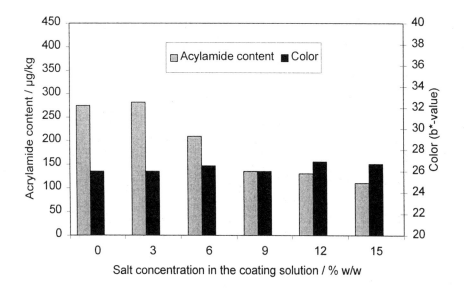

Figure 5. Acrylamide content of French fries coated in solutions with different salt contents.

Coating with salt concentrations of 6% and higher decreased acrylamide formation in French fries in a distinct way. At the highest level of 15%, a reduction from about 270 µg/kg to 120 µg/kg was obtained. Coating with a salt solution of 6% resulted in an acrylamide content of about 210 µg/kg. However, coating with salt solutions of 9% and higher lead to very salty French fries, which could not be marketed. French fries coated with the 6% salt solution had an acceptable taste and quality.

The results confirm the possibility of reducing acrylamide levels in French fries by lowering water activity at the surface of the product. The controlled design of interface processes during frying can be used to minimize acrylamide formation in fried foods.

5. INFLUENCE OF pH ON QUALITY AND ACRYLAMIDE FORMATION

The pH-value at the surface of the heated product seems to have an influence on the acrylamide formation during frying (Jung et al., 2003). Lower pH-values lead to a reduction of acrylamide content. To obtain more information about these effects, we investigated par-fried French fries coated before frying with aqueous solutions of different pH values.

5.1 Materials and methods

The coating procedure was carried out as described above (4.1), with a new batch of commercial frozen and par-fried French fries. Organic acids at a concentration of 2% w/w were used as coating solutions and the pH was adjusted using a phosphate buffer. The following acids and pH-values were examined:
- citric acid solution at a pH of 2.1 (without buffer)
- lactic acid solution:
    pH 2.1 (similar to the citric acid)
    pH 3.0
    pH 5.8 (natural pH-value of potatoes without pre-treatment)

After coating and freezing, the French fries were fried the next day using standard conditions, but for 3.5 min (see 3.1 above). The longer frying time was selected to obtain acrylamide values for all settings, which were in the validated measurement range. Additionally, the color of the French fries and the pH-value of the whole fries were measured after frying.

5.2 Results and discussion

Table 1. Influence of coating of French fries in solutions of different pH-values on acrylamide contents, pH-value and color after frying

Coating solution	Acrylamide content µg/kg	pH-value of the French fries after frying	Color b*-value
without coating	728 ± 7	5.5	22.0
only water	739 ± 105	5.4	20.9
citric acid	292 ± 65	4.6	21.8
lactic acid (pH 3.1)	225 ± 37	4.4	22.6
lactic acid (pH 3.0)	350 ± 94	4.8	24.2
lactic acid (pH 5.8)	696 ± 49	5.0	25.2

The results of these coating experiments are summarized in Table 1. As expected, the pH-values measured after frying were influenced somewhat by the pH-value of the coating solution. This is due to the fact that the coating only affected the outer layers; the pH-value in the core remained at the initial value. Regarding the color of the French fries (b*-value), a slight increase of browning (higher b*-values) was observed at the higher pH-values. The acrylamide content was much more affected, however. At the lowest pH-value, it was less than one third that of untreated French fries. The results for lactic acid and citric acid experiments at the same pH-level were similar. This means that the pH-influence dominates; possible chelating effects of citric acid can be neglected. The French fries coated in water at neutral pH did not show any decrease in acrylamide content indicating that leaching effects during coating for 5 min did not influence acrylamide formation in par-fried French fries. However, there was a distinct sensory influence of the coating on taste of the French fries. Products coated with pH solutions of low pH were sour and were refused by the panelists.

6. MINIMIZING ACRYLAMIDE CONTENT BY PRE-DRYING AND SALTING (SIMULATION)

Based on the successful results of pre-drying and coating with salt on lowering the acrylamide contents in French fries, further investigations were carried out on a possible overlap of these treatments. For this purpose, salt-solution-coated French fries (see section 4 above) were pre-dried before frying according to the procedure described above in 3.1. The resulting acrylamide content in the fried products were used to calculate the coefficients in a multiple regression approach according to the response surface method (RSM), where salt content and degree of pre-drying were the independent variables. The results are shown in Fig. 6 as the calculated surface.

The acrylamide contents were related to the acrylamide content of untreated French fries of this batch fried at standard conditions. The acrylamide concentration representing 100% was about 260 µg/kg.

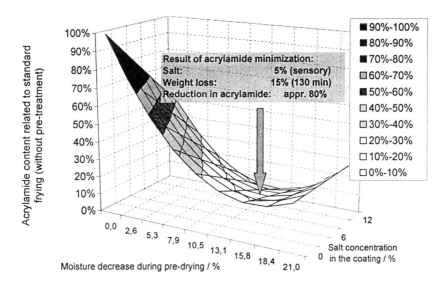

Figure 6. Calculated relative acrylamide content of French fries coated in salt solutions and pre-dried before frying including minimization approach.

As can be observed, both pre-drying and salting contributed to lower acrylamide values. However, the effect of pre-drying was more pronounced, especially at shorter drying times and, therefore, gave less of a decrease in moisture. For prolonged drying it seemed that the acrylamide content of the pre-treated French fries increased again. The influence of salt was more distinct without pre-drying (left side) and became lower at higher drying degrees.

Based on these results, an approach was simulated to minimize acrylamide concentration. A salt content of 5% in the coating was assumed because French fries coated with such a solution lead to an acceptable taste in the final product. For this salt level, the optimized degree of pre-drying was about 15% weight loss and the calculated acrylamide content was about 20% of the untreated French fries (Fig. 6). It is worth noting that all these French fries had a similar quality with respect to browning and crispness as those obtained by the standard frying procedure.

7. CONCLUSIONS

Since the beginning of the acrylamide discussion, French fries are one of the food products, which have been in focus. Due to the relationship between Maillard reaction and acrylamide formation, shorter frying times or lower oil temperatures have been recommended, often resulting in paler and less crispy fries. The challenge is to minimize acrylamide content while maintaining the product quality expected by the consumer. Such an approach requires knowledge about the whole process and its influencing variables, in order to separate the desired Maillard reaction pathways, e.g. browning, from acrylamide formation.

We have demonstrated that acrylamide is formed in the outer layers of the product. The interface between product and heating medium therefore plays an important role. Possible controlling parameters of these processes are the reduction of water activity at the outer layers (salting or pre-drying) and lowering pH-values. Applying these measures makes it possible to maintain desired product quality while lowering acrylamide content. Combined salt coating and pre-drying of French fries enable further reduction in acrylamide to about 20% compared to the untreated product, as demonstrated by simulation experiments. For a sustainable reduction in acrylamide levels of heated foods, this simple model has to include other variables, especially acrylamide precursor levels in the raw material and the technical parameters of the frying equipment.

ACKNOWLEDGEMENTS

The research project is supported by the FEI (Forschungskreis der Ernährungsindustrie e.V., Bonn), the AiF and the Ministry of Economics and Labour. AiF-Project-No.: 108 Z BG.

REFERENCES

AOAC, 2002, *Official Methods of Analysis of The Association of Official Analytical Chemists* 17th Edition.

Franke, K., Kreyenmeier, F., and Reimerdes, E. H., 2003, Ganzheitlicher Ansatz: Acrylamid- das gesamte Geschehen um die Bildung ist entscheidend, *Lebensmitteltechnik* 35(3):60-62.

Franke, K., and Reimerdes, E.H., 2004, Possibilities in simulating frying processes with respect to minimizing acrylamide contents, Hagen: 4th International Symposium on Deep Frying - Tastier and Healthier Fried Foods.

Friedman, M., 2003, Chemistry, biochemistry, and safety of acrylamide: a review, *J. Agric. Food Chem.* 51:4504-4526.

Grob, K., Biedermann, M., Biedermann-Brem, S., Noti, A., Imhof, D., Amrein, T., Pfefferle, A., and Bazzocco, D., 2003, French fries with less than 100 µg/kg acrylamide. A collaboration between cooks and analysts, *Eur. Food Res. Technol.* **217**(2): 185-194.

Jung, M.Y., Choi, D.S., and Ju, J.W., 2003, A novel technique for limitation of acrylamide formation in fried and baked corn chips and in French fries, *J. Food Sci.* **68**: 1287-1290.

Krokida, M.K., Oreopoulou, V., Maroulis, Z.B., and Marinos-Kouris, D., 2001, Colour changes during deep fat frying, *J. Food Eng.* **48**(3): 219-225.

Moreira, R.G., Castell-Perez, M.E., Barrufet, M.A., and Castwell-Perez, M.E., 1999, *Deep Fat Frying: Fundamentals and Applications*, Gaithersburg: Aspen Publishers.

Rosen, J., and Hellenäs, K.-E., 2002, Analysis of acrylamide in cooked foods by liquid chromatography tandem mass spectroscopy, *Analyst* **127**:880-882.

Stadler, R.H., Blank, I., Varga, N., Robert, F., Hau, J., Guy, P.A., Robert, M.-C., and Riediker, S., 2002, Acrylamide from Maillard reaction products, *Nature* **419**: 449.

Vitrac, O., Trystram, G., and Raoult-Wack, A.L., 2000, Deep-fat frying of food: heat and mass transfer, transformations and reactions inside the frying material, *Eur. J. Lipid Sci. Technol.* **102**(8/9):529-538.

GENETIC, PHYSIOLOGICAL, AND ENVIRONMENTAL FACTORS AFFECTING ACRYLAMIDE CONCENTRATION IN FRIED POTATO PRODUCTS

Erin M. Silva[1] and Philipp W. Simon[2]
[1]Department of Agronomy and Horticulture, New Mexico State University, Las Cruces, NM 88003; [2]USDA-ARS, Department of Horticulture, University of Wisconsin, Madison, WI 53706; e-mail: psimon@wisc.edu

Abstract: The discovery of acrylamide in processed potato products has brought increased interest in the controlling Maillard reaction precursors (reducing sugars and amino acids) in potato tubers. Because of their effects on nonenzymatic browning of fried potato products, reducing sugars and amino acids have been the focus of many potato research and breeding programs. This study focused on changes in sugars and amino acids in diploid potatoes selected for their storage qualities and their effect on acrylamide formation in the fried product. In addition, a second study was performed using cultivated lines that evaluated the effect of nitrogen fertilization on amino acid levels in tubers. Glucose, fructose, sucrose, and asparagine concentrations in tubers increased upon storage at 2°C. Glucose and fructose concentrations in the tubers were significantly and positively correlated with subsequent acrylamide formation in the products. Tuber sucrose and asparagine concentrations did not have an effect on acrylamide levels. Acrylamide levels in the products were significantly reduced if tubers were preconditioned before being placed in storage at 2°C. Higher rates of nitrogen fertilization resulted in increased amino acid concentrations in the tubers.

Key words: Acrylamide, amino acids, nitrogen, potato, storage, sugar

1. INTRODUCTION

The discovery of acrylamide in processed potato products resulting from the Maillard reaction has brought increased interest in the components of raw potatoes which contribute to this chemical process. The Maillard reaction begins with the condensation of an α- or ε-amino group with a carbonyl group, both of which can come from a wide range of compounds. In potatoes, this usually involves the reaction between reducing sugars and free amino acids, although other compounds such as ascorbic acid and phenolics may be involved (Rodriguez-Saona and Wrolstad, 1997).

The products resulting from the Maillard reaction include not only acrylamide, but also a complex mixture of flavor and color compounds that affect the sensory attributes of potato chips and French fries (Schallenberger et al., 1959). The "nonenzymatic" brown color, which is characteristic of heat processed potato products, is in part due to the brown melanoidin pigments resulting from the Maillard reaction (Eskin, 1990; Smith, 1987). While some degree of browning is desirable for the visual and taste appeal of the product, excessive browning due to high levels of Maillard reaction precursors produces unacceptable color and bitter taste (Danehy, 1986; Roe et al., 1990). In most circumstances, the major contributor and limiting factor in color development in potato products is reducing sugar concentration (Roe, 1990; Sowokinos, 1978). Sucrose can also be involved in the Maillard reaction since it is hydrolyzed during frying (Leszkowiat et al., 1990); however, its role in color development has been shown to be minimal (Roe and Faulks, 1991; Schallenberger, 1959).

Because of their effects on browning, Maillard reaction precursors in potatoes have been studied by many researchers over the past century. Several early studies found that reducing sugar concentrations have consistent and significant positive correlations with browning during processing (Butler, 1913; Denny and Thornton, 1940; Hoover and Xander, 1961, 1963; Wright et al., 1932). Perhaps because of this, most subsequent research regarding nonenzymatic browning in potato focused on changes in concentrations of reducing sugars and sucrose during both tuber development and postharvest storage. Studies involving free amino acid concentrations in the tuber and their effect on browning demonstrated inconsistent correlations of amino acid levels with chip color (Hoover and Xander, 1961; Schwimmer et al., 1957). This may be because sugars, which are in relatively low concentrations in the tuber (at least 5 times lower than amino acid concentrations), are the limiting factor in the first step of this reaction.

1.1 Sugars in potato tubers

The reducing sugar concentrations of potato tubers decrease over the course of the growing period. At the end of the growing season, the sugar levels in the tubers reach a minimum level, at which time chemical maturity is reached (Pritchard and Adam, 1992). It has been recommended that potatoes for chip production have a maximum reducing sugar concentration of 0.30 to 0.35 mg/g fresh weight (Sowokinos and Preston, 1988). Processors have known for more than a century that postharvest storage of potatoes in temperatures below 10°C results in a phenomenon called "cold-sweetening" (Müller-Thurgau, 1882). This term describes the marked increase in both reducing and non-reducing sugar concentrations at low temperature storage resulting from the breakdown of starch. Several factors affect the degree to which cold-sweetening occurs. Immature tubers with higher sucrose concentrations at harvest accumulate reducing sugars faster than mature tubers in storage (Clegg and Chapman, 1962; Sowokinos, 1978). Because of this, potato processors remedy high sugar levels by raising the storage temperatures and holding tubers above 10°C before processing. This process of "reconditioning' (Watada and Kunkel, 1955) converts some free sugars back to starch and allows for the production of chips and fries of acceptable quality. However, reconditioning is time-consuming and costly with often erratic results (Burton, 1963). Furthermore, while reconditioning lowers sugar levels, they rarely drop to harvest levels.

Since sugars have a marked effect on the processing quality of potato, breeding programs throughout the world have focused on developing potato varieties that accumulate low levels of sugar in storage. Changes in tuber reducing sugar and sucrose concentrations in storage are genetically determined, in combination with significant environmental effects (Ehlenfeldt et al., 1990; Loiselle et al., 1990). Several genes influencing sugar accumulation have been identified in both cultivated potatoes and their wild relatives (Jakuczun et al., 1995; Oltmans and Novy, 2002; Thill and Peloquin, 1993; 1994) and significant progress has been realized in new cultivars which can be stored at below 10°C (Douches et al., 1996; Love et. al., 1998).

1.2   Amino acids in potato tubers

The primary nitrogen transport compounds in plants are the amino acids glutamine and asparagine. Both of these amino acids play crucial roles in plant growth and development (Urquhart and Joy, 1981). These compounds are the major products of nitrate assimilation in the roots and can be transported to the shoots via the transpiration stream through the xylem. In light-grown plants, glutamine transports nitrogen and serves as the nitrogen donor in essentially all metabolic reactions (McGrath and Coruzzi, 1991). However, in dark grown plants, asparagine is used as the nitrogen transporter and donor because its higher nitrogen:carbon ratio results in a more efficient use of nitrogen. Glutamine and asparagine are synthesized at high levels in plant development when nitrogen must be re-mobilized and re-assimilated in different organs.

The key enzymes involved in nitrogen assimilation are glutamine synthetase (GS), asparagine synthetase (AS), glutamate synthase (GOGAT) and aspartate aminotransferase (AspAT). GS catalyzes the assimilation of ammonium into glutamine via glutamate. This enzyme has been well characterized and its activity has been found in the majority of plant organs (McGrath and Coruzzi, 1991). Different isoforms with distinctive functions have been characterized; GS2, a chloroplastic form, is involved in re-assimilation of photorespiratory ammonia (Freeman et al., 1990). Alternatively, GS3A functions to generate glutamine for intercellular nitrogen transport (Edwards et al., 1990). Thus, the different forms may have non-overlapping roles in nitrogen metabolism (Edwards et al., 1990).

The major route for the biosynthesis of asparagine in plants involves AS. This enzyme catalyzes the transfer of an amide group from glutamine to aspartate, generating glutamate and asparagine. Unlike GS, AS is extremely unstable and has proven difficult to characterize *in vitro* because of asparaginase and non-protein inhibitors (Joy et al., 1983; McGrath and Coruzzi, 1991; Streeter, 1977). However, several studies have evaluated the effect of light on AS gene expression and have demonstrated that AS1 and AS2 transcripts increase in dark grown plants (Tsai and Coruzzi, 1990, 1991). Other studies have found that the increase in asparagine synthetase activity which is responsible for the asparagine accumulation in dark adapted plants is a result of carbon deprivation (Brouquisse et al.,1992; King et al., 1990; Stulen and Oaks, 1977). Chevalier et al. (1996) isolated a root-specific AS transcript that was highly inducible when root tips were subjected to carbon deprivation. While research on plant amino acid biosynthesis and metabolism is an active research area, many details remain obscure.

In potato tubers, asparagine and glutamine are the predominant amino acids, often accounting for up to 90% of the total free amino acid composition (Brierley et al., 1997). An extensive study of the free amino acids of potato tubers grown in England and Ireland found great variation in the levels of amino acids depending on location, weather conditions, soil conditions, and other factor (Talley and Porter, 1970). Asparagine levels varied nearly tenfold from 3.71 – 34.90 mg/g dry weight. It appeared that some amino acids such as glutamine, proline, alanine, valine, tyrosine, and histidine were particularly variable. Davies (1977) also found that amounts of amino acids present in potato tubers may be quite different in samples from different growing locations and from different years. Most studies involving the effect of nitrogen fertilization on amino acid status of the tubers have found that increasing nitrogen rates lead to higher amino acid concnetrations in the tubers (Mulder and Bakema. 1956; Ogato and Ishizuka, 1967). In a greenhouse study using four different concentrations of nutrient solutions, asparagine increased about 2.5X and glutamine increased 6X over the range of nitrogen fertilization levels (Mack and Shjoerring, 2002). Similarly, a field study using four different levels of nitrogen fertilization found that total free amino acids approximately doubled when the rate of nitrogen application was increased from 36 lbs (40 kg/ha) to 336 lbs (376 kg/ha) and asparagine and glutamine reached maximal levels at the highest rates of fertilization (Hoff et al., 1971). A more recent study looking at the effects of nitrogen fertilization on tuber amino acids also showed a positive effect of nitrogen fertilization on asparagine (Amrein et al., 2003), with asparagine levels varying widely with cultivar.

Although nitrogen fertilization affects the level of amino acids in the tubers, the direct effect of nitrogen levels on chip color is not straightforward. A study evaluating the effects of nitrogen fertilization on chip color at four sites was highly variable, ranging from significantly positive to significantly negative relationships between nitrogen levels and chip color at the sites (Dahlenburg et al, 1990). The same type of results had been observed by Kunkel and Holstad (1972).

Studies concerning the changes of amino acids in tubers during storage also have yielded inconsistent results. Brierley et al. (1996; 1997) using tubers of two cultivars cured at 15°C and then stored at 5°C and 10°C demonstrated extremely variable patterns of changes in amino acids. During one year, amino acid concentrations peaked after approximately 12 weeks of storage and then declined; in another year, a steady increase in amino acid levels was observed from 12 weeks of storage to 24 weeks. Short periods of reconditioning did not change amino acid levels, but higher storage temperature (10°C) generally resulted in higher asparagine concentration than 5°C storage. The researchers hypothesized that the changes in free amino acids were due to changes in the nitrogen balance of the tubers during storage depending on their dormancy status. Talley et al. (1970) also found asparagine levels increased in storage. A more recent study observed that storage temperature did not affect the levels of asparagine or glutamine and their levels remained steady throughout the storage period (Olsson et al, 2004).

While the interplay between sugars and starch levels in potato tubers is fairly well-understood, the metabolic processes leading to changes in concentrations of asparagine and glutamine in storage are unclear. Patatin, the major storage protein in potato tubers, is comprised of only 4.40% asparagine and 3.63% glutamine (Rosahl et al., 1986). Furthermore, patatin concentrations change very little during storage. It has been hypothesized that free amino acids resulting from patatin degradation are probably subject to amidation in order to mobilize them for transport (Brierley et al. 1997). This is supported by the fact that high levels of cytosolic glutamine synthase, the enzyme involved in the synthesis of glutamine for export to the sprouts, are present at the onset of tuber sprouting (Pereira et al., 1996).

There are no published reports on selection of potatoes for specific amino acid content. However, long-term selection studies in maize for high and low protein content suggests that it should be possible to significantly alter potato storage protein content (Dembinski et al., 1991; Lohaus et al., 1998).

1.3 Crop production system effects on acrylamide concentration in potato products

Several recent studies have addressed the relationship between the concentration of Maillard reaction precursors in the tubers and subsequent acrylamide formation. A study by Haase et al. (2003) found that total reducing sugars correlated with the acrylamide concentration in fried products ($r^2=0.64$). Glucose concentrations ($r^2 = 0.60$) and fructose concentrations ($r^2 = 0.56$) alone also correlated with acrylamide formation, but not as strongly as both compounds combined. No correlation was found between acrylamide formation and sucrose concentrations in the tubers ($r^2=0.24$). Similarly, Amrein et al. (2003) discovered that in practice, fructose and glucose concentrations determine acrylamide formation and that little improvement of the predictive equation develops if asparagine levels are included. They conclude that the differences in the acrylamide forming potential of different cultivars reflect differing levels of reducing sugars. No correlation with sucrose was found in their study.

2. MATERIALS AND METHODS

2.1 Tuber composition effects on acrylamide concentration in potato products

Several potato families developed by the USDA Potato Germplasm Enhancement Laboratory were used in this study. These crosses involved diploid *Solanum tuberosum* plants and several wild relatives of cultivated potato, including *S. phureja*, *S. raphanifolium*, and *S. stenotomum*. The genotypes were selected due to their ability to be stored at low (2°C) temperatures and maintain their chipping quality. Potato plants were grown in Rhinelander, WI at the Lelah Starks Potato Breeding Farm during the summer of 2002. The tubers were harvested in August through October.

Tubers were suberized for 1 week at room temperature and placed at 2°C for three months. At this time, a subset of the tubers was removed from storage to evaluate their chipping quality directly from cold storage. Tubers were sliced in half longitudinally from bud end to apical end. One half of the tuber was frozen in liquid nitrogen and freeze-dried for subsequent amino acid and sugar analysis. The second half of the tuber was sliced, rinsed in tap water, and fried in 180°C vegetable oil until bubbling ceased (ca. 2 min). Chip color was rate on a scale from 1-10 using the Potato Chip Institute International Color Chart with a score of 1 (light) to 10 (dark), with 1-5 rating acceptable, 6-7 rating marginal, and 8-10 rating unacceptable. The remaining tubers were reconditioned at room temperature for 2 weeks, after which time chipping quality was once again evaluated.

2.1.1 Sugar extraction

Freeze-dried potato tuber was homogenized and a 1-g sample was weighed to the nearest 0.01g. The samples were mixed with 25 ml of boiling 80% ethanol and shaken for 10 min. The supernatant was decanted and collected. The process was repeated four times and the volume of supernatant brought to 100 ml in a volumetric flask. One ml of the supernatant was evaporated to dryness under vacuum at 40°C. The sample was re-suspended in 100 μL H_2O for analysis using High Performance Liquid Chromatography (HPLC).

2.1.2 Amino acid extraction

Amino acids were extracted in the same manner as sugars (described above). This is similar to the methodology used by Jaswal (1973) and Brierley et al. (1996, 1997) for extraction of free amino acids from potato tubers. Samples were evaporated to dryness under vacuum at 40°C and resuspended in 20 mM HCL prior to analysis. Amino acid analysis on the HPLC was performed using the Waters AccQ-tag kits following the instructions provided by the manufacturer.

2.1.3 Acrylamide analysis

Acrylamide analyses were done by Dr. Yeonwha Park (Food Research Insititute, University of Wisconsin). Two grams of potato chips were extracted in 20 ml water with 1 μg internal standard ($^{13}C3$-Acrylamide) added to the sample. The samples were placed on mechanical shaker for 10 min followed by centrifugation at 10,000 g for 30 min. The samples were brominated as per EPA method and incubated on ice for one hour. Excess bromine was decomposed with sodium thiosulfate. The samples were extracted twice with 6 ml ethyl acetate, shaken, centrifuged at 2000 rpm for 10 min, concentrated with roto-vap, and filtered with syringe filter (0.45 µm). Analysis was performed using Gas Chromatography-Mass Spectrometry.

2.1.4 HPLC analysis of sugars and amino acids

Sugars: Glucose, fructose, and sucrose were separated using a HPLC (Waters Corporation, Milford, MA, USA) equipped with a 10 µm 300 mm x 4.1 mm Alltech carbohydrate column (Alltech Associates Inc., Deerfield, IL, USA). The mobile phase was a mixture of 80% acetonitrile and 20% water. The injected volume was 50 µL at a flow rate of 2 ml/min for a total run time of 16 min. Sugars were detected using a Waters 401 Refractive Index Detector. Chromatograms were analyzed using Shimadzu Client/Server software version 7.2 SP1 Build 9 (Shimadzu, Tokyo, Japan).

Amino Acids: Amino acids were separated using a Waters HPLC equipped with a Waters AccQ-tag column. The column temperature was held at 37°C using a Fiatron CH-3 column heater. The mobile phase was a mixture of (A) 12.5 mM sodium phosphate (pH 6.3) and (B) 12.5 mM sodium phosphate (pH 6.3) and acetonitrile (70:30). The gradient was as follows: 98% A, 2% B for 15 min; 93% A, 7% B for 4 min; 90% A, 10% B for 7 min; 68% A, 32% B for 9 min; 60% A, 40% B for 16 min; 100% B for 15 min; 100% A for 19 min. An injection volume of 30 µL at a flow rate of 1 ml/min for 76 min was used. Amino acids were detected on a Shimadzu RF-10A XL fluorescence detector (λ_{ex} 250nm, λ_{em} 395nm). Asparagine and glutamine were calculated in reference to the internal standard using Millenium 3.1 software (Waters Corporation).

2.2 Nitrogen fertilization and tuber amino acid concentration

Potato plants were grown at the University of Wisconsin Agricultural Experiment Station in Hancock, WI during the 2002 and 2003 growing seasons. Two varieties developed from the University of Wisconsin potato breeding program and two common cultivated varieties (Atlantic and Russet Burbank) were selected for this study. Three rates of nitrogen fertilization were used to cultivate the plants (140 lbs/ac, 220 lbs/ac, and 300 lbs/ac). Tubers were harvested in October and dried in a forced-air oven. Amino acid concentrations were determined as described above.

2.3 Storage effects on acrylamide concentration in potato products

Tubers from the diploid plants described above were harvested in October 2003. Tubers were held at 15°C for 1 week for suberization. Subsequently, a subset of tubers was placed directly into 2°C cold storage. The other subset of tubers was brought slowly to the 2°C storage temperature at a rate of 0.2°C/day ("preconditioned"). Tubers were held in storage for three months and evaluated for color and acrylamide levels as described above.

3. RESULTS

3.1 Tuber composition effects on acrylamide concentration in potato products

Due to the genetic diversity of the material and the unusually cold storage temperatures, the range of sugar and asparagine values varied widely. At harvest, glucose and fructose concentrations in the tubers averaged 0.32±0.10% and 0.29±0.089% dry weight (dw) and sucrose concentrations averaged 1.17±0.12% dw. The tubers of the cold-sweetening resistant diploid potatoes accumulated both sucrose and reducing sugars in cold storage. After storage at 2°C for 3 months, glucose concentrations rose to 2.64±0.20% dw, fructose to 2.81±0.21% dw, and sucrose to 7.53±0.56% dw. Reconditioning for two weeks at 15°C decreased the sugar concentrations of the tubers, but not to the original levels before storage (glucose: 0.58±0.14% dw; fructose 0.86±0.20 dw; sucrose 2.67±0.37 dw).

Genotypes varied for the amount of sugars that accumulated during storage. Certain genotypes, such as 3277 and 3287, had very little reducing sugars at harvest but significant increases in the concentrations of these compounds in the tubers after storage (Fig. 1). Other genotypes, such as 3230, had a much less marked rise in reducing sugar concentrations in storage (Fig. 1).

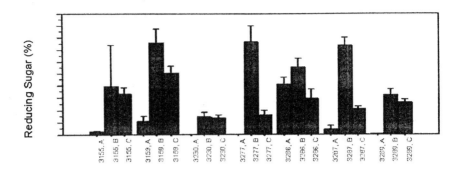

Figure 1. Changes in reducing sugar concentrations in different genotypes of cold-sweetening resistant potato tubers (A = at harvest; B = after 3 months storage at 2°C; C = after reconditioning at 15°C for 2 weeks).

Asparagine concentrations in the tubers increased throughout the storage period ($P = 0.010$). At harvest, the asparagine levels in the tubers ranged from 10.7 to 32.4 mg/g dw, averaging 21.2 mg/g dw. After 13 weeks of storage at 2°C, asparagine concentrations in the tubers averaged 34.9 mg/g dw (3.9-77.1 mg/g dw). After reconditioning at 15°C for two weeks, the asparagine concentrations in the tubers declined slightly (30.6 mg/g dw), but the change was not statistically significant.

The acrylamide levels of the chips produced from these tubers after reconditioning were high, ranging from 3,290 - 44,297 µg/kg. Significant positive correlations were found between reducing sugar concentrations and acrylamide formation in chips (glucose: $r^2 = 0.75$; fructose: $r^2 = 0.46$; total reducing sugar: $r^2 = 0.58$). No correlation was found between sucrose concentrations, asparagine concentrations, and acrylamide production (sucrose: $r^2 = 0.01$; asparagine: $r^2 = 0.01$). This is similar to the results of Haase et al. (2003) and Amrein et al. (2003), who found significant positive correlations between reducing sugar concentrations in tubers and acrylamide production in potato products, but no correlations between asparagine and acrylamide production or sucrose and acrylamide production. A relationship also was observed between chip color and acrylamide concentrations, with acrylamide levels in the chips increasing as the chip color darkened (Fig. 2).

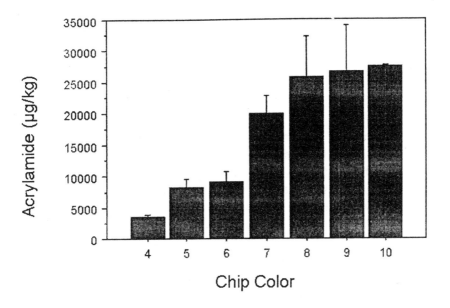

Figure 2. Relationship between chip color and acrylamide concentration from potato tubers fried at 180°C.

3.2 Nitrogen fertilization and tuber amino acid concentration

In general, tubers showed increasing asparagine concentrations in response to increased rates of nitrogen fertilization ($P = 0.05$). At the 140 lb/ac N fertilizer rate, asparagine concentrations in the tubers measured 4.15±0.43 mg/g dw. As the fertilizer rate increased to 220 lbs/ac, asparagine levels rose to 5.42±0.47 mg/g dw. The highest tuber asparagine concentrations (5.90±0.34 mg/g dw) occurred at the highest fertilizer rates of 330 lbs/ac. Generally, all of the cultivars used in the study showed similar trends. However, the extent of the response did differ depending on genotype.

3.3 Storage effects on acrylamide concentration in potato products

High levels of acrylamide were found in the both the chips of the tubers directly placed in 2°C and those that were slowly cooled to this temperature. However, acrylamide concentrations of the chips of the tubers directly from 2°C were 5X higher than those from the preconditioned tubers (35,004 ± 2,393 µg/kg vs.7,462±521 µg/kg) (P <0.001). In addition, the color of the preconditioned tuber chips was significantly lighter (chip color score = 4) than the chips from tubers placed directly in 2°C cold storage (chip color score = 10).

4. DISCUSSION

Management of non-enzymatic browning in processed potato products has been an issue for over a century. The degree to which this reaction occurs can be controlled to a significant extent through management of reducing sugar levels in the tubers. The concentration of reducing sugars are generally low in the tuber at harvest; thus, reducing sugars are the limiting precursor in this reaction. Upon storage below 10°C, tubers can accumulate high levels of these compounds and produce much higher levels of Maillard reaction products. Because of their limiting role in the initial steps of the Maillard reaction and the possibility of controlling their concentrations through storage practices and genetics, much of the work regarding non-enzymatic browning has focused on sugars. However, the discovery that high levels of acrylamide can be formed in tubers stored under current practices brought interest in rigorous control of the amounts of both precursors for acrylamide formation (reducing sugars and asparagine). The results of our preconditioning study indicate that that amount of acrylamide formed in potato products varies with storage practices. Insuring that correct storage practices are followed might minimize acrylamide levels.

This study, as well as previous studies, supports the fact that tuber sugar and asparagine levels may be affected by a number of factors including cultural practices, genetics, and storage. The increases in sugar concentrations in storage observed in this study are similar to those seen in previous studies. The extent of variation in asparagine concentrations in tubers depending on genetic, cultural, and storage factors is still unclear. This study suggests that amino acid concentrations in tubers increase upon cold storage. However, the degree of the change in asparagine concentrations in the tuber at different storage temperatures and lengths has not been investigated in depth. Similarly, although this study also suggests that tuber amino acid concentrations increase with higher nitrogen fertilization rates, other studies have shown that the amount of amino acids in tubers varies widely from year to year in the same location with the same management practices. Thus, if the possibility for control of acrylamide through cultural practices is to be brought to fruition, the nature of the variation in asparagine must be understood.

Breeding potatoes for low asparagine levels offers the possibility for a long-term solution to reducing acrylamide concentrations in processed potato products. Much progress has been made over the past century in the development of potato cultivars that are able to be stored at cold temperatures with less accumulation of sugars. The variation in asparagine concentrations in tubers allows for the possibility of selecting for low-asparagine genotypes. Screening of wild potato germplasm will offer another wealth of genetic diversity to help attain this goal.

ACKNOWLEDGEMENTS

The authors would like to thank Mr. Andy Hamernik, Dr. Keith Kelling, and Dr. A.J. Bussan for their technical assistance and the production of the potato tubers. We also thank Dr. Yeonwha Park for her assistance with the acrylamide analysis and Dr. Douglas Senalik for his assistance with the HPLC analysis.

REFERENCES

Amrein, T.M., Bachmann, S., Noti, A., Biedermann, M., Ferraz Barbosa, M, Biedermann-Brem, S., Grob, K., Keiser, A., Realini, P., Escher, F., and Amado, R., 2003, Potential of acrylamide formation, sugars, and free asparagine in potatoes: A comparison of cultivars and farming systems, *J. Agric. Food Chem.* **51**:5556-5560.

Brierley, E.R.,. Bonner, P.L.R., and Cobb, A.H., 1996, Factors influencing the free amino acid content of potato (*Solanum tuberosum* L.) tubers during prolonged storage, *J. Sci. Food Agric.* **70**:515-525.

Brierley, E.R., Bonner, P.L.R., and Cobb, A.H., 1997, Analysis of amino acid metabolism in stored potato tubers (cv. Pentland Dell), *Plant Sci.* **127**:17-24.

Brouquisse, R., James, F., Pradet, A., and Raymond, P., 1992, Asparagine metabolism and nitrogen distribution during protein degradation in sugar-starved maize tips, *Planta* **188**:384-385.

Burton, W.G., 1963, The basic principles of potato storage as practiced in Great Britain, *Eur. Potato J.* **6**:77-92.

Butler, O., 1913, A note on the significance of sugar in the tubers of *Solanum tuberosum, Torrey Botan. Club Bulletin* **40**:100-118.

Chevalier, C., Bourgeois, E., Just, D., and Raymond, D., 1996, Metabolic regulation of asparagines synthestase gene expression in maize (*Zea mays* L.) root tips, *Plant Physiol.* **9**:1-11.

Clegg, M.D., and Chapman, H.W., 1962, Post-harvest discoloration of chips from early summer potatoes, *Am. Potato J.* **39**: 176-184.

Dahlenburg, A.P., Maier, N.A., and Williams, C.M.J., 1990, Effect of nitrogen on the size, specific gravity, crisp colour, and reducing sugar concentration of potato tubers (Solanum tuberosum L.) cv. Kennebec, *Aust. J. of Exp. Agric.* **30**:123-130.

Danehy, J.P., 1986, Maillard reactions: Nonenzymatic browning in food systems with special reference to the development of flavor, *Advances Food Res.* **30**:77-138.

Davies, A.M.C., 1977, The free amino acids of tubers of potato varieties grown in England and Ireland, *Potato Res.* **20**:9-21.

Dembinski, E., Rafalski, A.,and Wisniewska, I., 1991, Effect of long-term selection for high and low protein content on the metabolism of amino acids and carbohydrates in maize kernel, *Plant Physiol. Biochem.* **29**:549-557.

Denny, F.E. and Thornton, N.C., 1940, Factors for color in the production of potato chips, *Boyce Thompson Inst. Contrib.* **11**:291-303.

Douches, D.S., Maas, D., Jastrzebski, K., and Chase, R.W., 1996, Assessment of potato breeding progress in the USA over the last century, *Crop Sci.* **36**:1544-1552.

Edwards, J.W., Walker, E.L., and Coruzzi, G.M., 1990, Cell-specifc expression in transgenic plants reveals nonoverlapping roles for chloroplast and cytosolic glutamine synthetase, *Proc. Natl. Acad. Sci.* **87**:3459-3463.

Eskin, N.M., 1990, Biochemistry of food processing: Browning reactions in foods, In: *Biochemistry of foods,* 2nd ed., Academic Press, Inc., USA.

Haase, N.U., Matthaus, B., and Vosmann, K., 2003, Minimierungsansatze zur Acrylamid-Bildung in pflanzlichen Lebensmittein – aufgezeight am Beispiel von Kartoffelchips, *Deutsche Lebensmittel-Rundshau* **3**:87-90.

Hoff, J.E., Jones, C.M., Wilcox, G.F., and Castro, M.D., 1977, The effect of nitrogen fertilization on the composition of the free amino acid pool of potato tubers, *Am. Potato J.* **48**: 391-394.

Hoover, E.F. and Xander, P.A., 1961, Potato composition and chipping quality, *Am. Potato J.* **38**:163-170.

Hoover, E.F. and Xander, P.A., 1963, Influence of specific compositional factors of potatoes on chipping color, *Am. Potato J.* **40**:17-24.

Jakuczun, H., Zgorska, K., and Zimnoch-Guzowksa, E., 1995, An investigation of the level of reducing sugars in diploid potatoes before and after cold storage, *Potato Res.* **38**:331-338.

Jaswal, A.S., 1973, Effects of various processing methods on free and bound amino acid contents of potatoes, *Am. Potato J.* **50**:86-95.

Johansen, H., 1999, Reducing sugar accumulation in progeny families of cold-chipping potato clones, *Am. Potato J.* **67**:83-91.

Joy, K.W., Ireland, R.J., and Lea, P.J., 1983, Asparagine synthesis in pea leaves, and the occurrence of an asparagine synthetase inhibitor, *Plant Physiol.* **73**:165-168.

King, G.A., Wollard, D.C., Irving, D.E., and Borst, W.E., 1990, Physiological changes in asparagus spear tips after harvest, *Plant Physiol.* **100**:1661-1669.

Kunkel, R., and Holstad, N., 1972, Potato chip colour, specific gravity, and fertilization of potatoes with N-P-K, *Am. Potato J.* **49**: 43-62.

Leszkowiat, M.J., Baricello, V., Yada, R., Coffin, R.H., Lougheed, E.C., and Stanley, D.W., 1990, Contribution of sucrose to nonenzymatic browning in potato chips, *J. Food Sci.* **55**:281-282.

Lohaus, G., Buker, M., Hußmann, M., Soave, C., Heldt, H.-W., 1998, Transport of amino acids with special emphasis on the synthesis and transport of asparagine in the Illinois Low Protein and Illinois High Protein strains of maize, *Planta* **205**:181-188.

Loiselle, F., Tai, G.C.C., and Christie, B.R., 1990, Genetic components of chip color evaluated after harvest, cold storage, and reconditioning, *Am. Potato J.* **67**:633-646.

Love, S.L., Pavek, J.J., Thompson-Johns, A., and Bohl, W., 1998, Breeding progress for potato chip quality in North American cultivars, *Am. J. Potato Res.* **75**:27-36.

Mack, G. and Schjoerring, J.K., 2002, Effect of NO_3^- supply on N metabolism of potato plants (*Solanum tuberosum* L.) with a special focus on the tubers, *Plant Cell Environ.* **25**:999-1009.

McGrath, R.B. and Coruzzi, G.M., 1991, A gene network controlling glutamine and asparagines biosynthesis in plants, *Plant J.* **1**:275-280.

Mulder, E.G. and Bakema, K., 1956, Effect of the nitrogen, phosphorus, potassium, and magnesium nutrition of potato plants on the content of the free amino acids and on the amino acid composition of the protein of tubers, *Plant Soil* **7**:136-166.

Müller-Thurgau, H., 1882. Über die Natur des süssen Kartoffeln sich vorfindenen Zuckers, *L. Jb.* **11**: 751.

Ogato, S. and Ishizuka, Y., 1967, Variation of nutritional and physical characteristics of crops by crossing. I. Response of cultivate, wild, and hybrid potatoes to nitrogen, *Nippon Dojo-Hiryogaku Zasshi* **38**:79-84.

Olsson, K., Svensson, R., and Roslund, C.-A., 2004, Tuber components affecting acrylamide formation and colour in fried potato: variation by variety, year, storage temperature, and storage time, *J. Sci. Food. Agric.* **84**:447-458.

Oltmans, S.M. and Novy, R.G., 2002, Identification of potato (*Solanum tuberosum* L.) Haploid x wild species hybrids with the capacity to cold-chip, *Am. J. Potato Res.* **79**:263-268.

Pereira, S., Pissarra, J., Sunkel, C., Salema, R., 1996, Tissue-specific distribution of glutamine synthetase in potato tubers, *Ann. Bot.* **77**:429-432.

Pritchard, M.K. and Adam, I.R., 1992, Preconditioning and storage of chemically immature russet Burbank and Shepody potatoes, *Am. Potato J.* **69**: 805-815.

Rodriguez-Saona, L.E. and Wrolstad, R.E., 1997, Influence of potato composition on chip color quality, *Am. Potato J.* **74**: 87-106.

Roe, M.A., Faulks, R.M., and Belsten, J.L., 1990, Role of reducing sugars and amino acids In: Fry color of chips from potatoes grown under different nitrogen regimes, *J. Sci. Food Agric.* **52**:207-214.

Roe, M.A. and Faulks, R.M., 1991, Color development in a model system during frying: Role of individual amino acids and sugars, *J. Food Sci.* **56**:1711-1713.

Rosahl, S., Schmidt, R., Schell, J., Willmitzer, L., 1986, Isolation and characterization of a gene from S. tuberosum encoding patatin, the major storage protein of potato tubers, *Mol. Gen. Genet.* **203**:214-220.

Schwimmer, S., Hendel, C.E., Harrington, W.O., and Olson, R.L., 1957, Interrelation among measurements of browning of processed potatoes and sugar components, *Am. Potato J.* **34**:119-132.

Shallenberger, R.S., Smith, O., and Treadway, R.H., 1959, Role of the sugars in the browning reaction in potato chips, *J. Agric. Food Chem.* **7**:274-277.

Smith, O., 1987, Potato chips, in: *Potato Processing*, 4th ed., Van Nostrand Reinhold Company Inc., USA.

Sowokinos, J.R., 1978, Relationship of harvest sucrose content to processing maturity and storage life of potatoes, *Am. Potato J.* **55**:333-343.

Sowokinos, J.R. and Preston, D.A., 1988, Maintenance of potato processing quality by chemical maturity monitoring (CMM), *Minn. Agric. Exp. Sta. Bull.* 586-1988 (Item No. AD-SB-3441).

Sowokinos, J.R., Orr, P.H., Knoper, J.A., and Varns, J.L., 1987, Influence of potato storage and handling stress on sugars, chip quality, and integrity of the starch (amyloplast) membrane, *Am. Potato J.* **64**:213-226.

Streeter, J.M., 1977, Asparaginase and asparagine transaminase in soybean leaves and root nodules, *Plant Physiol.* **60**: 680-683.

Stulen, I. and Oaks, A., 1977, Asparagine synthesis in corn roots, *Plant Physiol.* **60**:235-239.

Talley, E.A. and Porter, W.L., 1970, Chemical composition of potatoes. VII. Relationship of the free amino acid concentrations to specific gravity and storage time, *Am. Potato J.* **47**:214-224.

Thill, C.A. and Peloquin, S.J., 1993, Accelerated development of clones that chip from 4°C, *Abstracts of the 12th Triennial Conference of the EAPR, Paris, France, July 1993*, p. 229.

Thill, C.A. and Peloquin, S.J., 1994, Inheritance of potato chip color at the 24-chromosome level, *Am. Potato J.* **71**:629-646.

Tsai, F.-Y and Coruzzi, G.M., 1991, Light represses that transcription of asparagine synthetase genes in photosynthetic and non-photosynthetic organs of plants, *Mol. Cell Biol.* **11**:4966-4972.

Tsai, F-Y. and Coruzzi, G.M., 1990, Dark induced and organ specific expression of two aspargine synthetase genes in *Pisum sativum*, *EMBO J.* **9**:323-332.

Urquhart, A.A. and Joy, K.W., 1981, Use of phloem exudates technique in the study of amino acid transport in pea plants, *Plant Physiol.* **91**:702-708.

Watada, A.E. and Kunkel, R., 1955, The variation in reducing sugar content in different varieties of potatoes, *Am. Potato J.* **32**-132-140.

Wright, R.C., Peacock, W.M., and Whiteman, T.M., 1936, Cooking quality, palatability, and carbohydrate composition of potatoes influenced by storage temperature, *USDA Tech. Bull.* **507**:1-20.

ACRYLAMIDE REDUCTION IN PROCESSED FOODS

A.B. Hanley, C. Offen, M. Clarke, B. Ing, M. Roberts and R. Burch
Leatherhead Food International, Randalls Road, Leatherhead, KT22 7RY, UK; e-mail: bhanley@leatherheadfood.com

Abstract: The discovery of the formation of acrylamide in fried and baked foods containing high levels of starch and the amino acid asparagine, prompted widespread concern. Both processed and home cooked foods are affected and this has led to the increased study of variations in cooking and processing conditions to minimize formation. While changes in cooking protocols have been in part successful, particularly when lower frying and baking temperatures are used, pretreatments to reduce levels of acrylamide by prevention of formation or acceleration of destruction have been investigated. In this study, a range of pretreatments of grilled potato were investigated and compared with surface washing to remove asparagine and reducing sugars. Synergies were observed between different treatments, and reductions of up to 40% were achieved in a non-optimized system.

Key words: Acrylamide, processed food

1. INTRODUCTION

The formation of acrylamide in high starch foods was both unexpected and extensively reported. Initial studies concentrated upon confirming the initial results and, once this had been achieved, a range of questions remained to be answered. The first and foremost concerned the mechanism of formation. It quickly became clear that, in the main, acrylamide was formed from a single precursor – asparagine - with complete retention of all of the carbon backbone and the amide nitrogen (Zyzak *et al.*, 2003). It is presumed that the carbonyl oxygen is also retained. While the nature of the precursor was being established, other studies were taking place to determine the mechanism of the reaction. Considerable speculation took place including the potential role of other acrylates with exogenous addition of the amino functionality. This was effectively solved when the precursor was established and it became clear, in a series of model experiments, that Maillard type chemistry was involved, with the likelihood that the first step in the process was the addition of the amino group of asparagine to a carbonyl. Subsequent reactions lead to elimination of the carboxyl and amino functionalities and formation of acrylamide (Mottram *et al.*, 2002).

The discovery of the nature of the reaction pathway and the molecular requirements for the reaction to proceed enabled research to begin on developing proactive methods to prevent acrylamide formation. In effect, four strategies appeared possible.

- Prevention of formation of acrylamide by removal of the essential precursors (asparagine and a source of a carbonyl moiety – generally a reducing sugar).
- Inhibition of the reaction by introducing compounds (perhaps other amino acids) that would compete with asparagine for the carbonyl compounds in the Maillard reaction.
- Interruption of the reaction by addition of chemically reactive compounds that are able to react with intermediates in the Maillard reaction. These could include active thiol and amino groups.
- Removal of acrylamide after it has been formed.

There remains a further possibility. A number of researchers investigated changing the processing or cooking conditions to minimise acrylamide formation in products (e.g., Beidermann et al., 2002; Gertz and Klosterman, 2002; Grob et al., 2003). The major such approach was reduction of cooking temperature but this could have a negative effect on perceived product quality and may also be difficult to introduce into home cooked food processes.

Our strategy to reduce acrylamide formation in processed foods involved the development of quick and cheap methods that could reduce acrylamide in grilled potato slices. The reason for selecting this matrix was that it could be easily controlled, it allowed for simple solutions to be tested and it removed some potentially confounding factors. We could, of course, have spent time analyzing the kinetics of formation of acrylamide under a range of conditions in test systems. Because this was being carried out elsewhere and we felt that we should be able to develop simple treatments that could be applied cost effectively in an industrial setting, based upon current knowledge, that would demonstrate significant reductions. If such reductions were not observed then we would know that our procedures would not be applicable and more detailed mechanistic study may be necessary.

2. MATERIALS AND METHODS

2.1 Acrylamide analysis

Samples (25g) were extracted into water (200 mL) by homogenization followed by centrifugation and collection of the supernatant. Acrylamide was derivatised by bromination (bromine water in the presence of potassium bromide and hydrobromic acid), extracted into ethyl acetate and treated with triethylamine. Extracts were analyzed by GC-MS. ^{13}C labeled acrylamide, as internal standard, was added to the sample prior to extraction.

2.2 Sample preparation

Potatoes were peeled and sliced to 1 cm thickness. Initial experiments were carried out on random-shaped pieces (i.e. not cut any further). The remainder of experiments were carried out on standardized slices, where the slices were cut into squares of 9 cm x 5 cm. 'Control' potato slices were grilled without further treatment. 'Washed' slices were immersed in water, drained and grilled. Samples treated with 'herb oil' were sprayed with oil (sunflower) containing 1 % of each of rosemary, basil and thyme essential oils, before grilling. The control sample for this treatment consisted of potato slices sprayed with oil, without rosemary, basil and thyme oil added.

Other treated slices were immersed in solutions (detailed below) for 30 seconds, drained and grilled.

2.2.1 Treatment solutions

The following were prepared in water at the specified concentrations:
Aspartate (2 g/L)
Thiamine Hydrochloride (1 g/L)
Glutamine (1 g/L)
Lysine (0.5 g/L)
Citric Acid (1 g/L)
Glutathione / Cysteine (1 g/L each)
Glutamine/ Aspartic (1 g/L each)
Glutamine / Aspartic / Lysine (1 g/L each)

2.2.2 Grilling

Potato slices were grilled for 30 minutes at 220°C, except for initial experiments, for which three cooking times (19 min, 23.5 min and 28 min) were evaluated.

3. RESULTS AND DISCUSSION

3.1 Establishment of control conditions

The results from the first series of control samples are shown in Table 1. The aim of this series of experiments was to establish baseline conditions for the formation of acrylamide in the grilled potato slices.

Table 1. : Formation of acrylamide in randomly prepared grilled potato slices

Sample	Cook time	Temperature (°C)	Acrylamide (µg/kg)
1a	19 min	220	147
1b	19 min	220	36
2a	23.5 min	220	267
2b	23.5 min	220	738
3a	28 min	220	564

It is clear from these results that the use of randomly prepared potato slices did not give reproducible results. We had assumed that, with a sample size of 500g, any variation in the relative potato sizes within batches would average out the results between batches. Clearly, in order to obtain consistent results, it was necessary to standardize the volume/surface area ratio of the slices.

The second series of control experiments made use of 50g samples of potato in identical slices taken from the centre of the potato in order to minimize intra variation in reducing sugar or acrylamide levels. This gave levels of acrylamide of between 900 – 1038 µg/kg in the control grilled samples.

3.2 Trial conditions to reduce acrylamide

For the experimental samples, these were grilled after submersion in either plain water or water containing a range of treatments designed to minimize acrylamide formation. One set of samples was treated with oil soluble components (herb oil mixture) and in this case the herb oils were dissolved in sunflower oil before administration.

The results from these are shown in Tables 2 and 3.

Table 2. Levels of acrylamide after treatment with single components

Sample	Acrylamide (µg/kg)	Control (µg/kg)	Loss (%)
Surface washed	772	1068	28
Blanched	1038	1004	3
Citric acid	664	910	27
Thiamine	794	900	12
Rsmry/Bsl/Thyme	230	352	35

It is clear from these results that most of the treatments failed to result in a significant lessening in the formation of acrylamide over that achieved by surface washing and, indeed for the blanched sample, the level is comparable to that in the unwashed control. In this latter case, while the treatment may have removed some of the surface asparagine and reducing sugars, it seems likely that it has also opened up the surface allowing acrylamide to be formed deeper into the potato slice. The effect of citric acid may be related to the lowering of the surface pH thereby causing asparagine (which has a relatively low pKa compared to other amino acids) to be protonated therefore diminishing its preferential reaction with the carbonyl moiety at the start of the reaction (Rydberg et al., 2003). It is notable that, of the other amino acids, only cysteine has a lower pKa than asparagine and this may be part of the reason why others have reported a lowering of acrylamide formation after pretreatment with cysteine (i.e. it may be unrelated to the scavenging properties of the thiol group). The only other case which showed significant lessening of levels of acrylamide was in the oil treated samples. The levels of acrylamide in this case are lower even in the control (treated with sunflower oil) than potato slices not treated with oil and this may be due to more rapid cooking caused by the heat conductivity of the oil.

The next stage in the process was to consider potential additive effects. Mixtures of amino acids were used which could have a range of effects including lowering of pH, (aspartic acid), chelating effects (aspartic acid and lysine) and competition with asparagine. The results from this phase of the work are shown in Table 3.

Table 3. Treatment with amino acids to minimise acrylamide formation

Sample	Acrylamide (μg/kg)	Control (μg/kg)	Loss (%)
Glutamine	570	659	13
Lysine	478	659	27
Aspartic acid	469	659	29
Glutathione/cysteine	678	900	25
Glutamine/aspartate/lysine	543	900	40
Glutamine/aspartate	379	659	43

The combined treatments appear to have a greater potential to reduce acrylamide formation than those taken individually, however it is important to bear in mind that a range of concentrations would have to be tested to confirm if the effect is synergistic or simply additive. It is noteworthy that these results are not optimized and the treatments (immersion for only 30 seconds) are unlikely to lead to products with substantially different taste characteristics to those of the untreated samples, since the amount of each chemical absorbed is likely to be small.

It is not possible to determine the mechanistic basis for these reductions based upon the results thus far. However it is interesting to note that the most significant reductions occur when a combination of a basic and acidic amino acid is used which could both have multiple effects (competition, pH adjustment and chelation effects). To this end it would be useful to attempt to characterise the various products in the potatoes grilled in the presence of these amino acids to determine if this could provide a mechanistic basis and thereby aid in the development of optimal procedures.

4. CONCLUSION

We have studied a series of mild pretreatments designed to minimize the presence of acrylamide in processed potato products. The test system, which was designed to remove as much variation as possible from exogenous factors, was grilled potato. We have been able to demonstrate that mixtures of amino acids (possibly with complementary effects) show the greatest potential. Furthermore, these treatments were designed to be simple (dipping or spraying) and to use only very small amounts of material so as to minimize effects on organoleptic properties and costs of treatment.

We anticipate that, if such procedures are to be adopted in future, considerable optimization will be required both in the mixtures to be used and in the mode of application.

REFERENCES

Biedermann, M., Noti, A., Biedermann-Brem, S., Mozzetti, V., and Grob, K., 2002., Experiments on acrylamide formation and possibilities to decrease the potential of acrylamide formation in potatoes. *Mitt. Lebensm. Hyg.* **93**:668-687.

Gertz, C., and Klosterman, S., 2002, Analysis of acrylamide and mechanisms of its formation in deep-fried products. *Eur. J. Lipid Sci. Technol.* **104**:762-771.

Grob, K., Biedermann, M., Biedermann-Brem, S., Noti, A., Inhof, D., Amrein, T., Pfefferle, A., and Bazzocco, D., 2003, French fries with less than 100 µg/kg acrylamide. A collaboration between cooks and analysts, *Eur. Food Res. Technol.* **217**:185-194.

Mottram, D.S., Wedzicha, B.L., and Dodson, A.T., 2002, Acrylamide is formed in the Maillard reaction, *Nature* **419**:448-449.

Rydberg, P., Eriksson, S., Tareke, E., Karlsson, P., Ehrenberg, L., and Tornqvist, M., 2003, Investigation of factors that influence the acrylamide content of heated foodstuffs, *J. Agric. Food Chem.* **51**:7012-7018.

Zyzak, D.V., Sanders, R.A., Stojanovic, M., Tallmadge, D.H., Eberhart, B.L., Ewald, D.K., Gruber, D.C., Morsch, T.R., Strothers, M.A., Rizzi, G.P., and Villagran, M.D., 2003, Acrylamide formation mechanism in heated foods, *J. Agric. Food Chem.* **51**:4782-4787.

CHEMICAL INTERVENTION STRATEGIES FOR SUBSTANTIAL SUPPRESSION OF ACRYLAMIDE FORMATION IN FRIED POTATO PRODUCTS

Robert C. Lindsay and Sungjoon Jang
Department of Food Science, 1605 Linden Drive, University of Wisconsin-Madison, Madison, WI 53706; e-mail: rlindsay@wisc.edu

Abstract: Prototype processes were developed for the substantial suppression of acrylamide formation (40-95% compared to untreated controls) in cut surface fried potato products using potato chips (crisps) as the primary model. The most efficacious procedures employed sequentially both surface preparation and subsequent acrylamide precursor complexation and/or competitive inhibition processing steps. Surface preparation processing involved either various low-temperature (50–75°C) aqueous (5-30 min) or *ca.* 80% ethanol blanch solutions for various times (1-5 min) combined with aqueous leaching steps (1-10 min) to reduce concentration of acrylamide precursors in the critical frying zone of cut potato surfaces. Acrylamide precursor complexation and/or competitive inhibition processing strategies included immersion exposure of prepared cut potato surfaces to solutions or dispersions of various combinations of either calcium chloride, phytic acid, chitosan, sodium acid pyrophosphate, or N-acetylcysteine.

Key words: Acrylamide reduction, processing, phytate, calcium ions, chitosan, pectinmethylesterase, yeast, fried potatoes

1. INTRODUCTION

Fried potato products generally contain notable quantities of acrylamide (Livsmedelsverket, 2002; Becalski et al., 2003; FDA, 2004) because they contain abundant asparagine and reducing sugars that readily undergo the Maillard browning reaction under typical frying conditions (Mottram et al., 2002; Stadler et al., 2002; Biederman et al., 2002; Grob et al., 2003;

Yaylayan et al., 2003; Zyzak et al., 2003). A number of potential strategies have been identified for suppressing acrylamide formation in potato products, and some of these are shown in Figure 1. Even though the overwhelming majority of acrylamide formed in fried potatoes originates via the Maillard reaction, small amounts may arise under some conditions from other degradative pathways (Yashuhara et al., 2003; Yaylayan et al., 2004; Granvogl et al., 2004), and thus, some of these pathways also are accommodated in the overall schematic (Figure 1).

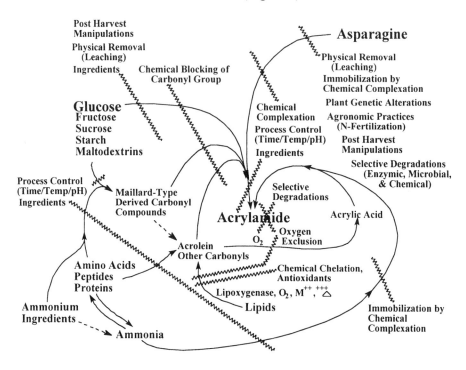

Figure 1. Schematic representation indicating some potential strategies and points of intervention for suppressing the formation of acrylamide in fried potato products. (Note: Points of intersection of wavy lines and reaction arrows denote potential intervention targets).

Basic strategies for suppression of acrylamide have included either lowering concentrations of precursors, controlling physical process conditions, or inhibiting the Maillard reaction via functional group blocking or competitive inhibition. Grob et al. (2003) have proposed the selection of raw potatoes containing low concentration of precursors, an avoidance of exposures of raw potatoes to low temperature storage, incorporation of aqueous pretreatment steps (rinsing, soaking, blanching of potato slices), and the avoidance over-cooking as means to lower acrylamide concentrations in potato products.

Others (Biederman et al., 2002; Rydberg et. al., 2003) have reported that levels of acrylamide in foods were elevated when samples were exposed to high temperatures for extended times. Also, Jung et al. (2003) have reported that the treatments with various concentrations of citric acid before frying suppressed acrylamide in both model French fries and potato chips, but there are concerns regarding sour flavors in resulting products. Zyzac et al. (2003) have reported nearly complete suppression of acrylamide in a fabricated potato chip model using asparaginase to remove asparagine in the dough before frying. In the present paper, we report some additional processing strategies for the substantial suppression of acrylamide formation in fried potato products.

2. MATERIALS AND METHODS

2.1 Supplies and Processing

The materials and methods employed were the same as described in detail in another chapter in this volume (Lindsay and Jang, 2004) except as noted below.

2.1.1 Low temperature (70°C) blanching

Cut potato pieces were held submersed in water at 70°C for 30 min, and then were cooled by submersion for 1 min in excess water (initially 15°C) before draining and frying.

2.1.2 Hot ethanol (*ca.* 80°C) blanching

Cut potato pieces were submersed in refluxing 80% ethanol (*ca* 80°C; grain alcohol; caution: potential fire hazard if improperly contained) for 1 min. Then, potato pieces were soaked for 3 min at 50°C in excess water or aqueous solutions containing phytate plus calcium chloride at 50°C for 3 min before draining and frying.

2.1.3 Low temperature (70°) blanching plus chitosan

Cut potato pieces were held submersed in 5% NaCl solution at 70°C for 1 min, and then were cooled by submersion for 1 min in excess water (initially 15°C) before transferring to a solution of 1000 ppm chitosan at 50°C. A 1% chitosan (Sigma; crustacean origin) stock solution was prepared by adding commercial chitosan powder (5.0 g) into 495 g of 10% acetic acid in deionized water. The chitosan powder suspension was stirred with a magnetic stirring bar until it was dispersed in solution (became transparent; about 3 hr at 21°C). Appropriate amounts of this stock solution were added to aqueous treatment solutions for application. Diluted treatment solutions were used without further adjustments (i.e., pH) with regard to the dissolved chitosan component.

2.1.4 Potato nugget model fabrication

Experimental nuggets were prepared with commercial dehydrated potato flakes and active commercial baker's yeast. Potato nuggets were prepared from basic combinations of 50 g potato flakes plus 100 g deionized water that were first mixed into a dough. Appropriate amounts of baker's yeast (Red Star Active Dry Yeast; Milwaukee, WI; wt basis) and other ingredients were then added, mixed, and incubated for 60 min to permit yeast metabolic activity on reducing sugars and asparagine in the dough. Finally, the dough was extruded at ambient temperature using hand-held chef's piping bag equipped with a plastic 10 mm i.d. plastic tip (Hutzler Manufacturing Co., Canaan, CT) into potato nuggets (solid rods, 10 mm diameter X 30 mm length) before frying at *ca* 180°C in food-service partially hydrogenated vegetable oil for 1 min.

2.2 Acrylamide analysis

Samples were analyzed using an adaptation of the US EPA procedure for GC-MS detection of brominated acrylamide (US EPA, 1996; Tareke et al., 2002; see also Lindsay and Jang, 2004 and Park et al., 2004).

3. RESULTS AND DISCUSSION

3.1 Pectinmethylesterase activation (70°C blanch and hold)

During studies with model systems, we found that the introduction of polyanionic substances, including various pectin derivatives, into glucose plus asparagine Maillard reactants substantially reduced acrylamide formation during frying (Lindsay and Jang, 2004). However, the direct introduction of pectic substances to required binding sites in potato pieces proved to be minimally effective in reducing acrylamide in fried potatoes.

On the other hand, firming of the texture of vegetable materials via *in situ* pectic acid formation through low temperature blanching (50 to 75°C) to

Intervention Strategies for Suppression of Acylamide

activate pectinmethylesterase is a familiar technology (*cf.*, Bartolome and Hoff, 1972). When applied to the processing of fried potato products, we found that low temperature blanching provided surprisingly effective inhibition of acrylamide formation (Table 1). We hypothesize that pectic acid-containing polymers, which are situated in or near tuber cell walls, surround the cell contents and capture ionic Maillard reactive substances as they attempt diffusion from the thermally injured cells. A schematic chemical representation portraying the proposed immobilization of asparagine by pectic acid-containing polymers is shown in Figure 3.

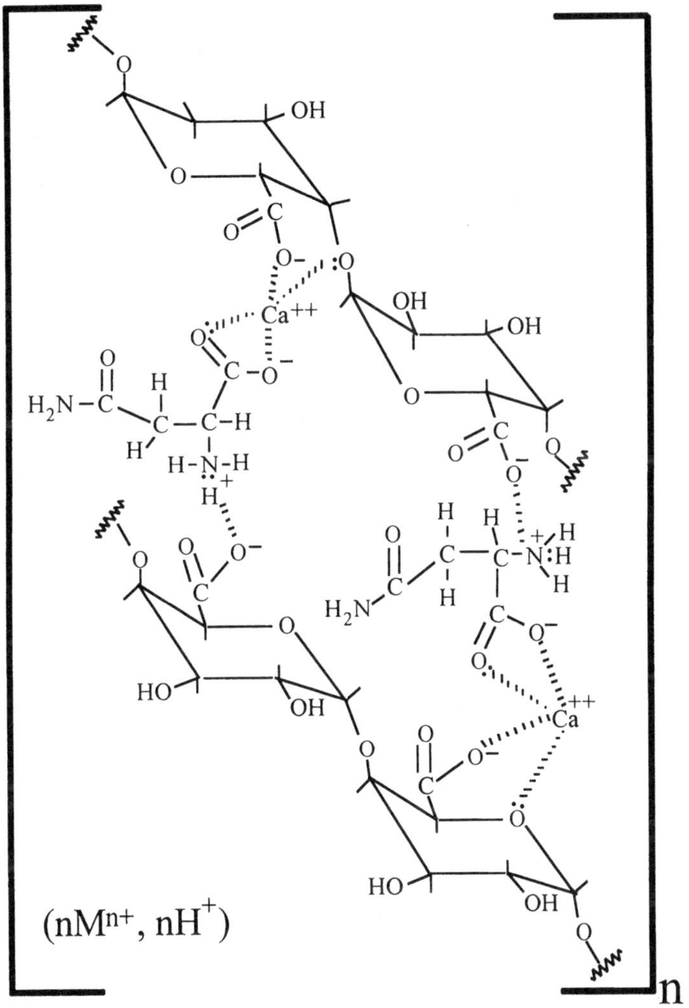

Figure 2. Schematic representation indicating proposed ionic complexations via salt formation, multivalent cation crosslinkings and chelation of asparagine and calcium ions by pectic acid moieties (polygalacturonate segments-αGal4C1) in or near potato tuber cell walls.

Table 1. Use of pectinmethylesterase activation (70°C blanch and 30 min hold) processing for the reduction of acrylamide in potato chips.

Sample Treatment	Acrylamide (µg/kg; ppb)	Reduction
Untreated [a]	661	–
Treated [b]	59	91%

[a] cut, wash, hold in deionized water at 21°C for 30 min
[b] cut, wash, blanch in deionized water at 70°C for 30 min, hold in deionized water at 21°C for 1 min

Water blanching (usually 80°C to 84°C for 1 to 4 min) of potato pieces during processing, especially for French fries, is commonly practiced for whitening of finished products through thermal inactivation of polyphenoloxidase and partial removal of sugars (Talburt and Smith, 1987). However, at these temperatures pectinmethylesterase is rapidly thermally inactivated (Bartolome and Hoff, 1972).

3.2 Hot ethanol (ca 80°C) blanching

In a search for possible alternative processing treatments that might provide reduction in acrylamide potential in processed potato, ethanol-water combinations were investigated as media for reactant (especially, reducing sugars) extractions, cell thermal injury effects, and thermal stabilization for inhibition of gelatinization of starch granules during processing. The results of a short blanch of potato slices in refluxing 80% ethanol combined with an ionic complexation treatment involving phytate and calcium chloride are shown in Table 2, and it can be seen that substantial (89%) reduction in acrylamide was obtained for this treatment.

Table 2. Use of ethanol blanch processing for damaging potato tuber cells followed by phytate plus calcium ion complexation processing for the reduction of acrylamide in potato chips.

Sample Treatment	Acrylamide (µg/kg; ppb)	Reduction
Untreated [a]	587	–
Treated [b]	64	89%

[a] cut, wash, and fry
[b] cut, wash in tap water, blanch at 80°C in 80% ethanol for 1 min, soak in 250 ppm phytate plus 1000 ppm CaCl$_2$ at 50°C for 3 min at pH 3 in tap water, fry

While food processing with ethanol is novel, we found that appropriate sequencing of process steps yielded products without detectable ethanol flavors that were very light in color and crisp in texture. A proposed schematic sequence for a novel ethanol blanching process of potato slices is shown in Figure 3.

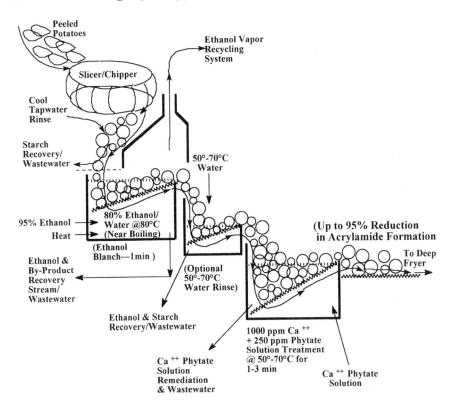

Figure 3. Schematic representation showing a prototype process utilizing 80°C ethanol blanching plus sodium chloride/sodium acid pyrophosphate exposure for suppression of acrylamide formation in potato chips a cap.

3.3 Pectinmethylesterase activation (70°C blanch) combined with chitosan application

Pectinmethylesterase activation technology effectively targets only ionic Maillard reactants (i.e., asparagine, glycosylasparagine, phosphorylated reducing sugars, etc.). Thus, we explored the possibility of using ammoniated polymers (e.g., chitosan) for blocking the carbonyl groups of neutral reducing sugars on cut potato surfaces. We found that chitosan treatments used alone effected modest acrylamide reductions, but they were more effective when used in combination with other acrylamide reduction technologies. Such an application is illustrated in Table 3 where potato slices were first briefly exposed to low temperature blanching (70°C for 1 min), and then were held in a chitosan solution (1000 ppm).

Table 3. Use of pectinmethylesterase activation (70°C blanch) and chitosan surface application (submersion) processing to block reducing sugar sugar carbonyl groups for the reduction of acrylamide in potato chips.

Sample Treatment	Acrylamide (µg/kg; ppb)	Reduction
Untreated [a]	317	–
Treated – blanch in 5% NaCl [b]	159	50%
Treated as above plus chitosan [c]	37	88%

[a] cut, wash, & fry
[b] cut, wash 5% NaCl in dionized water at 70°C for 1 min. hold in deionized water at 21°C for 1 min, fry
[c] blanch as above, plus hold additional 3 min in 1000 ppm chitosan at *ca* 50 °C

Again, substantial acrylamide reduction was observed in the finished chips, and acetic acid used in the solubilization of chitosan did not yield noticeable acid carryover flavors and were exceptionally crisp in texture. We hypothesize that the amino groups on chitosan monomer units react with reducing sugars providing a corresponding immobilization of sugar molecules.

Additional carbonyl-blocking compounds were investigated, and these included cysteine, N-acetylcysteine, and lysine. When used alone in modest concentrations (<1000 ppm in soak solutions), none of these were greatly effective in reducing acrylamide concentrations in finished fried potato products. Incorporation of these compounds into the glucose-asparagine (10 mM each) fried stacked-glass pad model (Lindsay and Jang, 2004) showed that low concentrations (<250 ppm) were ineffective, but very high concentrations were quite effective in reducing acrylamide formation (Table 4). Overall, the results of these trials indicated that these carbonyl blocker compounds entered the Maillard reaction in a competitive manner with asparagine and other amino acids, and thus reduced acrylamide notably when used in relatively very high concentrations.

Table 4. Effect of adding N-acetylcysteine into a solution of glucose and asparagine (10 mM each) upon acrylamide formation in a fried glass pad model system.

Treatment to Glass Pad	Acrylamide (µg/kg; ppb)	Reduction
Untreated [a]	337	–
Treated 250 ppm N-acetylcysteine [b]	383	Increase 14%
Treated 2500 ppm N-acetylcysteine [c]	50	85%

[a] 10 mM each glucose and asparagine at pH 3 with HCl
[b] plus 250 ppm N-acetylcysteine at pH 3.5
[c] plus 2500 ppm N-acetylcysteine at pH 3.5

3.4 Suppression of acrylamide in potato dough by metabolizing baker's yeast

These trials were carried out in a series of experiments to determine the role that yeast fermentations play in reducing acrylamide in finished cooked foods. As can be seen in Table 5, yeast metabolic activity in potato dough substantially reduced the acrylamide content of fried potato nuggets. We speculate that both asparagine degradation for nitrogen sourcing and reducing sugar metabolism contribute to the net reduction of acrylamide.

Table 5. Use of metabolizing yeast in the preparation of potato dough for suppression of acrylamide in deep fried potato nuggets.

Sample Treatment	Acrylamide (µg/kg; ppb)	Reduction
Untreated [a]	409	–
Treated by holding before frying [b]	152	63%
Treated 2500 ppm N-acetylcysteine [c]	127	70%

[a] Base potato dough (50% solids) + 1% baker's yeast (wt basis), not held, fried
[b] as above but held 60 min at 30°C before frying
[c] as above plus 1% pectic acid (wt basis) and held 60 min at 30°C before frying

3.5 Typical commercial (80°C blanch) French fry preparation replaced with 70°C blanch/hold processing

Water blanching (usually 80°C to 84°C for 1 to 4 min) of potatoes strips during processing of French fries substantially reduces acrylamide concentrations compared simply cutting, rinsing, and frying comparable potato strips (Table 6; in this case, 70%). As part of assessments of proposed commercial applications of 70°C blanch/hold technology for fried potato products, this approach was used in the typical commercial processing protocol, and about 95% reduction in acrylamide was realized. Moreover, the edibility characteristics of the resulting French fries were remarkably similar to commercial counterparts. Similar applications to an experimental potato chip processing protocol (Figure 4) also gave comparable notable reductions in acrylamide concentrations (data not shown).

Table 6. Use of 70°C blanch/hold plus sodium chloride/sodium acid pyrophosphate (NaCl/SAPP) processing for suppression of acrylamide in French fried potatoes.

French Fry Treatment	Acrylamide (μg/kg; ppb)	Reduction
Untreated [a]	489	–
Treated - typical commercial [b]	141	71%
Treated - Modified [c]	30	95%

[a] cut and wash, finish fry immediately
[b] cut, wash, blanch 80°C for 4 min; soak in 1.5% NaCl + 0.5% SAPP at 21°C for 3 min; par-fry 190°C for 30 sec; freeze; finish fry 190°C for 1.75 min
[c] cut, wash, blanch 70°C for 30 min; soak in 1.5% NaCl + 0.5% SAPP at 21°C for 3 min; par-fry 190°C for 30 sec; freeze; finish fry 190°C for 1.75 min

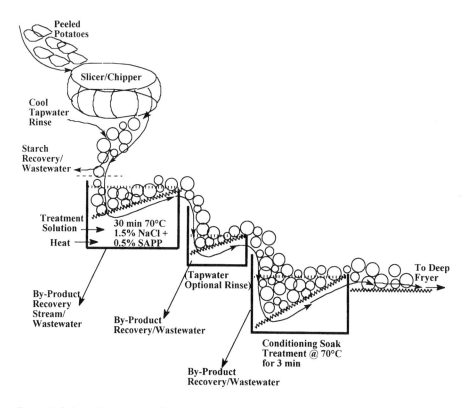

Figure 4. Schematic representation showing a prototype process utilizing 70°C blanching plus sodium chloride/sodium acid pyrophosphate exposure for suppression of acrylamide formation in potato chips.

3.6 Summary

The processing strategies for fried potato products reported herein join a variety of other processing options that have been identified by other researchers for reducing acrylamide in fried potato products. In many instances, laboratory prototype processes are ready for technology transfer to commercial processing operations. However, research is continuing on identifying improved acrylamide suppression technology for a variety of vulnerable foods.

REFERENCES

Bartolome, L. G., and Hoff, J. E., 1972, Firming of potatoes: Biochemical effects of preheating, *J. Agric. Food Chem.* **20**:266-270

Becalski, A., Lau, B. P.-Y., Lewis, D., and Seaman, S., 2003, Acrylamide in foods; occurrence and sources, and modeling, *J. Agric. Food Chem.* **51**:802-808.

Biedermann, M., Biederman, B. S., Noti, A., and Grob, K., 2002, Methods for determining the potential of acrylamide formation and its elimination in raw materials for food preparation, such as potatoes, *Mitt. Lebensm. Hyg.* **93**:638-652.

FDA, 2004, Exploratory Data on Acrylamide in Food; http://www.cfsan.fda.gov/~dms/acrydata.html.

Graf, E., 1983 Application of phytic acid, *J. Am. Oil Chem. Soc.* **60**:1861-1867.

Granvogl, M., Jezussek, M., Koehler, P., and Schieberle, P., 2004, Quantitation of 3-aminopropionamide in potatoes-A minor but potent precursor in acrylamide formation, *J. Agric. Food Chem.* **52**:4751-4757.

Grob, K., Biedermann, M., Biedermann, B. S., Noti, A., Imhof, D., Amrein, T., Pfefferle, A., and Bazzocco, D., 2003, French fries with less than 100ug/kg acrylamide. A colloboration between cooks and analysts, *Eur Food Res Technol.* **217**:185-194.

Jung, M. Y., Choi, D. S., and Ju, J. W., 2003, A novel technique for limitation of acrylamide formation in fried and baked corn chips and in French fries, *J. Food Sci.* **68**:1287-1290.

Lindsay, R.C. and Jang, S., 2005, Model systems for evaluating factors affecting acrylamide formation in deep fried foods, in: *Chemistry and Safety of Acrylamide in Food*, M. Friedman and D.S. Mottram, eds, Springer, New York, pp. 329-341.

Livsmedelsverket (The National Food Administration, Uppsala, Sweden), 2002, April 26, Individual acrylamide results for all tested samples; http://192.71.90.8/engakryltabell.htm

Mottram, D. S., Wedzicha, B. L., and Dodson, A. T., 2002, Acrylamide is formed in the Maillard reaction, *Nature* **419**:448-449.

Park, Y., Yang, H., Storkson, J.M., Albright, K.J., Liu, W., Lindsay, R.C., and Pariza, M.W., 2004, Controlling acrylamide in French fry and potato chip models and a mathematical model of acrylamide formation, in: *Chemistry and Safety of Acrylamide in Food*, M. Friedman and D.S. Mottram, eds, Springer, New York, pp. 343-356.

Rydberg et al., 2003, Investigations of factors that influence the acrylamide content of heated foodstuffs, *J. Agric. Food Chem.* **51**:7012-7018.

Stadler, R. H., Blank, I.., Varga, N., Robert, F., Hau, J., Guy P. A., Robert, M. C., and Riediker, S., 2002, Acrylamide from Maillard reaction products, *Nature* **419**:449-450.

Talburt, W. F., and Smith, O., 1987, *Potato Processing*, 4th ed., AVI Van Nostrand Reinhold, NY.

Tareke, E., Rydberg, P., Karlsson, P., Eriksson, S., and Tornqvist, M., 2002, Analysis of acrylamide, a carcinogen formed in heated foodstuffs, *J. Agric. Food Chem.* **50**:4998-5006.

US EPA, 1996, *SW 846, Test methods for evaluating solid waste, Acrylamide by gas chromatography, Method 8032,*: U.S. Environmental Protection Agency, Washington, DC.

Yasuhara, A., Tanaka, Y., Hengel, M., and Shibamoto, T., 2003, Gas chromatographic investigation of acrylamide formation in browning model systems, *J. Agric. Food Chem.* **51**:3999-4003.

Yaylayan, V. A., Wnorowksi, A., and Locas, C. P., 2003, Why asparagine needs carbohydrate to generate acrylamide, *J. Agric. Food Chem.* **51**:1753-1757.

Yaylayan, V. A., Locas, C. P., Wnorowski, A., and O'Brien, John., 2004, The role of creatine in the generation of N-Methylacrylamide: A new toxicant in cooked meat, *J. Agric. Food Chem.* **52**:5559-5565.

Zyzak, D. V., Sanders, R. A., Stojanovic, M., Talmadge, D. H., Eberhart, B.L., Ewald, D. K., Gruber, D. C., Morsch, T. R., Strothers, M. A., Rizzi, G. P., and Villagran, M. D., 2003, Acrylamide formation mechanism in heated foods, *J. Agric. Food Chem.* **51**:4782-4787.

ACRYLAMIDE IN JAPANESE PROCESSED FOODS AND FACTORS AFFECTING ACRYLAMIDE LEVEL IN POTATO CHIPS AND TEA

Mitsuru Yoshida[1], Hiroshi Ono[1], Yoshihiro Chuda[2], Hiroshi Yada[1], Mayumi Ohnishi-Kameyama[1], Hidetaka Kobayashi[1], Akiko Ohara-Takada[3], Chie Matsuura-Endo[3], Motoyuki Mori[3], Nobuyuki Hayashi[4], and Yuichi Yamaguchi[4]

[1]*National Food Research Institute, 2-1-12 Kannondai, Tsukuba, Ibaraki 305-8642, Japan*
[2]*Center for Food Quality, Labeling and Consumer Services, 2-1 Shintoshin, Chuo-ku, Saitama 330-9731, Japan;* [3]*National Agricultural Research Center for Hokkaido Region, National Agriculture and Bio-oriented Research Organization, Shinsei, Memuro-cho, Hokkaido 082-0071;* [4] *National Institute of Vegetable and Tea Science, National Agriculture and Bio-oriented Research Organization, 2769 Kanaya, Kanaya, Shizuoka 428-8501, Japan; e-mail: mitsuru@nfri.affrc.go.jp*

Abstract: Acrylamide concentrations in processed foods sold in Japanese markets were analyzed by LC-MS/MS and GC-MS methods. Most potato chips and whole potato-based fried snacks showed acrylamide concentration higher than 1000 μg/kg. The concentrations in non-whole potato based Japanese snacks, including rice crackers and candied sweet potatoes, were less than 350 μg/kg. Those in instant precooked noodles were less than 100 μg/kg with only one exception. The effect of storage condition of potato tubers on acrylamide concentration in potato chips after frying was also investigated. Sugar content in the tubers increased during cold storage, and the acrylamide concentration increased accordingly. The concentrations of asparagine and other amino acids, however, did not change during the cold storage. High correlations were observed between the acrylamide content in the chips and glucose and fructose contents in the tubers. This fact indicated that the limiting factor for acrylamide formation in potato chips is reducing sugar, not asparagine content in the tubers. Effects of roasting time and temperature on acrylamide concentration in roasted green tea are also described.

Key words: Acrylamide; LC-MS/MS; GC-MS, Japanese food; potato; cold storage; glucose; fructose; asparagine; roasting condition; tea

1. ACRYLAMIDE ANALYSIS OF PROCESSED FOODS IN JAPANESE MARKET

After the press release on the presence of acrylamide in common processed foods by University of Stockholm and National Food Administration of Sweden in April 2002, we started analysis of acrylamide in processed foods in Japanese market (National Food Research Institute, 2002; Ono et al., 2003).

Commercial processed foods were purchased at supermarkets in Tsukuba, Japan from 31 May to 5 June 2002. Our analytical sample preparation procedure is shown in Fig. 1. Food samples to be analyzed were pulverized by a food processor. After addition of the internal standard acrylamide-d_3 solution, the sample was shaken, and then water was added. The mixture was homogenized and centrifuged. The supernatant was subjected to a mixed-mode SPE cartridge, which had been conditioned with methanol and water, and eluted with water.

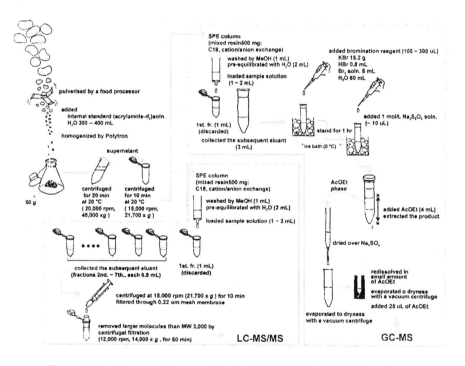

Figure 1. Sample preparation procedure for LC-MS/MS and GC-MS analyses.

For LC-MS/MS analysis, an aliquot of the eluate was centrifuged and passed through a syringe filter with a hydrophilic membrane. The filtrate was ultrafiltered and kept at -20°C under dark until LC-MS/MS analysis on an API 3000 triple quadrupole instrument equipped with a Turbo Ion Spray source (Applied Biosystems) coupled to a Nanospace SI-2 HPLC system (Shiseido). The sample (2 µl) was injected into the HPLC with an Atlantis dC18 column (2.1×150 mm, 3 µm, Waters), a C18 column for polar molecules, and eluted by 10 % methanol in water isocratically at a flow rate of 0.1 ml/min at 40°C. The eluate from 4.8 min to 10.2 min after the sample injection was introduced to the mass spectrometer operating in the positive-ion mode. Acrylamide was eluted at 5.4 min, and detected as a transition ion at m/z 72>55 by selected reaction monitoring. The internal standard was detected at m/z 75>58.

For GC-MS analysis, the SPE eluate was mixed with a bromination reagent (Castle et al., 1991) in an ice bath until the reddish color of bromine remained. The reaction mixture was stood for 1 hr, and then an aqueous sodium thiosulfate solution was added to destroy excess bromine until the solution became colorless. The brominated acrylamide was extracted with ethyl acetate and dried over anhydrous sodium sulfate. After most of the solvent was removed by centrifugal evaporation under reduced pressure at 30°C, the residue was transferred to a micro grass tube to be kept at 4°C till GC-MS analysis. The sample (1 µl) was introduced into the GC-MS-QP2010 system (Shimadzu) with a CP-Sil 24 CB Lowbleed/MS column (50% phenyl, 50% dimethylpolysiloxane (equivalent to OV-17), 0.25 mm i.d. ×30 m, 0.25 µm film thickness, Varian) by splitless injection method. Injector temperature was 120°C, and the temperature program used was as follows: isothermal for 1 min at 85°C, increased at a rate of 25°C/min to 175°C, isothermal for 6 min, increased at a rate of 40°C/min to 250°C, and isothermal for 7.52 min. Fragment ion peaks of 2,3-dibromopropionamide derived from acrylamide at m/z 150 $[C_3H_5NO^{79}Br]^+$ and 152 $[C_3H_5NO^{81}Br]^+$, and those from acrylamide-d_3 at m/z 153 $[C_3H_2D_3NO^{79}Br]^+$ and 155 $[C_3H_2D_3NO^{81}Br]^+$ were detected by selected ion monitoring.

The limit of detection and limit of quantification of acrylamide were 0.2 ng/ml (6 fmol) and 0.8 ng/ml (22 fmol), respectively, on the LC-MS/MS, and those of 2,3-dibromopropionamide derived from acrylamide were 12 ng/ml (52 fmol) and 40 ng/ml (170 fmol), respectively, on the GC-MS. Repeatability given as RSD was <5% and <15% for the LC-MS/MS and the GC-MS methods, respectively. High correlation (r^2=0.946, n=74) was observed between values obtained by the two methods.

Fig. 2 shows the results of our analysis of Japanese processed foods (National Food Research Institute, 2002; Ono et al., 2003). The ranges of acrylamide concentration in potato chips, cookies, and breakfast cereals observed in this survey were within those reported from Western countries in the FAO/WHO consultation meeting (FAO/WHO, 2002). Most potato chips and whole potato based fried snacks showed acrylamide concentration

>1000 μg/kg. Rice crackers, which are processed by grilling at 200-300°C for several minutes or frying at 160-260°C for about a few minutes, contained 20 to 300 μg/kg acrylamide. Non-whole potato based snacks, made from potato starch, wheat flour, or corn whole meal, shaped, fried, and seasoned, contained similar level of acrylamide as did rice crackers'. Candied sweet potatoes, which are cooked by frying at around 160°C for several minutes followed by syrup coating, also showed lower acrylamide concentrations than did potato chips. Instant noodles and won-tons contained about 10 to 60 μg/kg acrylamide, with one exception with a concentration >500 μg/kg. Roasted barley grains used as a tea substitute or herb tea blend contained 200-600 μg/kg acrylamide, comparable to the concentration in coffee powder (FAO/WHO, 2002). Other Japanese and Asian type foods measured here contained around 100 μg/kg acrylamide or less.

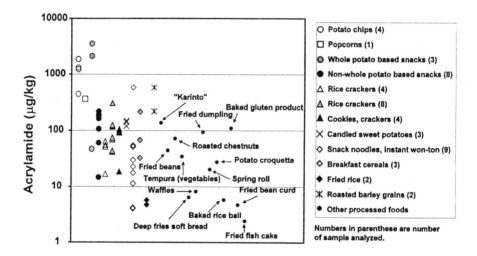

Figure 2. Acrylamide concentration in processed foods in Japanese market.

2. EFFECT OF STORAGE CONDITIONS OF POTATO TUBERS ON ACRYLAMIDE FORMATION DURING FRYING

A large variation has been found in acrylamide concentration in potato based foods. This cannot be accounted for only by difference of processing conditions. It is known that sugar content in potato tuber increases during cold storage by starch degradation, which is known as "low temperature sweetening." This sugar level change during the storage should affect acrylamide formation during cooking because reaction of reducing sugars with asparagine is thought to be a major rout for acrylamide formation in starchy foods. We then investigated the change of sugar and amino acid contents during storage of potato tubers and its effect on acrylamide concentration after frying (Chuda et al., 2003).

Potato tubers (cultivar Toyoshiro) with similar weight and starch values were selected for the storage experiment. Toyoshiro is the most popular cultivar for potato chips in Japan. Three tubers stored under each condition were cut in half longitudinally, with a half used for chipping and the other for amino acid and sugar analyses by HPLC. For chipping, 4 slices 1.3 mm thick were prepared from each of 3 tubers (total 12 slices). Slices were washed for 100 sec in water and fried at 180°C for 90 sec in cottonseed oil. The acrylamide content of the chips was determined by the GC-MS method.

The concentrations of asparagine and other amino acids did not change during the cold storage. On the other hand, sugars increased during cold storage (Fig. 3). Acrylamide concentration after frying increased accordingly (Fig. 4). The acrylamide content in chips showed a high correlation with glucose content in the tubers ($r^2=0.884$, n=20). A similar correlation was observed for fructose ($r^2=0.884$, n=20). No correlation was observed with asparagine content ($r^2=0.056$, n=20). This fact indicated that the limiting factor for acrylamide formation in fried potato-based foods is the reducing sugar and not the asparagine content in the tubers.

We investigated the sugar contents of tubers stored at various temperatures for 2 weeks. When they were stored at 6°C or below, the sugar content was higher than when stored at 8°C or above. Acrylamide concentration in the chips also increased when the tubers were stored at 6°C or below. Therefore, refrigeration of potatoes for chips and French fries at 6°C or below should be avoided to suppress the acrylamide formation. Avoiding the cold storage also suppresses browning during frying, probably by inhibition of accumulation of reducing sugars in the tubers.

There should be cultivar difference in sugar content and sugar accumulation pattern during the cold storage. Thus we are now investigating difference in accumulation of acrylamide by frying into potato chips among cultivars and breeding lines for selection of material with low potential of acrylamide formation even after storage.

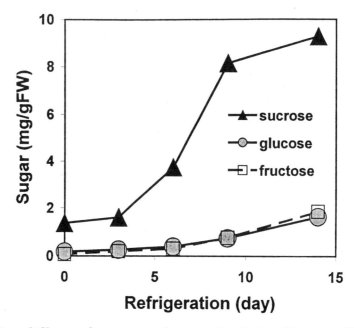

Figure 3. Changes of sugar contents in potato tubers during refrigeration at 2°C.

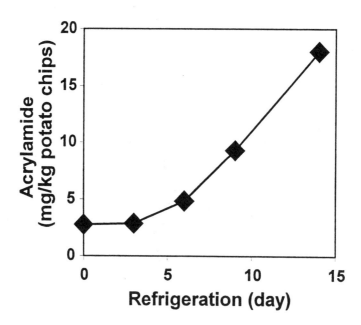

Figure 4. Effect of refrigeration period of tubers on acrylamide content in potato chips.

3. ACRYLAMIDE ANALYSIS OF ROASTED GREEN TEA LEAVES

Various types of tea (Fig. 5) are consumed in Asian countries. In Japan, non-fermented green teas are popular, and among green teas, Sencha is consumed most. Hojicha, roasted green tea, is often served in Japanese restaurants. Recently Chinese type semi-fermented Oolong tea has become popular. Bottled Oolong tea is now sold in stores.

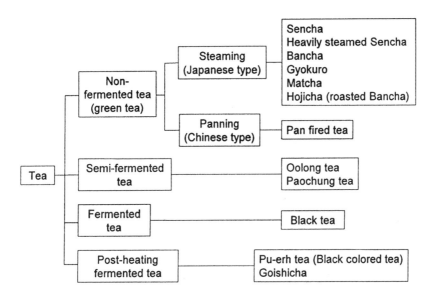

Figure 5. Types of tea.

Results of analyses of tea and tea substitutes reported by the Ministry of Agriculture, Forestry and Fisheries of Japan (2003) (Table 1) show that the acrylamide level in green tea is not as high as in roasted products such as Houjicha and roasted cereal grains used as tea substitutes or in herb teas. Based on the analytical data, it appears that steeping in boiling or hot water resulted in extraction of most of the acrylamide in the infusions. We also investigated the effect of roasting conditions on acrylamide accumulation in the tea products. In the course of the analysis, it was found that epigallocatechin in the tea samples inhibited the formation of the brominated derivative. We then fractionated the SPE eluate and used only a 1.5 ml portion after discarding 1 ml early eluate used for bromination to avoid contamination of epigallocatechin eluted in the later fractions.

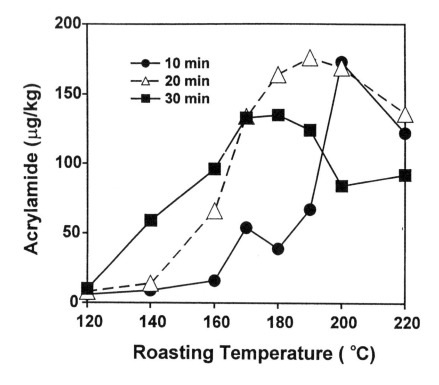

Figure 6. Effect of roasting temperature on acrylamide content in tealeaves.

Green tea samples prepared by the conventional method were roasted in a laboratory oven for 10, 20, or 30 min, and the acrylamide content measured by the GC-MS method was plotted against roasting temperature (Fig. 6). For each roasting time, a maximum of acrylamide content was observed. The maximum shifted at lower temperatures with increased roasting times. In this graph, the maximum acrylamide concentration is about 180 µg/kg, and this seems to be the maximum potential of acrylamide accumulation of this tea sample. But some roasted green tea samples showed values above 500 µg/kg (Table 1). Therefore, not only roasting time and temperature but also components of tealeaves should affect the acrylamide level in tea products. Effect of the leaf components on the acrylamide formation is the subject to be studied next.

Table 1. Acrylamide in tea* (The Ministry of Agriculture, Forestry and Fisheries of Japan, 2003)

	Sample No.	Acrylamide (ng/g) Product	Acrylamide (ng/g) Infusion	Condition of infusion
Green tea	1	nd	nd	10 g / 90°C 430 ml
(Sencha)	2	20	nd	1 min
	3	30	nd	
Green tea	1	70	nd	10 g / 90°C 430 ml
(Pan fired tea)	2	50	nd	1 min
	3	100	3	
Hojicha	1	520	10	15 g / 90°C 650 ml
(roasted green tea)	2	260	4	0.5 min
	3	570	11	
	4	190	4	
Black tea	1	20	nd	5 g / boiling water 360 ml
	2	20	nd	1.5 - 4 min
	3	nd	nd	
Oolong tea	1	60	nd	15 g / 90°C 650 ml
	2	90	2	0.5 min
Roasted barley grains	1	320	14	50 g / hot water 1500 ml
(Mugicha)	2	290	11	5 min standing after boil up
	3	180	5	
Roasted pearl barley	1	130	nd	30 g / hot water 1000 ml
grains (Hatomugicha)	2	130	nd	15 min boiling,
	3	120	7	5 min standing

*Limit of detection: 20 ng/g (product), 2 ng/g (infusion).

REFERENCES

Castle, L., Campos, M., and Gilbert, J., 1991, Determination of acrylamide monomer in hydroponically grown tomato fruit by capillary gas chromatography-mass spectrometry, *J. Sci. Food Agric.* **54**:549-555.

Chuda, Y., Ono, H., Yada, H., Ohara-Takada, A., Matsuura-Endo, C., and Mori, M., 2003, Effects of physiological changes in potato tubers (*Solanum tuberosum* L.) after low temperature storage on the level of acrylamide formed in potato chips, *Biosci. Biotechnol. Biochem.* **67**:1188-1190.

FAO/WHO, 2002, Health implications of acrylamide in food. Report of a Joint FAO/WHO Consultation, World Health Organization, Geneva, pp. 8-9.

The Ministry of Agriculture, Forestry and Fisheries of Japan, 2003, Press release, Result of analysis of acrylamide in tea (June 27, 2003); http://www.maff.go.jp/www/press/cont/20030627press_7b.pdf (in Japanese).

National Food Research Institute, 2002, Analysis of acrylamide in processed foods in Japan; http://aa.iacfc.affrc.go.jp/en/.

Ono, H., Chuda, Y., Ohnishi-Kameyama, M., Yada, H., Ishizaka, M., Kobayashi H., and Yoshida, M., 2003, Analysis of acrylamide by LC-MS/MS and GC-MS in processed Japanese foods, *Food Addit. Contam.* **20**:215-220.

THE FORMATION OF ACRYLAMIDE IN UK CEREAL PRODUCTS

Peter Sadd and Colin Hamlet
RHM Technology Ltd, Lincoln Road, High Wycombe, Bucks, HP12 3QR, UK; e-mail: pasadd@rhmtech.co.uk

Abstract: Many bakery products sold in the UK such as crumpets, batch bread and Naan might be expected to show high levels of acrylamide because they have strong Maillard colours and flavours. However, analysis of commercial products has shown that the highest levels of acrylamide are seen in dry biscuit type products. With the exception of spiced products such as ginger cake, moist high sugar products (e.g. cakes and fruit loaves) show relatively low levels of acrylamide, even in darkly browned crusts. This is in contrast to bread where acrylamide levels in excess of 100 µg/kg are common in the crust region, but are diluted by low levels in the crumb. Acrylamide levels in bread are significantly raised by domestic toasting, but other products such as crumpets and Naan bread have been found to be less sensitive. A mathematical model has been developed (and validated against tests on model dough) which shows that once obvious recipe differences are allowed for, the key factor limiting acrylamide levels is crust moisture. Chemical decay of acrylamide and depletion of amino acids are also limiting factors at higher temperatures.

Key words: Acrylamide, asparagine, biscuits, bread, cake, crust, decay, dough, formation, mathematical model, morning goods

1. INTRODUCTION

Once acrylamide was identified as an issue in foods, potato products rapidly emerged as the product group with the highest levels, and subsequent research has tended to focus on these rather than cereal based products. While some data on acrylamide has been reported for breads (e.g. Ahn et al., 2002; Konings et al., 2003; Surdyk et al., 2004; Svensson at al, 2003; Tareke at al., 2003), bread recipes and processes are very diverse, so results from different workers vary widely and can be misleading.

UK bakery products can be divided into four broad sectors: bread, cake, biscuits, and morning goods (croissants, crumpets, muffins, rolls, pancakes, scones, waffles etc.). Of these, only bread has received significant attention, largely because it is a dietary staple. However the other three market sectors are each comparable in sales value to bread, and so there was a need to assess acrylamide levels in products in all four sectors. The levels of acrylamide in many UK specific bakery goods had never been measured at all, so the UK Food Standards Agency funded work the work presented here on UK cereal products.

This chapter begins with results from a scoping survey of acrylamide and precursor levels in commercial cereal products. Then a mathematical model of acrylamide formation is developed, validated against data from controlled cooking experiments with model doughs, and used to understand the variation in acrylamide levels seen in commercial products.

2. METHODS

2.1 Food samples, materials and reagents

Food samples were obtained from UK retail outlets (supermarkets). Flour and bakery ingredients were obtained from commercial suppliers. Acrylamide and amino acid reference standards were obtained from Sigma (Poole, UK). Acrylamide-1,2,3-$^{13}C_3$ was obtained from Cambridge Isotopes (Andover, MA, USA).

2.2 Preparation of experimental dough samples

Model bread dough comprising white flour + salt + water (1000:20:600 w/w) was vacuum mixed on a 1 kg (flour) scale by the Chorleywood baking process (Chamberlain et al., 1962). Low moisture dough samples were obtained by freeze-drying over 16-48 h in a Supermodulyo 12K freeze dryer, (Edwards, Crawley, UK). These samples were reduced to a fine homogeneous powder using a BL 300 domestic blender (Kenwood, Havant, UK).

2.3 Controlled cooking experiments

Dough samples were cooked using a custom built pressure-cooking apparatus used to simulate the crust region of baked cereal products (Hamlet et al., 2002, 2004a, 2004b). Briefly, the dough samples were contained in stainless steel HPLC tubes of 75 mm length with an internal diameter (I.D.) of 7.75 mm (Hichrom Ltd, Reading, UK) and secured with stainless steel end caps modified to accept 1.6 mm O.D. calibrated type K thermocouple probes (Labfacility, Bognor Regis, UK). The samples were cooked for 20 minutes in a Mega series (Carlo Erba, Milan, Italy) GC oven equilibrated to a preset temperature. The temperature from each of two tube/dough thermocouples together with a further oven air thermocouple (certified reference, type AK28M) was recorded at 4 s intervals using a Squirrel 1200 series data logger (Grant Instruments Ltd, Cambridge, UK).

2.4 Analysis

Amino acid analysis. Free amino acids were analyzed by HPLC with fluorescence detection following aqueous extraction and automated on-line derivatization with o-phthalaldehyde (Henderson et al., 2003).

Acrylamide. Samples (5 g) were extracted into deionised water (50 ml). Centrifuged and clarified (Carrez) extracts were analyzed by GC/MS following bromination (Ahn et al., 2002) and/or LC/MS after cleanup with multi-mode solid phase extraction cartridges (Tareke et al., 2002). Acrylamide was quantified by a stable isotope internal standard method using acrylamide-1,2,3-$^{13}C_3$ to measure acrylamide generated by the cooking experiments.

Moisture. The moisture content of samples was determined gravimetrically following heating overnight at 105°C.

3. LEVELS OF ACRYLAMIDE IN COMMERCIAL BAKERY GOODS

The aim of the scoping survey presented here was to identify where the higher levels of acrylamide were to be found in bakery products, particularly products which had received less attention than bread because of their lower sales volumes.

It is now well established that acrylamide is a side product of the Maillard reaction, in particular, the reaction of asparagine with reducing sugars (Mottram, et al., 2002, Stadler et al., 2002). Hence high acrylamide levels would be expected to be present in products with added sugar, fruit, malted grains or heavy crust colouration. In addition, ginger has been suggested as an acrylamide promoter (Konings et al., 2003) and products with very thick crusts or little crumb to dilute the acrylamide formed in their crust would also be expected show high levels. Accordingly, the products chosen were deliberately biased towards those which met some or all of these criteria.

3.1 White bread

The levels of acrylamide found in commercial white breads are shown in Table 1. Most products showed low levels, typically 20 µg/kg; however unusual oven profiles which gave dry and highly coloured crusts produced levels as high as 91 µg/kg.

Table 1. Measured acrylamide levels in white bread (in µg/kg, as consumed)

Product	As is	Crumb	Crust
White loaf 800g	15-91		
Scottish batch bread 800g		6.4	127.7, 155.6, 174.7
Organic loaf (in store bakery)		6.5	7.7
French rolls (in store bakery)	8.8		
Ready to bake baguettes	<10 (<10 when baked)		
White pitta bread	7.4		
Oregano focaccia	8.4		
Pizza bases 300g	10.0		
Tandoori Naan bread	30.8 (40.8 after toasting)		

Levels in bread crumb were consistently low, as would be expected from the low cooking temperatures involved.

The high (up to 175 µg/kg) levels seen in the crust of Scottish batch bread are consistent with the dark thick crust distinctive of this product. The variation in crust acrylamide level from sample to sample is not surprising given the natural variation in crust colour from loaf to loaf.

Flat breads such as pizza bases and Naan had surprisingly low levels of acrylamide, despite their larger proportion of crust to crumb. This may be because only a small proportion of the crust surface is actually strongly coloured during baking.

3.2 Non-white bread

The pattern for non-white bread broadly followed that of white (Table 2). The one exception was the multigrain loaf, where added malted grains raised the free sugar and amino acid levels, so that significant levels of acrylamide were formed even in the crumb.

Table 2. Measured acrylamide levels in non-white bread (in µg/kg, as consumed)

Product	As is	Crumb	Crust
Brown loaf 800g		6.1	15.5
Extra wheatgerm loaf 800g		7.7	99.9
In store multigrain loaf		36.3	142.6
Wholemeal loaf 800g		6.0	84.1
Wholemeal pitta bread	52.5		

3.3 Morning goods

Many of the products in this sector have high levels of added sugar in their recipes, and some (like the Scotch pancakes and crumpets) have dark crusts from being baked on hotplates. In practice, these factors do not seem to lead to high acrylamide levels (Table 3). One possibility is that this is because the cook times are relatively short. However these products also have quite moist surface crusts, and this will inhibit acrylamide formation too (see Fig. 4).

Table 3. Measured acrylamide levels in morning goods (in µg/kg, as consumed)

Product	As is	After toasting
In store bakery doughnut rings	8.4	
Raisin loaf 400g	7.8	
Sliced fruit loaf 500g	15.1	
Scotch pancakes	41.6	
Croissants (from in store bakery)	29.2	
Crumpets	10.1	11.5

3.4 Cake

All the products selected for this section had well coloured dark brown crusts, but only the ginger cake showed acrylamide levels above those seen in bread. Interestingly, the crumb actually showed more acrylamide than the crust, suggesting that an earlier stage in manufacture than baking was key in forming acrylamide or its precursors.

Table 4. Measured acrylamide levels in cake (in µg/kg, as consumed)

Product	As is	Crumb	Crust
Ginger cake		69.9	54.7
All butter Madeira cake		8.3	10.7
Farmhouse fruit cake 300g	11.8		
Blueberry muffin (from in store bakery)	13.8		
Chocolate brownies	22.8		

The generally low levels of acrylamide might reflect the commercial practice of using lower oven temperatures in cake baking to avoid burning the crusts (cake crust colours much more readily than bread crust because of its high sugar content).

Another possibility was that the heat treatment used to make commercial cake flours in the UK was removing asparagine. To check this, the amino acid levels in flours subjected to the full range of commercial heat treatments were measured. The seven flours with no or minimal heat treatment actually had lower levels of asparagine than the eight which had received moderate to high treatment levels, though the difference was not significant (61.9 v. 73.1 mg/kg respectively). A similar effect was seen for the sum total of sixteen free amino acids. Again the less treated flours showed lower levels (455.8 *versus*. 530.8 mg/kg respectively).

However the asparagine levels found in the cake flours were typically less than a third of those seen in bread flour (~235 mg/kg). This is because cake flours contain a lower proportion of the outer parts of the grain, and hence lower protein, and this is undoubtedly a significant factor in the low levels of acrylamide seen in cakes.

3.5 Biscuits

Acrylamide levels in biscuits cover a very wide range (Table 5). However the results shown here agree well with the limited literature data for biscuits: Leung et al. (2003) reported 12 µg/kg (shortbread), 110-340 µg/kg (digestives), and a maximum level of 1100 µg/kg (unspecified).

Table 5. Measured acrylamide levels in biscuits (in µg/kg, as consumed)

Product	As is	Type of raising agent used
Organic shortbread fingers (retailer own label)	13	Ammonium
Pure butter stem ginger shortbread	49	Sodium
Ginger thins	79	Sodium
Bourbon (retailer own label)	55	Mixed
Baby rusks	63	Mixed
Gingerbread (from in store bakery)	124	Mixed
Amaretti 250g	303	Ammonium
Rich tea (retailer own label)	371	Mixed
Rich tea	463	Mixed
Digestive (retailer own label)	707, 733	Mixed
Digestive (brand 1)	177	Sodium
Digestive (brand 2)	393	Mixed
Extra wheatgerm crackers	317	Mixed
Black treacle biscuits	810	Mixed
Ginger nuts	1646	Mixed

Most of the biscuits were declared as using a mix of sodium and ammonium based raising agents, but three were entirely sodium based and two used only ammonium salts.

The biscuits in the first half of the table are typically pale (e.g. shortbread), or obtain their colour from coloured ingredients such as chocolate (e.g. bourbon). All the biscuits where significant colour was generated during baking had high levels of acrylamide.

In the case of rich tea biscuits, darker surface colour and higher acrylamide levels went hand in hand. However, for the digestive samples, the darkest biscuit (brand 1) actually had the lowest acrylamide content, so colour is not always a good indicator of acrylamide level. It is possible that this is evidence of acrylamide loss or decay. However brand 1 was the only digestive biscuit to use sodium rather than ammonium based raising agents and this may account for the difference as ammonium has been reported to promote acrylamide formation (Biedermann and Grob, 2003). The three sodium based biscuits all had acrylamide levels towards the lower end of the range, but it is not possible to be certain about this effect without detailed process and recipe information about the different biscuits. All that can be concluded at this stage is that neither ginger nor ammonium are necessarily fatal ingredients if present in a commercial product, but they can be associated with higher acrylamide levels.

4. MODELLING ACRYLAMIDE FORMATION

It is now accepted that acrylamide (Acr) is primarily formed from asparagine (Asn) reacting with a dicarbonyl intermediate (I), which itself is formed from a reaction between amino acids (AA) and reducing sugars (S). Given that all the reactants and acrylamide itself are subject to loss due to competing reactions, the simplest possible reaction scheme for the formation of acrylamide needs six chemical reactions:

$$AA \xrightarrow{k_1} X_1 \tag{1}$$

$$Asn \xrightarrow{k_2} X_2 \tag{2}$$

$$AA + S \xrightarrow{k_3} I \tag{3}$$

$$AA + I \xrightarrow{k_4} X_3 \tag{4}$$

$$Asn + I \xrightarrow{k_4} Acr \tag{5}$$

$$Acr \xrightarrow{k_5} X_4 \tag{6}$$

where X_1 to X_4 are by-products which are of no interest here.

The levels of seventeen amino acids (alanine, arginine, asparagine, aspartate, cystine, glutamate, glutamine, glycine, isoleucine, lysine, methionine, phenylalanine, serine, threonine, tryptophan, tyrosine and valine) were measured in cooked model doughs over a range of temperatures and moistures, and the data used to determine k_1 and k_2 (by fitting first order rate equations to the data).

Fig. 1 shows the effect of dough moisture on these two rate constants, and Figs. 2 and 3 compare measured and predicted amino acid levels. In the case of asparagine, the Arrhenius constant was 3.2×10^9 (1-7.0 mc^3) 1/s (where mc is the fractional moisture content) with an activation energy of 100 kJ/mol. For the total amino acid levels, the Arrhenius constant 4.0×10^7 (1-1.89 mc) 1/s and the activation energy was 85 kJ/mol.

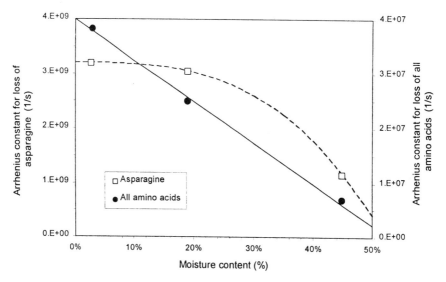

Figure 1. Effect of dough moisture content on the Arrhenius constants for the loss of just asparagine and for all amino acids together (i.e. for k_2 and k_1 respectively).

The rate of amino acid loss was faster at lower moistures, but not as fast as might have been expected given the increase in concentration of the reactants. For asparagine, the rate constants at 2.7 and 19 % moisture were little different.

Some evidence was seen of free amino acids being released by protein breakdown at 45 % moisture and temperatures above 170°C, but the effect was small.

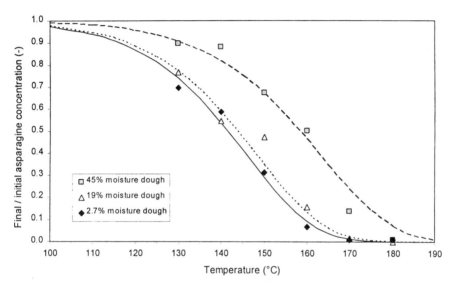

Figure 2. Measured loss of free asparagine in flour salt water doughs after 20 minutes heating at different temperatures and moistures, and model fit.

The proportion of asparagine in the residual amino acids after cooking was the same irrespective of moisture, but strongly temperature dependent. At temperatures up to 150°C, asparagine was lost at the same rate as the other amino acids, but above this temperature it became a steadily smaller fraction of the amino acid population: doughs cooked at 130-150°C contained ~19% of their amino acids as asparagine, while those cooked at 170-180°C had only ~3%. This drop in asparagine availability helps to limit acrylamide formation at the higher temperatures.

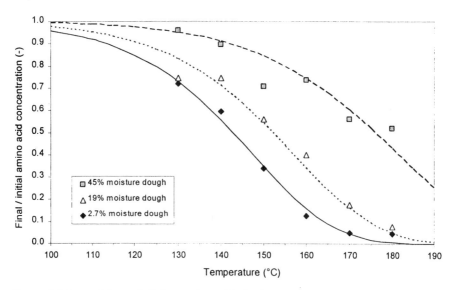

Figure 3. Measured loss of all free amino acids in flour salt water doughs after 20 minutes heating at different temperatures and moistures, and model fit.

The rate constant for decay of acrylamide (k_5) was obtained from data given in Biedermann and Grob (2003). They added C_{13} labeled acrylamide to flour and measured the fraction of acrylamide eliminated after 30 minutes heating. Assuming first order kinetics and fitting this data gave an Arrhenius constant of A=1.42 1/s and an activation energy of 30.3 kJ/mol. The later figure was lower than would be expected on physical grounds, but was kept in the absence of better data.

In theory, this left two rate constants, k_3 and k_4 to be determined by fitting to experimental data. However, when the model was fitted to data on acrylamide formation in model doughs, it turned out that k_4 was essentially infinite and added nothing to the fit. This left k_3 as the only fitted parameter required, and Eqs. (3)-(5) could be condensed to just

$$AA + S \xrightarrow{k_3 \frac{Asn}{AA}} Acr. \qquad (7)$$

Equation (7) assumes that as soon as any intermediate is formed it immediately reacts with the nearest amino acid; hence the rate of acrylamide formation is proportional to the asparagine content of the amino acid population.

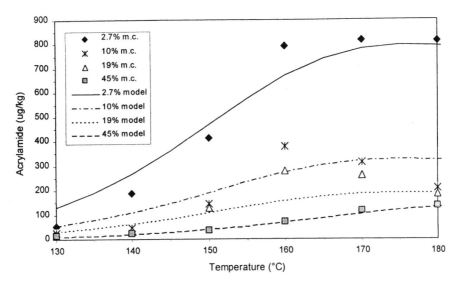

Figure 4. Comparison of predicted acrylamide levels with those measured in 20 minute cooked doughs over 2.7-45% moisture and 130-180°C.

To provide the data for k_3, 2.7, 10, 19 and 45 % moisture doughs were cooked for 20 minutes and the resulting acrylamide levels measured. Fig. 4 shows the experimental data and the model fit. Agreement is reasonable given the complexity of the underlying chemistry that the model is attempting to represent.

In order to minimize the effect of moisture on k_3, concentrations were expressed in terms of kg of reactant per m^3 of moisture in the dough. This gave an Arrhenius constant for k_3 of 2.4×10^7 ($mc^{0.3}$) m^3/kg/s (where mc is the fractional moisture content) with an activation energy of 120 kJ/mol.

It is clear from the experimental data that while low moisture promotes acrylamide formation, it is not essential. In addition, the decline in acrylamide levels seen here at the higher temperatures must be due to chemical rather than physical loss as all the experiments were conducted in a sealed system. Of course in real food systems, evaporative loss of acrylamide may be important as well.

The decay of acrylamide once it has formed is clearly important, but exhaustion of the asparagine supply also limits the amount of acrylamide which can form. The model shows that acrylamide production would still tail off at higher temperatures, even in the absence of any decay (Fig. 5).

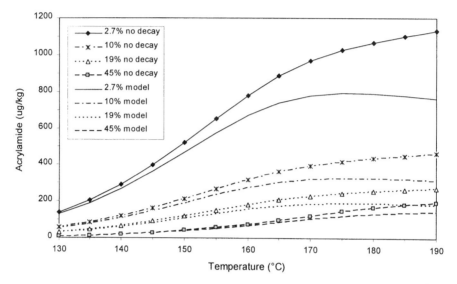

Figure 5. The effect of decay on predicted acrylamide levels for different dough moisture contents.

The most interesting finding from the model, however, is that contours of constant acrylamide are dominated by moisture, whereas contours of equal Maillard browning follow a quite different pattern. This is illustrated in Figs. 6 and 7, which contrast the degree of colour generated in cooked model dough samples to the predicted acrylamide levels for the same cooking conditions. (The 45 % moisture samples are not shown as they were very similar to the 19 % ones in appearance).

The Formation of Acrylamide in UK Cereal Products 427

Figure 6. Colour generation in dough samples which have been cooked and then freeze dried. Left to right: 130, 140, 150, 160, 170 and 180°C. Bottom row: 2.7 % moisture, middle row: 10%, top row: 19 %.

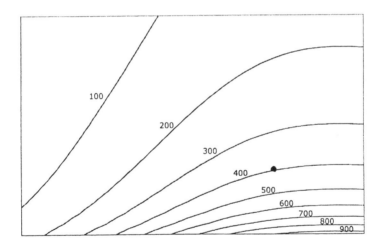

Figure 7. Predicted acrylamide levels (µg/kg) for the same range of temperatures and moistures as the cooked doughs shown in Fig. 6.

This difference explains why commercial biscuits show such high levels of acrylamide while other equally sweet products such as cakes or pancakes do not. A typical biscuit dough starts off drier than a cake batter or bread dough, and the whole of the finished product is dried to a very low moisture content during baking. In contrast, the crust region in cake or bread never dries out to the same extent because the center of the cake or bread is constantly releasing steam. Hence the crust never develops the levels of acrylamide observed in biscuits.

5. CONCLUSIONS

The highest levels of acrylamide in UK cereal products are seen in browned biscuits. Some strongly coloured products show surprisingly low levels for their degree of colour, so the degree of browning is not a good guide to acrylamide level.

Levels of acrylamide in cakes are low, even in the crust, partly due to the low levels of asparagine in cake flours. Bread crusts can contain up to 175 µg/kg of acrylamide, but levels in the whole product as eaten are much lower, typically only 20 µg/kg.

There is some evidence that recipes which incorporate ginger and ammonium salts are prone to higher acrylamide formation. However this is by no means universally true for all products in practice.

A mathematical model has been developed to predict acrylamide levels in bakery products and validated against tests on model dough. The model shows that levels of acrylamide are limited by amino acids running out as well as direct decay or loss of the acrylamide produced.

It is also clear that the levels of Maillard products do not necessarily align with acrylamide levels; low moisture content is a more important promoter of acrylamide than temperature, so crust moisture is a key factor controlling acrylamide levels, and this accounts for much of the variation seen between different types of bakery products.

6. RESEARCH NEEDS

There is a need for more information on the rate of decay of acrylamide in bakery products, and particularly the effects of dough composition and moisture.

There is little understanding of what role minor ingredients such as flavourings (e.g. ginger), raising agents (e.g. ammonium salts) and preservatives (e.g. vinegar, and so pH) have on acrylamide formation in real product formulations.

In the case of fermented yeasted products, the yeast may well affect both sugar and amino acid levels. If practical conditions which promote the yeast consuming acrylamide precursors can be found, then this may offer a route to controlling acrylamide levels while maintaining product quality.

ACKNOWLEDGEMENTS

The authors would like to thank Li Liang for assistance with amino acid analysis and Ian Slaiding of Brewing Research International for assistance with GC-MS analysis.

The support of the UK Foods Standards Agency for this work is gratefully acknowledged.

REFERENCES

Ahn, J.S., Castle, L., Clarke, D.B., Lloyd, A.S., Philo, M.R., Speck, D.R., 2002, Verification of the findings of acrylamide in heated foods *Food Add. Contam.*, **19**:1116-1124.

Biedermann, M., and Grob, K., 2003, Model studies on acrylamide formation in potato, wheat flour and corn starch; ways to reduce acrylamide contents in bakery ware, *Mitt. Lebensm. Hyg.* **94**:406-22.

Chamberlain, N., Collins, T.H., Elton, G.A.H., 1962, The Chorleywood bread process, *Bakers Digest* **36**:52-53.

Hamlet, C.G., Sadd, P.A., Gray, D.A., 2002, Influence of composition, moisture, pH and temperature on the formation and decay kinetics of monochloropropanols in wheat flour dough, *European Food Research and Technology* **216**:122-128.

Hamlet, C.G., Sadd, P.A., Gray, D.A., 2004a, Generation of monochloropropanols (MCPDs) in model dough systems. 1. Leavened doughs, *J. Agric. Food Chem.* **52**:2059-2066.

Hamlet, C.G., Sadd, P.A., Gray, D.A., 2004b, Generation of monochloropropanols (MCPDs) in model dough systems. 2. Unleavened doughs *J. Agric. Food Chem.* **52**:2067-2072.

Konings, E.J.M., Baars, A.J., van Klaveren, J.D., Spanjer,M.C., Rensen, P.N., Hiemstra, M., van Kooij, J.A., Peters, P.W.J. 2003, Acrylamide exposure from foods of the Dutch population and an assessment of the consequent risks *Food Chem. Tox.* **41**:1569-1579.

Henderson, J.W., Ricker, R.D., Bidlingmeyer, B.A., Woodward, C., Sept. 2003, Rapid, accurate, sensitive, and reproducible HPLC Analysis of Amino Acids, *Agilent technical report No. 5980-1193EN, 2001.* Available at: <www.chem.agilent.com/temp/radE9216/00023867.pdf>

Leung, K.S., Lin, A., Tsang, C.K., and Yeung, S.T.K., 2003, Acrylamide in Asian foods in Hong Kong, *Food Add. Contam.* **20**:1105-1113.

Mottram, D.S., Wedzicha, B., and Dodson, A.T., 2002, Acrylamide is formed in the Maillard reaction, *Nature* **419**:448-449.

Stadler, R.H., Blank, I., Varga, N., Robert, F., Hau, J., Guy, P.A., Robert, M.-C., and Riediker, S., 2002, Acrylamide from Maillard reaction products, *Nature* **419**:449-450.

Surdyk, N., Rosen, J., Andersson, R., Aman, P., 2004, Effects of asparagine, fructose and baking conditions on acrylamide content in yeast-leavened white bread *J. Agric. Food Chem.* **52**:2047-2051.

Svensson, K., Abramsson, L., Becker, W., Glynn, A., Hellenäs, K.-E., Lind, Y., Rosen, J., 2003, Dietary intake of acrylamide in Sweden *Food Chem. Tox.* **41**:1581-1586.

Tareke, E., Rydberg, P., Karlsson, P., Eriksson, S., Törnqvist, M., 2002, Analysis of acrylamide, a carcinogen formed in heated foods, *J. Agric. Food Chem.* **50**:4998-5006.

FACTORS INFLUENCING ACRYLAMIDE FORMATION IN GINGERBREAD

Thomas M. Amrein, Barbara Schönbächler, Felix Escher, and Renato Amadò
Institute of Food Science and Nutrition, Swiss Federal Institute of Technology (ETH), ETH-Zentrum, CH-8092 Zurich, Switzerland; e-mail: thomas.amrein@ilw.agrl.ethz.ch

Abstract: The influence of ingredients, additives, and process conditions on the acrylamide formation in gingerbread was investigated. The sources for reducing sugars and free asparagine were identified and the effect of different baking agents on the acrylamide formation was evaluated. Ammonium hydrogencarbonate strongly enhanced the acrylamide formation, but its N-atom was not incorporated into acrylamide, nor did acrylic acid form acrylamide in gingerbread. Acrylamide concentration and browning intensity increased both with baking time and correlated with each other. The use of sodium hydrogencarbonate as baking agent reduced the acrylamide concentration by more than 60%. Free asparagine was a limiting factor for acrylamide formation, but the acrylamide content could also be lowered by replacing reducing sugars with sucrose or by adding moderate amounts of organic acids. A significant reduction of the acrylamide content in gingerbread can be achieved by using sodium hydrogencarbonate as baking agent, minimizing free asparagine, and avoiding prolonged baking.

Key words: Acrylamide, gingerbread, free asparagine, reducing sugars, baking agent, ammonium hydrogencarbonate, sodium hydrogencarbonate, citric acid, glycine, asparaginase

1. INTRODUCTION

The detection of acrylamide in heated foodstuffs (Tareke et al., 2002) led to a world-wide concern because acrylamide is a known neurotoxin and is classified as "probably carcinogenic to humans" (group 2A) (IARC, 1994). Concentrations exceeding 1000 µg/kg were found in heated potato products such as French fries and potato crisps (Tareke et al., 2002). It was shown that acrylamide is formed in the Maillard reaction (Mottram et al., 2002; Stadler et al., 2002). Acrylamide is eliminated concurrently to its formation, particularly at high temperatures (Biedermann et al., 2002a; Mottram et al., 2002). Asparagine is the main precursor for acrylamide formation in foods and it delivers the backbone of the acrylamide molecule (Becalski et al., 2003; Zyzak et al., 2003). A reducing sugar is needed for the formation of the Schiff's base of asparagine which is transformed via an oxazolidin-5-one intermediate to a decarboxylated Amadori product that releases acrylamide (Yaylayan et al., 2003). In analogy to the formation of acrylamide from asparagine, the release of vinylogous compounds from other amino acids was assumed and it was shown that acrylic acid is formed in copyrolysates of aspartic acid and sugars (Stadler et al., 2003).

Gingerbread may contain up to 1000 µg of acrylamide per kg. In Germany, acrylamide contents in gingerbread ranged from < 20 µg/kg to more than 8000 µg/kg (average 481 µg/kg) (Holtmannsspötter, 2003). Dutch gingerbread products contained acrylamide from 260 to 1410 µg/kg (average: 890 µg/kg; median: 1070 µg/kg). Since gingerbread is consumed frequently during the entire year in the Netherlands, these products were estimated to contribute 16% of the total acrylamide exposure of the Dutch population (Konings et al., 2003). In our laboratory preliminary analyses of typical Swiss gingerbread of the 2003 Christmas season showed acrylamide contents from 100 to 800 µg/kg. These relatively high concentrations are in contrast to the low content of free asparagine in cereal flours (Prieto et al., 1990) which contain about 100 times less free asparagine than do potatoes (Amrein et al., 2003). Therefore the question of an alternative mechanism of acrylamide formation in gingerbread was raised. It was shown that ammonium hydrogencarbonate (i.e. the typical baking agent for gingerbread in Switzerland) strongly enhanced acrylamide formation in a model system for bakery products (Biedermann and Grob, 2003). Since acrylic acid is formed upon pyrolysis of aspartic acid with sugars (Stadler et al., 2003) and it can form high levels of acrylamide with ammonia (Yasuhara et al., 2003), a mechanism other than the thermal degradation of asparagine was imaginable.

In this study, we first investigated the hypothesis of ammonia incorporation from the baking agent into acrylamide. Second, we investigated ways to reduce the acrylamide content in gingerbread and the influence of the type and the amount of baking agent and other additives on acrylamide formation. Critical factors for acrylamide formation such as ingredients and process conditions were identified and discussed. Several possible ways are suggested to significantly reduce the acrylamide content in gingerbread.

2. MATERIALS AND METHODS

2.1 Preparation of gingerbreads

All ingredients were obtained from JOWA AG (Volketswil, Switzerland): flour (a 70/30 mixture of spelt and wheat, corresponding to a flour type 720), inverted sugar syrup, powdered sucrose, American honey, water, spices, caramel coloring, whole egg, ammonium hydrogencarbonate (baking agent), and lezirol. Wheat flour type 400 was purchased from a supermarket. ^{15}N-labeled baking agent was a mixture of ^{15}N-ammonium sulfate (CIL, Andover, Massachusetts, USA) and sodium hydrogencarbonate (Fluka, Buchs, Switzerland) containing the same amounts of NH_4^+ and HCO_3^- ions as the normal baking agent. Acrylic acid, amino acids, and other organic acids were obtained from Fluka. Asparaginase (islated from *E. coli*) was purchased from Fluka and Sigma-Aldrich (Steinheim, Germany), diluted, and directly added to the dough during kneading (4 units per kg dough). Preparation of gingerbreads was performed as close to the industrial process as possible (Amrein et al., 2004).

2.2 Analytical procedures

For each gingerbread sample color (L*, a*, b*-system), dry weight, and pH (in a 10% aqueous slurry) were determined. All ingredients were analyzed for free asparagine as described in (Lebet et al., 1994) and for glucose, fructose, and sucrose using an enzymatic kit form Scil diagnostics (Martinsried, Germany). For details, see (Amrein et al., 2004). Analysis of acrylamide was done with the GC-MS method described by Biedermann et al. (Biedermann et al., 2002b). $^{13}C_3$-acrylamide (CIL) and methacrylamide (Fluka) were used as internal standards.

3. RESULTS AND DISCUSSION

3.1 Influence of baking agent

Different amounts of ammonium hydrogencarbonate (NH_4HCO_3) were added to the dough (standard: 0.8 g/100 g dough) and samples were baked under standard conditions (3 min at 180°C followed by 7 min at 190°C). Acrylamide concentrations are shown in **Figure 1**. Gingerbread prepared according to the recipe contained 501 µg acrylamide per kg with a relative standard deviation of 16% (n = 9). This is well within the range of reported acrylamide levels in gingerbread products (Holtmannsspötter, 2003; Konings et al., 2003). Separate analysis of the crumb and crust of a standard gingerbread showed that 54% of the total acrylamide is contained in the crumb (441 µg/kg) and 46% in the crust (698 µg/kg). Thus, acrylamide formation in gingerbread is not uniquely related to the formation of dry crust but also occurs in the more humid environment of the crumb.

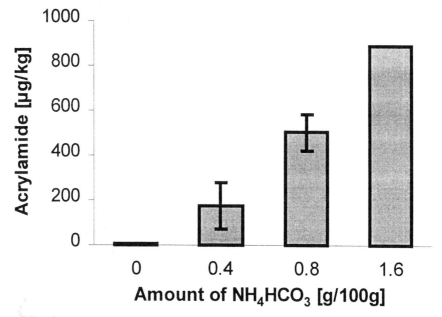

Figure 1. Acrylamide content in gingerbreads produced with different amounts of ammonium hydrogencarbonate as baking agent (error bars are ± standard deviation, n = 2, except for 1.6 g/100g).

The amount of NH_4HCO_3 had a very strong influence on the acrylamide formation in gingerbread. In the absence of this baking agent, almost no acrylamide was formed (about 10 µg/kg), but the product was unsatisfactory because it lacked browning and leavening. If 0.4 g/100 g was added, the acrylamide content decreased to one third (170 µg/kg) and the color was too bright, whereas 1.6 g/100 g increased the acrylamide content to 880 µg/kg and enhanced browning. Apparently, NH_4HCO_3 strongly promotes the formation of acrylamide in gingerbread which was also observed in a bakery model system if this baking agent was present (Biedermann and Grob, 2003). The amount of NH_4HCO_3 correlated with the pH and the color (L-value) since the darkness of the product was directly related to the amount of baking agent and pH.

To check if promotion of acrylamide formation was due to the incorporation of ammonia into acrylamide and, therefore, due to an another mechanism, gingerbread was produced with ^{15}N-labeled baking agent ((^{15}NH$_4$)$_2$SO$_4$ + NaHCO$_3$). In a control experiment with unlabeled (NH$_4$)$_2$SO$_4$ and NaHCO$_3$, a normal product with slightly increased acrylamide content was obtained. Gingerbreads with the labeled baking agent were baked under standard (3 min at 180°C and 7 min at 190°C) and more drastic process conditions (12 min at 200°C) in order to enhance temperature-dependent effects. No ^{15}N-acrylamide was detected in any of the products. Addition of L-aspartic acid (1000 mg/kg), a precursor of acrylic acid or of acrylic acid (1000 mg/kg) resulted in no detectable ^{15}N-acrylamide in gingerbread prepared with labeled baking agent. In standard samples and samples with added L-aspartic acid, only trace amounts of acrylic acid were detected. Addition of L-aspartic acid or acrylic acid to normal dough (1000 mg/kg) did not lead to a significant increase in the acrylamide content: 551 and 573 µg/kg, respectively (the standard product of same batch contained 562 µg/kg). Thus, ammonia is not incorporated into acrylamide and the amidation of acrylic acid by ammonia from the baking agent does not contribute to the high acrylamide content in gingerbread. Therefore, a mechanism for acrylamide formation other than by thermal degradation of asparagine is unlikely in gingerbread. On the other hand, formation of acrylic acid from aspartic acid, and reaction of ammonia with acrylic acid leading to acrylamide was shown in model systems (Stadler et al., 2003; Yasuhara et al., 2003). However, the thermal conditions in these studies were clearly different from the temperatures in gingerbreads during baking. Formation of acrylic acid from aspartic acid and glucose or fructose started at 150°C (Stadler et al., 2003), and reaction of ammonia with acrylic acid was performed at 180°C for 30 min (Yasuhara et al., 2003). For the amidation of aliphatic acids by ammonia heating at 160°C for 5 h is needed for a yield of about 80% (Mitchell and Reid, 1931). The temperature within gingerbreads during the baking process stayed below 100°C in the first 6 min due to water evaporation and did not exceed 110°C until the end of the baking process. Therefore, the formation of acrylamide via amidation of acrylic acid is unlikely to take place in gingerbread. The promoting effect of NH$_4$HCO$_3$ on the acrylamide formation could be explained by the reaction of asparagine with reactive carbonyls. Glyoxal and methylglyoxal are formed from reducing sugars in Maillard reaction models (Hollnagel and Kroh, 1998) and have been shown to react faster with amino acids than glucose or fructose (Piloty and Baltes, 1979). In addition, many other α-dicarbonyls and α-hydroxy-carbonyls are formed from reducing sugars in the Maillard reaction (Ledl and Schleicher, 1990). The sum of these reactive carbonyls might be responsible for the high yield of acrylamide. This hypothesis is supported by the finding that glyoxal and glyceraldehydes formed more acrylamide from asparagine than did glucose (Zyzak et al., 2003). Thus, the promoting effect of NH_4HCO_3 on acrylamide formation might be indirect; the bicarbonate provides more reactive carbonyls originating from the reaction of ammonia with reducing sugars. The fact that almost no acrylamide is formed in gingerbread prepared without NH_4HCO_3 supports this hypothesis.

If $NaHCO_3$ was used as an alternative baking agent, the acrylamide content was reduced to one third (**Figure 2**). However, only 1.67 g $NaHCO_3$ led to a product with a color comparable to that of the standard product (L-value = 47.3): L-values were 45.1 (1.67 g $NaHCO_3$ /100 g) and 51.3 (0.83 g $NaHCO_3$ /100 g). The pH was significantly higher in the samples with $NaHCO_3$(8.2 and 8.8, respectively), when compared to the standard product (pH = 6.9), causing an alkaline taste. Thus, $NaHCO_3$ allows the preparation of gingerbreads with a substantially less acrylamide, acceptable browning and sensory properties (taste, volume). A more alkaline pH does not necessarily imply a higher acrylamide content in gingerbread. Other factors such as the presence of ammonia have a stronger impact. Experiments with K_2CO_3 as baking agent were somewhat less effective. If NH_4HCO_3 was replaced by 0.53 g K_2CO_3, an acrylamide content of 416 µg/kg was found, only marginally below the average content of the standard product (501 µg/kg). The product was less brown and the pH was slightly higher (pH 7.6).

Since $NaHCO_3$ and K_2CO_3 as baking agents lead to products with an increased pH, an alkaline taste and lesser leavening compared to NH_4HCO_3, they were applied together with some organic acid. The combination of 1.67 g $NaHCO_3$ and 1.47 g tartaric acid per 100 g dough led to an acrylamide content of 30 µg/kg. Gingerbread prepared with 0.53 g K_2CO_3 and 0.53 g lactic per 100 g dough contained 37 µg/kg acrylamide. However, the browning was also less pronounced for both products. This shows that a combination of organic acid with $NaHCO_3$ or K_2CO_3 as baking agents is a way to significantly reduce the acrylamide content. However, the use of this strategy is limited because of the inhibiting effect of acids on the browning reaction. Only small amounts of organic acids allow the preparation of sufficiently browned products.

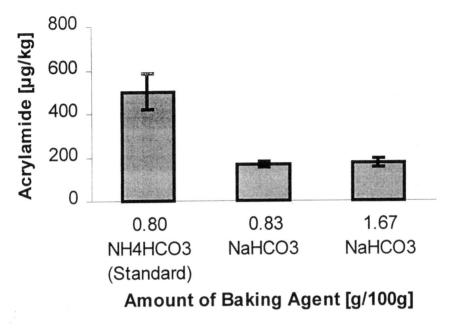

Figure 2. Acrylamide contents in gingerbreads prepared with different baking agents (error bars are ± standard deviation, n = 9 for NH_4HCO_3; n = 2 for $NaHCO_3$).

3.2 Sources of reducing sugars and free asparagine

Sugars and free asparagine were determined in all of the ingredients, and their contribution to the total amount in the gingerbread dough was calculated (**Table 1**). Flour was clearly the main source for free asparagine in the gingerbread dough. It contributed 87.5% of the total free asparagine while honey contributed less than 10%. The contribution from the spice mixture was only marginal (3.2%), although it contained almost twice as much free asparagine as the flour.

Table 1. Concentrations of sugars and free asparagine in the ingredients of gingerbread (referring to fresh weight; n = 2)

Ingredient	Glucose [g/100g]	Fructose [g/100g]	Sucrose [g/100g]	Free Asparagine [mg/kg]
Flour	0.03	0.03	0.44	139
Inverted sugar syrup	34.52	31.74	4.54	n.d.
Powdered sugar	n.d.	n.d.	96.4	n.d.
Honey	30.69	39.66	3.37	64
Caramel coloring	31.71	23.77	5.79	n.d.
Spices	0.77	0.81	0.57	212
Whole egg	0.39	n.d.	n.d.	31

(n.d.: not detected)

In contrast to free asparagine, the contribution to the total reducing sugar content from the flour was negligible (0.2%). The inverted sugar syrup, honey, and caramel coloring contributed 99% of the total amount of glucose and fructose with inverted sugar being the main source (52.3%), followed by the honey (38.8%). Interestingly, honey was a source for all of the compounds that are involved in acrylamide formation. The literature reports very similar values for the sugar and free amino acid contents in flours (MacArthur and D'Appolonia, 1976; Prieto et al., 1990) and honeys (Speer and Montag, 1986; Sporns et al., 1992) .In one experiment, honey, inverted sugar syrup, and caramel coloring were replaced by sucrose solutions containing the same amount of sucrose instead of glucose and fructose. This virtual depletion of reducing sugars reduced the acrylamide content by a factor of 20: Gingerbread prepared with sucrose contained only 25 µg/kg. However, browning of these samples was clearly insufficient. This can be explained by the fact that a reducing sugar (or a reactive carbonyl) is needed for acrylamide formation (Yaylayan et al., 2003; Zyzak et al., 2003) and Maillard reaction in general (Ledl and Schleicher, 1990). "White gingerbread" is a specialty in some regions of Switzerland, prepared with NH_4HCO_3 and sucrose instead of honey and inverted sugar syrup, and baked at 230°C for 15 minutes. It contains very little acrylamide (< 10 µg/kg). Thus, ammonium bicarbonate is only a critical factor if reducing sugars are present; its promoting effect is not related to free asparagine, and sucrose does not contribute to the formation of acrylamide (Amrein et al., 2003; Biedermann and Grob, 2003).

One kg of fresh gingerbread dough contains about 80 mg of free asparagine, and after standard baking 500 µg acrylamide per kg. This corresponds to a yield of 0.6% acrylamide based on free asparagine which is higher than the usually reported yields of about 0.1% (Becalski et al., 2003; Mottram et al., 2002; Yasuhara et al., 2003). This demonstrates that NH_4HCO_3 promotes the formation of acrylamide. Even higher yields (3.5

and 4.8%) were found in bakery model systems consisting of wheat flour, fructose, and $(NH_4)_2CO_3$ (Biedermann and Grob, 2003). These high yields are probably due the high sixfold molar ratio of fructose to asparagine. In our study, the molar concentration of free asparagine in the dough was about 500 times smaller compared to glucose or fructose. Thus, free asparagine is a limiting factor for acrylamide formation in gingerbread and in other bakery products (Springer et al., 2003; Weisshaar, 2004). Addition of free asparagine to the flour before preparing the dough led to drastic increases in the acrylamide contents of baked gingerbreads. Addition of only 250 mg or 500 mg of asparagine to 1 kg dough increased the acrylamide contents to 1950 and 4160 µg/kg, respectively. When 1000 mg/kg were added, the acrylamide concentrations rose to > 8000 µg/kg. The color of the product was only somewhat darker than was standard gingerbread. This shows that free asparagine governs the acrylamide formation in gingerbread.

A similar effect was observed in bakery products prepared with different flour types; the higher the extraction during milling, the more free asparagine present in the flour, the higher the acrylamide content in the product (Springer et al., 2003). Apparently, acrylamide formation in food is largely determined by the precursor with the lowest concentration. Thus, in gingerbread, free asparagine is limiting, whereas in potatoes glucose and fructose are the limiting parameters (Amrein et al., 2003).

As a consequence, the decomposition of free asparagine prior to baking is should result in a decrease in acrylamide formation. This hypothesis was tested by adding an asparaginase during dough preparation in order to hydrolyze the amide group of asparagine. Gingerbreads from this dough contained 228 µg acrylamide per kg (n = 3) which corresponds to a decrease of 55% of the normal acrylamide content. Taste and color were virtually identical to the normal product which is a clear advantage of this approach. The fresh dough treated with asparaginase still contained 22 mg free asparagine per kg. Therefore, about 75% of the total free asparagine had been degraded, which explains why the acrylamide formation was not fully inhibited. The incomplete hydrolysis was probably due to the limited mobility of the enzyme and the substrate within the dough. It has been shown that incubation of a potato matrix with an asparaginase can be an effective way to reduce the formation of acrylamide (Zyzak et al., 2003). Thus, the application of this enzyme to different food matrices prior to heating should be further investigated.

The use of a flour type 400 (67 mg free asparagine per kg) instead of the standard flour (139 mg/kg) did not decrease the acrylamide content (540 µg/kg), although the dough contained 50% less free asparagine compared to the standard dough. Thus, acrylamide formation in gingerbread may not be substantially reduced by only choosing ingredients with low content of free asparagine, probably because of the high yield of acrylamide per free asparagine.

3.3 Influence of process conditions

Gingerbreads were baked under various temperature-time conditions to investigate the influence of temperature and time on acrylamide formation. **Figure 3** shows the acrylamide concentrations measured during baking at 180°C and 200°C. Although the temperature stayed below 100°C in the first 6 min, some acrylamide was already formed in this period. Even in the raw dough, traces of acrylamide were detected. It seems that the presence of ammonia allows the formation of acrylamide at temperatures < 100°C. In model systems, acrylamide formation was observed at 60°C and even at room temperature, provided that $(NH_4)_2CO_3$ and fructose were present (Biedermann and Grob, 2003). The acrylamide concentrations increased steadily in the first 20 min of the baking process. A linear rather than an exponential correlation between acrylamide concentration and time could be assumed. In contrast to French fries, where most of the acrylamide is formed in the last minute of the frying process (Grob et al., 2003), acrylamide was formed almost evenly and over a broader period of the baking process. At 200°C, the acrylamide contents were slightly higher than at 180°C during the first 15 min.

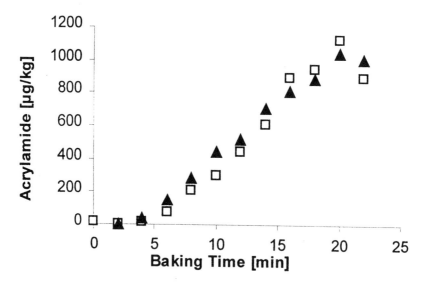

Figure 3. Influence of temperature and time on the acrylamide formation during baking (□: 180°C, ▲: 200°C).

Extension of the baking process beyond the necessary time (10 min) resulted in a further increase of the acrylamide content. Thus, prolonged baking should be avoided. A lower temperature combined with a prolonged baking time did not result in lower acrylamide contents if the same browning of the product was to be achieved. It was generally observed that a prolonged baking at lower temperatures resulted even in higher acrylamide contents. For instance, gingerbread baked at 160°C for 20 min exhibited the same color as a sample prepared at 200°C for 10 min, but with acrylamide contents of 910 and 440 µg/kg, respectively. Thus, a shorter baking at higher temperatures is more suitable to contain the formation of acrylamide in gingerbread. At all temperatures tested (160°C, 180°C, and 200°C), the acrylamide content decreased after 20 min baking indicating the elimination of acrylamide. To determine the extent of elimination, $^{13}C_3$-acrylamide (500 µg/kg) was added to the dough during mixing, and gingerbreads were baked at 180°C for different lengths of time. Analysis was performed by using only methacrylamide as internal standard. A linear elimination of acrylamide was observed all over the baking process, but its extent was limited. Only about one third of the added $^{13}C_3$-acrylamide was eliminated after 10 min; and 50% was still present after 28 min. Furthermore, the elimination of acrylamide is of no practical importance because baking exceeding 10 min results in unacceptably dark and dry products with high acrylamide content.

During baking, a change in brightness of the gingerbreads (determined as L-value) was observed. In the first 2 to 3 min of the baking process the brightness increased, afterwards a steady decrease occurred. This decrease in brightness strongly correlated with the acrylamide content ($R^2 = 0.9883$): The darker the product, the higher the acrylamide concentration. This was to be expected since browning and acrylamide both originate from the Maillard reaction (Martins et al., 2000; Mottram et al., 2002). Dry matter content increased steadily during baking and reached about 85% after the standard baking process. Since dry matter and acrylamide content both increased steadily during baking (e.g. at 180°C) they were strongly correlated with each other ($R^2 = 0.9874$). However, addition of some extra water (5 g / 100 g dough) during preparation of the dough did not reduce the acrylamide formation, but instead led to a 25% increase. This shows that the observed correlation between dry matter and acrylamide content is coincidental and no causal. A similar phenomenon was observed for the pH effect. After 2 to 3 min baking a slight increase in pH was observed followed by a steady decrease. These changes in pH were probably due to the formation, evaporation and reaction of NH_3 and CO_2 released from the baking agent. Accordingly, a strong correlation between decreasing pH and acrylamide content was observed ($R^2 = 0.9725$). However, results from experiments with citric acid (see 3.4) demonstrated that this correlation is also coincidental and not causal.

3.4 Organic acids reduce acrylamide content

Various experiments with addition of different organic acids were performed to check if some of them are able to reduce the acrylamide content in gingerbread (**Table 2**). In all of these experiments gingerbreads were prepared with NH_4HCO_3 as baking agent. A moderate addition of 0.25 g citric acid per 100 g dough decreased the acrylamide content by about 40%, whereas the browning was not affected. Although the pH was decreased to about 6, the product did not have an acidic taste. Addition of 0.5 and 1.0 g citric acid per 100 g dough resulted in a drop of pH to 5.6 and 5.0, and in a reduction of the acrylamide concentration by a factor of 4 and 40, respectively. At the same time browning was also substantially reduced. Addition of 0.5 g/100 g citric acid led to a slightly acidic product, while gingerbread with 1 g citric acid per 100 g had a clearly acidic taste and its leavening was not sufficient, probably due to the protonation of NH_3, whereby the gas volume was reduced during baking. The reduction of acrylamide formation by citric acid has also been reported in French fries and various model systems (Gama-Baumgartner et al., 2004; Jung et al., 2003; Rydberg et al., 2003). The protonation of the α-amino group of asparagine hinders the formation of the N-substituted glycosylamine which is an important early intermediate in the formation of acrylamide and in the Maillard reaction in general (Ledl and Schleicher, 1990; Yaylayan et al., 2003). Furthermore, NH_4^+ is not easily de-protonated in the presence of citric acid. Thus, the reaction of NH_3 with reducing sugars leading to reactive carbonyls was limited.

Table 2. Effect of added organic acids on the acrylamide content, color, and pH (n = 1).

Organic acid	Amount [mg/kg]	Acrylamide [μg/kg]	L- Value [-]	pH
No addition	-[a]	501	47.3	6.9
Citric acid	2500	286	45.2	6.0
	5000[b]	133	56.0	5.6
	10000[b]	12	55.3	5.0
Glycine	2000	430	41.2	7.0
	10000	151	38.4	6.5
L-Cysteine	500	368	48.7	6.7
	2000	380	42.7	6.4
L-Glutamine	2000	587	41.9	6.8
L-Lysine	2000	542	41.7	7.1

(a: n = 9; b: n = 2)

Addition of L-glutamine, L-lysine, or glycine (2000 mg/kg dough) did not lower the acrylamide contents, but enhanced browning. This might be due to the higher number of available α-amino groups which undergo Maillard reaction resulting in formation of additional melanoidins. However, addition of a large amount of glycine (10000 mg/kg dough) reduced the acrylamide content to one third. Browning was even stronger and the pH slightly lower. Glycine is known to strongly enhance browning (Ashoor and Zent, 1984) and to react readily with α-dicarbonyls (Piloty and Baltes, 1979). Thus, the observed effect could be due to the competition of the amino acids for the reactive carbonyls and/or the elimination of formed acrylamide by a reaction with glycine. Acrylamide is known to react with NH_2- and SH-groups through a Michael-type nucleophilic addition reaction (Friedman, 2003), a possible explanation for its elimination in foods. L-Cysteine showed a tendency to reduce the acrylamide content and the pH. However, these samples had an unpleasant taste and odor, presumably caused by S-containing decomposition products of cysteine.

The acrylamide content in a potato model system was reduced up to 92% following the addition of amino acids (Rydberg et al., 2003). However, the authors used far larger amounts of amino acids (5000 to 20000 mg/kg), which might explain the pronounced effect. A possible explanation for the observed effect of added glycine (10000 mg/kg dough) could be the competition of the amino acids for the reactive carbonyls, and/or a higher reactivity of glycine towards the carbonyls. Glycine caused more browning with reducing sugars than did asparagine (Ashoor and Zent, 1984), and was very reactive towards α-dicarbonyl compounds in model systems (Piloty and Baltes, 1979).

4. CONCLUSIONS

The formation of acrylamide in gingerbread is strongly promoted by the baking agent NH_4HCO_3. Its N-atom is not incorporated into acrylamide nor is the amidation of acrylic acid a relevant mechanism for acrylamide formation in gingerbread. The acrylamide concentrations are correlated with the extent of browning; both increase with baking time. Thus, a prolonged baking should be avoided to minimize the acrylamide content. $NaHCO_3$ is a valuable alternative baking agent and can reduce the acrylamide content by over 60%. In gingerbread, asparagine is a limiting factor for acrylamide formation and mainly comes from flour and honey. Therefore, a further reduction of the acrylamide content can be achieved by choosing ingredients with a lower content of free asparagine, or by adding the enzyme asparaginase. Citric acid can strongly reduce the acrylamide formation, but its use is limited by the acidic taste, poor leavening, and insufficient browning of the product. Addition of amino acids enhances browning. However, a reduction of acrylamide concentration is only observed with high amounts. The use of $NaHCO_3$ as baking agent, control of the baking conditions (i.e. time and temperature), moderate addition of organic acids, and reduction of free asparagine in the dough by more than 50% are the most promising ways to achieve a significant reduction of acrylamide content in gingerbreads.

ACKNOWLEDGEMENTS

We thank Horst Adelmann for pilot plant support, Maurus Biedermann and Koni Grob (Official Food Control Authority of the Canton of Zurich) for valuable support and cooperation in the acrylamide analysis, and JOWA AG for supplying raw materials. Financial support was provided by the Federation of Swiss Food Industries (FIAL), the Swiss Federal Office for Public Health (BAG), COOP Switzerland, and Cooperative Migros.

Part of this study has been already published elsewhere. Reprinted with permission from *Journal of Agricultural and Food Chemistry*, June 30, 2004, *52*, 4282-4288. Copyright 2004 American Chemical Society.

REFERENCES

Amrein, T. M., Bachmann, S., Noti, A., Biedermann, M., Barbosa, M. F., Biedermann-Brem, S., Grob, K., Keiser, A., Realini, P., Escher, F., and Amadò, R., 2003, Potential of acrylamide formation, sugars, and free asparagine in potatoes: A comparison of cultivars and farming systems, *J. Agric. Food Chem.* **51**:5556-5560.

Amrein, T. M., Schönbächler, B., Escher, F., and Amadò, R., 2004, Acrylamide in gingerbread: Critical factors for formation and possible ways for reduction, *J. Agric. Food Chem.* **52**:4282-4288.

Ashoor, S. H., and Zent, J. B., 1984, Maillard browning of common amino acids and sugars, *J. Food Sci.* **49**:1206-1207.

Becalski, A., Lau, B. P. Y., Lewis, D., and Seaman, S. W., 2003, Acrylamide in foods: Occurrence, sources, and modeling, *J. Agric. Food Chem.* **51**:802-808.

Biedermann, M., Biedermann-Brem, S., Noti, A., and Grob, K., 2002a, Methods for determining the potential of acrylamide formation and its elimination in raw materials for food preparation, such as potatoes, *Mitt. Geb. Lebensm. Unters. Hyg.* **93**:653-667.

Biedermann, M., Biedermann-Brem, S., Noti, A., Grob, K., Egli, P., and Mändli, H., 2002b, Two GC-MS methods for the analysis of acrylamide in foods, *Mitt. Geb. Lebensm. Unters. Hyg.* **93**:638-652.

Biedermann, M., and Grob, K., 2003, Model studies on acrylamide formation in potato, wheat flour and corn starch; ways to reduce acrylamide contents in bakery ware, *Mitt. Geb. Lebensm. Unters. Hyg.* **94**:406-422.

Friedman, M., 2003, Chemistry, biochemistry, and safety of acrylamide. A review, *J. Agric. Food Chem.* **51**:4504-4526.

Gama-Baumgartner, F., Grob, K., and Biedermann, M., 2004, Citric acid to reduce acrylamide formation in French fries and roasted potatoes?, *Mitt. Geb. Lebensm. Unters. Hyg.* **95**:110-117.

Grob, K., Biedermann, M., Biedermann-Brem, S., Noti, A., Imhof, D., Amrein, T., Pfefferle, A., and Bazzocco, D., 2003, French fries with less than 100 µg/kg acrylamide. A collaboration between cooks and analysts, *Eur. Food Res. Technol.* **217**:185-194.

Hollnagel, A., and Kroh, L. W., 1998, Formation of alpha-dicarbonyl fragments from mono- and disaccharides under caramelization and Maillard reaction conditions, *Z Lebensm. Unters. Forsch. A-Food Res. Technol.* **207**:50-54.

Holtmannsspötter, H., 2003. Bayer. Landesamt für Gesundheit und Lebensmittelsicherheit, Erlangen, Germany.

IARC, 1994, Acrylamide, in: *Monographs on the Evaluation of Carcinogenic Risks to Humans: Some Industrial Chemicals*, International Agency for Research on Cancer, Lyon, France, pp. 389-433.

Jung, M. Y., Choi, D. S., and Ju, J. W., 2003, A novel technique for limitation of acrylamide formation in fried and baked corn chips and in French fries, *J. Food Sci.* **68**:1287-1290.

Konings, E. J. M., Baars, A. J., van Klaveren, J. D., Spanjer, M. C., Rensen, P. M., Hiemstra, M., van Kooij, J. A., and Peters, P. W. J., 2003, Acrylamide exposure from foods of the Dutch population and an assessment of the consequent risk., *Food Chem. Toxicol.* **41**:1569-1579.

Lebet, V., Schneider, H., Arrigoni, E., and Amadò, R., 1994, A critical appreciation of the protein content determination by Kjeldahl's method based on the amino acid analysis, *Mitt. Geb. Lebensm. Unters. Hyg.* **85**:46-58.

Ledl, F., and Schleicher, E., 1990, New aspects of the Maillard reaction in foods and in the human body, *Angew. Chem.-Int. Edit. Engl.* **29**:565-594.

MacArthur, L. A., and D'Appolonia, B. L., 1976, Carbohydrates of various pin-milled and air-classified flour streams. 1. Sugar analyses, *Cereal Chem.* **53**:916-927.

Martins, S., Jongen, W. M. F., and van Boekel, M., 2000, A review of Maillard reaction in food and implications to kinetic modelling, *Trends Food Sci. Technol.* **11**:364-373.

Mitchell, J. A., and Reid, E. E., 1931, The preparation of aliphatic amides, *J. Am. Chem. Soc.* **53**:1879-1883.

Mottram, D. S., Wedzicha, B. L., and Dodson, A. T., 2002, Acrylamide is formed in the Maillard reaction, *Nature* **419**:448-449.

Piloty, M., and Baltes, W., 1979, Investigations on the reaction of amino-acids with alpha-dicarbonyl compounds. 1. Reactivity of amino-acids in the reaction with alpha-dicarbonyl compounds (in German), *Z. Lebensm.-Unters.-Forsch.* **168**:368-373.

Prieto, J. A., Collar, C., and Debarber, C. B., 1990, Reversed phase high performance liquid chromatographic determination of biochemical changes in free amino acids during wheat flour mixing and bread baking, *J. Chromatogr. Sci.* **28**:572-577.

Rydberg, P., Eriksson, S., Tareke, E., Karlsson, P., Ehrenberg, L., and Törnqvist, M., 2003, Investigations of factors that influence the acrylamide content of heated foodstuffs, *J. Agric. Food Chem.* **51**:7012-7018.

Speer, K., and Montag, A., 1986, Distribution of free amino acids in honeys - considering particularly German and French heath honeys (in German), *Dtsch. Lebensm.-Rundsch.* **82**:248-253.

Sporns, P., Plhak, L., and Friedrich, J., 1992, Alberta honey composition, *Food Res. Int.* **25**:93-100.

Springer, M., Fischer, T., Lehrack , A., and Freund, W., 2003, Acrylamidbildung in Backwaren (in German), *Getreide Mehl und Brot* **57**:274-278.

Stadler, R. H., Blank, I., Varga, N., Robert, F., Hau, J., Guy, P. A., Robert, M. C., and Riediker, S., 2002, Acrylamide from Maillard reaction products, *Nature* **419**:449-450.

Stadler, R. H., Verzegnassi, L., Varga, N., Grigorov, M., Studer, A., Riediker, S., and Schilter, B., 2003, Formation of vinylogous compounds in model Maillard reaction systems, *Chem. Res. Toxicol.* **16**:1242-1250.

Tareke, E., Rydberg, P., Karlsson, P., Eriksson, S., and Törnqvist, M., 2002, Analysis of acrylamide, a carcinogen formed in heated foodstuffs, *J. Agric. Food Chem.* **50**:4998-5006.

Weisshaar, R., 2004, Acrylamide in bakery products - Results from model experiments (in German), *Dtsch. Lebensm.-Rundsch.* **100**:92-97.

Yasuhara, A., Tanaka, Y., Hengel, M., and Shibamoto, T., 2003, Gas chromatographic investigation of acrylamide formation in browning model systems, *J. Agric. Food Chem.* **51**:3999-4003.

Yaylayan, V. A., Wnorowski, A., and Locas, C. P., 2003, Why asparagine needs carbohydrates to generate acrylamide, *J. Agric. Food Chem.* **51**:1753-1757.

Zyzak, D. V., Sanders, R. A., Stojanovic, M., Tallmadge, D. H., Eberhart, B. L., Ewald, D. K., Gruber, D. C., Morsch, T. R., Strothers, M. A., Rizzi, G. P., and Villagran, M. D., 2003, Acrylamide formation mechanism in heated foods, *J. Agric. Food Chem.* **51**:4782-4787.

EFFECTS OF CONSUMER FOOD PREPARATION ON ACRYLAMIDE FORMATION

Lauren S. Jackson[1] and Fadwa Al-Taher[2]
[1]*U.S. Food and Drug Administration, National Center for Food Safety and Technology (NCFST), 6502 S. Archer Rd., Summit-Argo, IL 60501;* [2]*Illinois Institute of Technology, NCFST, 6502 S. Archer Rd., Summit-Argo, IL 60501USA; e-mail: Lauren.Jackson@cfsan.fda.gov*

Abstract: Acrylamide is formed in high-carbohydrate foods during high temperature processes such as frying, baking, roasting and extrusion. Although acrylamide is known to form during industrial processing of food, high levels of the chemical have been found in home-cooked foods, mainly potato- and grain-based products. This chapter will focus on the effects of cooking conditions (e.g. time/temperature) on acrylamide formation in consumer-prepared foods, the use of surface color (browning) as an indicator of acrylamide levels in some foods, and methods for reducing acrylamide levels in home-prepared foods. As with commercially processed foods, acrylamide levels in home-prepared foods tend to increase with cooking time and temperature. In experiments conducted at the NCFST, we found that acrylamide levels in cooked food depended greatly on the cooking conditions and the degree of "doneness", as measured by the level of surface browning. For example, French fries fried at 150-190°C for up to 10 min had acrylamide levels of 55 to 2130 μg/kg (wet weight), with the highest levels in the most processed (highest frying times/temperatures) and the most highly browned fries. Similarly, more acrylamide was formed in "dark" toasted bread slices (43.7-610.7 μg/kg wet weight), than "light" (8.27-217.5 μg/kg) or "medium" (10.9-213.7 μg/kg) toasted slices. Analysis of the surface color by colorimetry indicated that some components of surface color ("a" and "L" values) correlated highly with acrylamide levels. This indicates that the degree of surface browning could be used as an indicator of acrylamide formation during cooking. Soaking raw potato slices in water before frying was effective at reducing acrylamide levels in French fries. Additional studies are needed to develop practical methods for reducing acrylamide formation in home-prepared foods without changing the acceptability of these foods.

Key words: Acrylamide; consumers; cooking; frying; toasting; bread; potato; browning

1. INTRODUCTION

In April 2002, researchers at the Swedish National Food Administration (NFA) and the University of Stockholm reported the presence of acrylamide in variety of heat-treated, carbohydrate-rich foods such as potato- and grain-based products. This report caused worldwide concern since acrylamide has been found to induce tumors in experimental animals (Johnson et al., 1986; Friedman et al., 1995), to be neurotoxic in humans and in laboratory animals (Bachmann et al, 1992; Callemen et al., 1994; Towell et al., 2000; LoPachin, 2004) and to be genotoxic in *in vivo* and *in vitro* toxicity tests (Dearfield et al., 1995). The International Agency on Research on Cancer (IARC) has classified acrylamide as a probable human carcinogen (IARC, 1994).

Acrylamide forms in foods that are subjected to high-temperature (>120 °C) processes such as frying, baking and extrusion. It is not present in raw food (i.e. before cooking or processing) or in foods that are processed at lower temperatures (e.g. boiled foods). In model systems and in foods, acrylamide content increases in the temperature range of 120-175°C, then decreases when the material is heated at higher temperatures (Mottram et al., 2002; Rydberg et al., 2003; Taubert et al., 2004). The mechanism(s) by which acrylamide degrades at temperatures >175°C is not known at this time, but there are reports that the compound decomposes and polymerizes at high temperatures (US EPA, 2004).

Research to-date suggests that acrylamide forms in foods mainly through Maillard reactions between reducing sugars and specific amino acids. Using model systems, four different laboratories (Becalski et al., 2003; Mottram et al., 2002; Stadler et al., 2002; and Zyzak et al., 2003) demonstrated that asparagine is the major amino acid precursor. These results explain the occurrence of acrylamide in cereals and potato-based foods which are particularly rich in free asparagine (Mottram et al., 2002). Similar to the Maillard reaction, acrylamide formation in model systems has been shown be at a maximum at about pH 8 (Rydberg et al., 2003).

The finding of acrylamide in food staples in Europe prompted surveys on acrylamide levels in foods consumed in the U.S. (U.S. FDA, 2004) and elsewhere (SNFA, 2002; Leung et al., 2003; Ono et al., 2003; Health Canada, 2003). According to such surveys, acrylamide may be present at concentrations exceeding 2000 µg/kg, especially in fried potato products (potato chips, French fries, hash browns). The surveys indicate that acrylamide levels can vary considerably between brands of a particular food category (i.e. potato chips) and also within lots of a single brand (U.S. FDA,

2004). It is likely that the wide variations of acrylamide concentration in foods are at least partially caused by different levels of acrylamide precursors in various batches of raw materials (Becalski et al., 2003; Roach et al., 2003). They also suggest that acrylamide levels in food are influenced by the method in which they are cooked or processed.

Although acrylamide is known to form during industrial processing of food, high levels of the chemical have been found in home-cooked foods, including baked, fried and roasted potato- and grain-based foods (Biedermann-Brem et al., 2003; US FDA, 2004). One research gap that has been identified at several international meetings on acrylamide (Food and Agriculture Organization/World Health Organization, 2002; JIFSAN, 2002) is the needed to determine the effects of home food preparation on acrylamide formation and to assess the relative contribution of home-cooked foods to dietary exposure to acrylamide. Since acrylamide may have detrimental effects on public health, methods need to be identified for the consumer to reduce acrylamide formation during home preparation of food.

This chapter will focus on research conducted at the FDA-National Center for Food Safety and Technology (NCFST) and elsewhere on 1) the effects of cooking conditions (e.g. time/temperature) on acrylamide formation in consumer-prepared foods, 2) the use of surface color (browning) as an indicator of acrylamide levels in some foods, and 3) methods for reducing acrylamide levels in home-prepared foods.

2. EFFECTS OF COOKING CONDITIONS ON ACRYLAMIDE FORMATION

2.1 Introduction

Research-to-date has shown that the manner in which heat is transmitted to a food (e.g. frying, baking, roasting, grilling, etc.) appears to have a negligible impact on the rate of acrylamide formation (Taubert et al., 2004). However, processing and cooking conditions such as temperature and length of time of heat exposure are important factors affecting the formation and degradation of acrylamide in model systems and in foods (Mottram et al., 2002; Stadtler et al., 2002; Biedermann et al., 2002a,b; Becalski et al., 2003; Leung et al., 2003; Tareke et al., 2002; Taubert et al., 2004). This section will focus on cooking conditions and their effect on acrylamide formation in potato and cereal-based foods, two of the major dietary sources of acrylamide.

2.2 Potato products

2.2.1 Introduction

U.S. (DiNovi and Howard, 2004), Norwegian (NFCA, 2002), Dutch (Konings et al., 2003) and Swedish (Svennsson et al., 2003) authorities have estimated that over one-third of total dietary exposure to acrylamide in Western countries is due to fried, baked and roasted potato products. Cooked or processed potatoes are a significant source of intake of acrylamide since raw potato tubers contain substantial quantities of acrylamide precursors (free asparagine, fructose and glucose) (Amrein et al., 2003). Since the potato is a dietary staple in much of the world, research is needed to study the effects of cooking conditions on acrylamide formation in potato products and to identify conditions that reduce its generation.

2.2.2 Effects of cooking times and temperatures

Work at the NCFST focused on studying acrylamide formation in French fries as a function of cooking (frying, baking) conditions. In frying experiments, commercially available frozen French fries were deep-fried in corn oil at temperatures of 150-190°C for 0-10 min. For baking experiments, the fries were oven baked at 232°C for 16-24 min. These conditions were in the range of cooking conditions recommended by the manufacturer of this product. Acrylamide levels in the French fries were determined by LC-MS using the method of Zyzak et al. (2003).

As shown in Fig. 1, acrylamide content of the fries increased with frying time and temperature. The presence of acrylamide in this product before frying (55 µg/kg) indicates that they were pre-fried (par-fried) by the manufacturer before they were frozen and packaged. Acrylamide levels for deep-fried French fries ranged from 265 µg/kg for samples fried at 150°C for 6 min to 2130 µg/kg for French fries prepared at 190°C for 5 min. The acrylamide levels we report for this French fry product are within the range of levels reported by Becalski et al. (2003) and the US FDA (2004) for restaurant-prepared French fries. It is interesting to note that if this product was prepared according to the maximum frying times/temperatures recommended by the manufacturer (204°C for 5 min), acrylamide levels would have likely exceeded 3000 µg/kg.

Fig. 1 indicates that at higher frying temperatures (180-190 °C), acrylamide levels increased exponentially at the end of the frying process. Similar results were reported by Grob et al. (2003) for French fries prepared from fresh cut potatoes. This phenomenon is likely due to the fact that acrylamide formation occurs at the surface of food during cooking. At the end of the frying process, this surface becomes sufficiently dry as to allow the temperature to rise to >120°C, the temperature above which acrylamide is believed to form (Mottram et al., 2002, Tareke et al., 2002; Taubert et al., 2004).

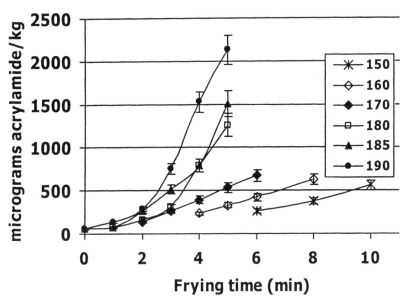

Figure 1. Effects of frying time and temperature on acrylamide formation in French fries that were fried in corn oil for 0-10 min at temperatures of 150-190°C. Acrylamide levels were measured by the LC-MS method of Zyzak et al. (2003). All frying trials were done in triplicate and acrylamide analyses were done four times. Error bars represent one standard deviation of the mean (Jackson et al., 2004).

We (Jackson et al., 2004) as well as others (Tareke et al., 2002; Biedermann et al., 2002a; Grob et al., 2003; Rydberg et al., 2003) studied the effects of baking on acrylamide formation in French fries. In our experiments, acrylamide levels in frozen French fries baked at 232°C increased with baking time and ranged from 198-725 µg/kg (Fig. 2). These values are within the range reported by Grob et al. (2003) for a similar product. Similar to the frying experiments, acrylamide levels increased exponentially during the baking run. This was likely due to a higher rate of acrylamide formation in the dry crust of the French fries.

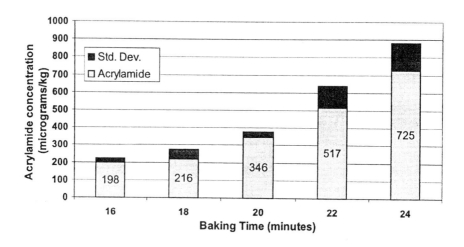

Figure 2. Effect of baking time on acrylamide formation in French fries (frozen) baked at 232°C for 16-24 min (as per manufacturer's cooking suggestions). Acrylamide levels were measured by the LC-MS method of Zyzak et al. (2003). All baking trials were done in triplicate and acrylamide analyses were done four times (Jackson et al., 2004).

2.2.3 Effects of other cooking conditions

Besides time and temperature, other cooking factors may alter acrylamide formation in fried potato products. Tauber et al. (2004) reported that the surface area of raw potato pieces influenced acrylamide formation during frying. In shapes with lower surface to volume ratios (SVRs), such as potato slices (potato chips), acrylamide levels increased with increasing frying times and temperatures. However, in shapes with high SVRs (i.e. shredded potato), acrylamide levels were greatest at 160-180°C, then decreased with higher frying temperatures and more prolonged frying times. These results suggest that the surface of potato pieces with large SVRs may reach 175°C, resulting in degradation of acrylamide. In contrast, the temperature at the surface of potato pieces with small SVR never approaches 175°C, even when frying oil temperatures reach 220°C (Taubert et al., 2004).

There is conflicting information on the effects of frying oil type and age on formation of acrylamide. Becalski et al. (2003) reported that potatoes pan-fried in olive oil had 60% more acrylamide than those fried in corn oil, while Gertz and Klostermann (2002) found higher acrylamide concentrations in potatoes fried in palm oil than other cooking oils. Roach et al. (2003) found more acrylamide in potato chips fried in thermally aged oil than those fried in fresh cooking oil. Possible explanations for these results include differences in the heat transfer characteristics for the oils (Grob et al., 2003) and the formation of carbonyls (Maillard browning precursors) in oils that are less thermally stable. In contrast to the above studies, we found that commonly used frying oils (peanut, canola, corn, safflower, olive and hydrogenated soybean) had no significant effect on acrylamide levels in French fries deep-fried at 180°C for 4 min (Jackson et al., 2004).

Grob et al. (2003) reported that the ratio of the frying oil: potato may affect acrylamide formation during deep-frying. Acrylamide levels were minimized at oil: potato ratios of 1.5 L oil: 100 g potato since at this ratio, the oil temperature remained over 140°C throughout the frying process, resulting in decreased frying times.

2.2.4 Surface browning and acrylamide formation

Since acrylamide formation increases exponentially towards the end of the frying or baking process, an important factor for minimizing acrylamide formation is to determine the proper cooking end-point. This may involve the use of visual clues such as surface browning as an indicator of product "doneness". Since acrylamide and the brown color of cooked foods are formed in similar reactions and from similar precursors, it is likely that acrylamide formed in parallel with browning (Amrein et al., 2003).

We studied surface browning of the French fries as a function of frying conditions and acrylamide formation. Not unexpectedly, the surface of fries became visibly browner as frying times and temperatures increased. A colorimeter (Hunterlab Labscan XE, Reston, VA) was used to obtain a more objective assessment of surface color as a function of frying conditions. As the fries were fried for longer periods of time and at higher temperatures, the "L" component of color (a measure of the white/black component of color) decreased while the "a" values (degree of redness) increased. Statistical analysis (regression analysis) indicated that the "a" and "L" components of color correlated highly (r^2=0.8558 and r^2=0.8551, respectively) with the log of acrylamide levels in the French fries (Fig. 3). In contrast, the "b" color values (a measure of the yellow/blue component of color) for the samples correlated poorly with acrylamide levels ($r^2 = 0.089$; data not shown). Visual examination of the fries indicated that acrylamide levels tended to be lower in fries that were golden in color with light browning at the edges. More extensive browning of the surface resulted in acrylamide levels of >1000 µg/kg.

We also studied surface browning as a function of acrylamide levels in baked French fries (Jackson et al., 2004). Visual inspection of the fries indicated more extensive surface browning as baking time increased. However, the degree of surface browning between replicate baked samples was much more variable than between replicate fried samples. Similarly colorimeter measurements of surface color for baked samples were more variable than for fried samples. Colorimeter determination of surface color for baked fries indicated a weak positive correlation of "L" and "a" values with acrylamide ($r^2=0.3425$ and $r^2=0.3502$; data not shown). As with fried samples, the "b" color values correlated poorly with acrylamide levels ($r^2 = 0.039$). The greater variability in the results for baked samples than fried samples is not unexpected since deep frying produces more even heating of the food surface than baking.

A high degree of correlation between the brown color of the fries and acrylamide levels is not unexpected since acrylamide is formed in similar reactions responsible for the development of flavor and brown color in cooked foods. Taubert et al. (2004) studied the impact of browning level on acrylamide formation in potato slices of different surface areas. They found a close linear correlation between browning levels and acrylamide concentration for fried potato slices that had lower surface-to-volume ratios (e.g shapes that approximated potato chips). However, color did not correlate well with acrylamide levels in potato pieces with high surface area (shredded potatoes). These authors suggested that the lack of correlation in shredded potatoes is due to degradation of acrylamide at the end of the frying process.

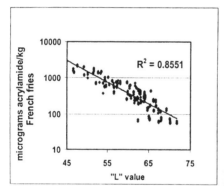 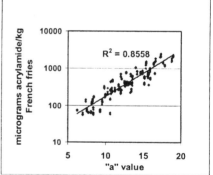

Figure 3. Correlation between "L" and "a" color values and acrylamide levels in deep-fried French fries cooked for 0-10 min in corn oil at 150-190°C (Jackson et al., 2004).

Color measurements (both visual and instrumental) indicate that regardless of frying conditions, it is the degree of surface browning that determines final acrylamide content of French fries. A lower frying temperature combined with a prolonged frying time did not result in lower acrylamide content if the product had similar amounts of surface browning. Therefore, determining the degree of surface browning appears to be a good method for estimating acrylamide formation during cooking (Tareke et al.,

2002). All studies point to the need to control the cooking process to achieve the desired food qualities such as proper flavor and texture development, yet prevent excessive browning. They also stress the importance of using surface browning rather than cooking times and temperatures to determine the degree of doneness of potato products.

2.3 Bread products

According to several estimates (US FDA, 2004; Svensson et al., 2003; Konings et al., 2003), bread, crackers, cookies, breakfast cereals and other grain-based products are major dietary sources of acrylamide. At present, little has been published on the effects of cooking conditions on acrylamide formation in grain-based foods, especially those prepared at home. Surdyk et al. (2004) examined baking conditions on acrylamide content of yeast-leavened wheat bread. In experiments with wheat flour naturally deficient in asparagine as well as asparagine supplemented flour, over 99% of acrylamide in bread was found in the crust (Surdyk et al., 2004). They also reported that temperature, but also time, increased acrylamide levels in the bread crust. When baked with the same recipe but with different baking temperatures and times, Surdyk et al. (2004) found a strong correlation between bread crust color and acrylamide content. These results indicate that color could be used as a gauge of acrylamide formation during bread-making. They also indicate that if at all possible, baking conditions should allow proper crumb formation, but yet prevent excessive browning in the crust.

Since bread is frequently toasted before it is consumed, work at the NCFST focused on studying the effects of toasting conditions on acrylamide formation in six different types of store-bought bread. "Light", "medium", and "dark" degrees of doneness were achieved by toasting the bread slices for 10, 13 and 15 min, respectively at the broil/bake setting in a toaster oven. Infrared (IR) thermometry measured the surface temperature of bread slices during toasting.

In general, acrylamide levels increased with toasting time (Fig. 4). However, in most cases, toasting to a "light" degree of doneness resulted in a negligible increase in acrylamide levels as compared to the untoasted bread. Similarly, Tareke et al. (2002) reported only slight increases in acrylamide levels as white bread was toasted (unspecified degree of doneness). As measured by IR thermometry, average surface temperatures of the bread slices at the end of the toasting runs were 120-143°C, 139-178°C, and 143-223°C for "light", "medium", and "dark" toasted bread, respectively. These findings are in accordance with previous studies (Tareke et al., 2002; Taubert et al., 2004; Becalski et al., 2003) that formation of acrylamide requires the food surface to reach temperatures >120°C. The data indicate that for most types of bread, toasting to a "medium" degree of doneness results in small to moderate acrylamide levels (<200 µg/kg). However, "dark" toast made from potato bread may have substantial (>600 µg/kg) acrylamide levels. Overall, potato bread tended to form more acrylamide during toasting than other bread types. This is likely due to higher concentrations of asparagine in breads containing potato flour than those without.

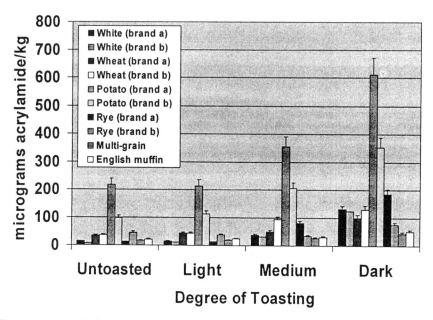

Figure 4. Degree of toasting affects acrylamide levels in bread. Six different varieties of bread (up to two brands of each variety) were toasted to light, medium and dark degrees of doneness. Three toasting trials were done for each type of bread and toasting condition. Acrylamide levels were measured as described previously. Error bars represent one standard deviation of the mean (Jackson et al., 2004).

In a separate experiment, we determined that scraping the surface of darkly toasted potato bread to remove the browned portions reduced acrylamide levels from 483 µg/kg to 181 µg/kg. These findings support data that indicate that over 99% of acrylamide in bread is found in the crust (Surdyk et al., 2004). Overall these results indicate that consumers should avoid toasting bread to a "dark" degree of doneness to reduce their exposure to acrylamide. In addition, scraping the toast surface to remove excessively browned portions could further reduce acrylamide levels.

The only other publication that describes cooking conditions and their effects on acrylamide formation in cereal-based foods besides bread is by Amrein et al. (2004). In their study, the effects on baking times and temperature on acrylamide formation in gingerbread were investigated. Both at 180°C and 200°C, acrylamide formation increased linearly over the 20 min baking period. This contrasts with acrylamide formation in potato products where acrylamide formation increased exponentially with time in this temperature range. Similar to our work with French fries, Amrein et al. (2004) reported that the degree of browning in baked gingerbread was an excellent predictor of acrylamide levels. The overall message from this work is that prolonged baking or excessive browning should be avoided to minimize acrylamide formation in gingerbread or other baked goods.

3. METHODS FOR REDUCING ACRYLAMIDE LEVELS IN FOOD

3.1 Raw ingredient selection and storage

Current research indicates that some simple measures may reduce acrylamide formation during home-preparation of foods. One method is to use raw ingredients such as potatoes or grain products that contain lower levels of acrylamide precursors. For example, an active area of research has been to identify potato cultivars that have reduced amounts of asparagine and reducing sugars. Amrein et al. (2003) and Becalski et al (2004) found that reducing sugar levels varied by a factor of 32 among potato cultivars, while free asparagine contents varied only within a narrow range. Both reported a linear relationship between sugar content of the raw tuber and the potential for acrylamide formation. Amrein et al. (2002) reported that neither the farming system (organic vs. conventional) nor extent of nitrogen fertilization influenced precursor levels and acrylamide forming potential in potatoes. These authors concluded that acrylamide content in cooked/processed potato products can be substantially reduced by selecting cultivars with low levels of reducing sugars.

A high degree of variation in reducing sugar content among potatoes of the same cultivar suggest that other factors such as storage conditions may have an even stronger influence on sugar content than cultivar. We studied acrylamide formation in French fries prepared from two different varieties of fresh potatoes (Russet and Klondike Rose) stored at room temperature (22-26 °C) or under refrigeration conditions (6-8 °C) for up to 28 days (Jackson et al., 2004). These conditions were an attempt to mimic conditions used by consumer to store fresh potatoes. After storage for 0, 1, 7, 14, 21, and 28 days, the potatoes were sliced into strips, soaked in water for 15 min, and then fried in corn oil (180°C, 3 min). The French fries were analyzed for acrylamide content using methods described earlier.

French fries prepared from Klondike Rose potatoes before storage contained more acrylamide (1132 µg/kg) than those made with Russet potatoes (677 µg/kg) (Fig. 5). The acrylamide forming potential of both potato cultivars did not significantly change during the 28 days of storage under refrigeration conditions. In contrast, acrylamide levels in fries made from potatoes stored at room temperature decreased during the 28 day study (Fig. 5). Fries made from Klondike Rose potatoes showed approximately a 73% decrease in acrylamide concentration while those from Russet potatoes had about a 50% decrease in acrylamide levels. Noti et al. (2003) and Biedermann et al. (2002b) reported that even short-term storage of potatoes at 4°C markedly increases the potential for acrylamide formation. Collectively, these data suggest that the potatoes used in our study were stored at refrigeration temperatures in the supermarket from which they were purchased.

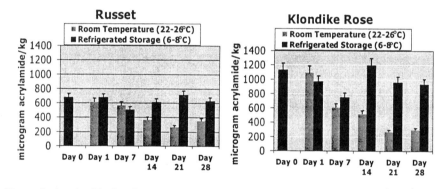

Figure 5. Acrylamide forming potential in potatoes stored at room temperature and under refrigeration conditions. Russet and Klondike Rose potatoes were stored at 6-8°C or 22-26°C for up to 28 days. French fries were prepared by frying (180 °C; 3 min) slices of stored potato in corn oil (Jackson et al., 2004).

Work by Noti et al. (2003), Biedermann et al., (2002b), as well as data generated in our laboratory, indicate that cooling potatoes to temperatures <10°C causes an increase in reducing sugars and thus, an increase in the potential for acrylamide formation. The phenomenon by which potatoes accumulate sugars during storage at low temperatures is known as "cold sweetening" and is believed to be a response by the potato plant to protect the tuber from freezing (Amrein et al., 2003). Biedermann et al. (2002b) stored potatoes of the cultivar Erntestolz for 15 days at 4°C and showed an increase in reducing sugars and the acrylamide forming potential by a factor of 28. Our work (Jackson et al., 2004) as well as reports by Amrein et al. (2003) and Biedermann et al. (2002b), show a high degree of correlation (r^2 >0.89) between reducing sugar content (glucose + fructose) and the acrylamide forming potential in potatoes Biedermann-Brem et al. (2003) determined that potatoes used for roasting and frying should contain less than 1 g/kg fresh weight of reducing sugar to minimize acrylamide formation during cooking.

To minimize acrylamide formation in potatoes cooked at high temperatures (fried, baked, roasted), it is imperative that raw potatoes not be stored at temperatures <10°C. This may be difficult especially during the late spring months when the majority of potatoes sold are from the previous year's harvest. During this time period, low temperature storage is needed to inhibit sprout formation and to prevent softening of the potato tissue and spoilage. More work is needed to identify or develop potato cultivars that are resistant to the "cold sweetening" phenomenon.

In baked grain-based products, there are several ingredients that may enhance acrylamide formation during baking. Amrein et al. (2004) found that use of ammonium hydrogen carbonate (ammonium bicarbonate) rather than sodium hydrogen carbonate (sodium bicarbonate) as a leavening agent in gingerbread augmented levels by three-fold. The promoting effect of ammonium hydrogen carbonate on the formation of acrylamide might be indirect by providing more reactive carbonyls originating from the reaction of ammonia with reducing sugars (Amrein et al., 2004). Amrein et al. (2004) also reported that using sucrose instead of honey or inverted sugar syrup in a gingerbread recipe reduced acrylamide content by a factor of 20. These results are explained by the fact both honey and inverted sugar syrup are a rich source of reducing sugars, precursors of acrylamide. Clearly more work is needed to identify ingredient or recipe changes that result in reduced acrylamide levels in home-prepared foods and to determine the effects of the changes on the sensory properties and acceptability of these foods.

3.2 Cooking conditions

Several investigators have made attempts at identifying cooking conditions that result in reduced acrylamide formation in home-prepared foods (Biedermann et al., 2002b; Grob et al., 2003; Taubert et al., 2004). Conditions that minimize acrylamide in French fries involve frying or baking potato pieces only as long as necessary to get the surface golden in color and the texture crispy (Grob et al., 2003). In fried potato products,

Grob et al. (2003) found that keeping the oil temperature at 170°C or lower was effective at minimizing acrylamide formation if the cooking process was terminated once the potatoes were golden in color. Using raw potatoes with low amounts of reducing sugars, a 15 min soak to remove surface precursors, frying temperatures of 165-170°C, and a oil: potato ratio 1L: 100 g, Grob et al. (2003) was able to consistently produce French fries with 40-70 µg acrylamide/kg. Suggestions made by Grob et al. (2003) to limit acrylamide formation in baked or roasted potatoes included the use of fresh potatoes with reduced amounts of reducing sugars and lower cooking temperatures <250°C. In grain-based foods such as breads and cookies, baking should proceed until the proper moisture levels are obtained and minimum browning in the crust or surface occurs. More work is needed to determine the flavor, texture and acceptability of products cooked at conditions that reduce acrylamide formation.

3.3 Additives

Some work has been published on additives that modulate acrylamide formation in food. In experiments aimed at finding conditions that might reduce or prevent the formation of acrylamide, Tareke et al. (2002) found that addition of oils, antioxidants, or argon gas during frying of beef had minor or nonsignificant acrylamide-reducing effects. Vattem and Shetty (2003) and Rydberg et al. (2003) found that the presence of sodium ascorbate, ascorbyl palmitate, or the phenolic antioxidants from cranberry and oregano extract had no effect or slightly enhanced acrylamide formation in potatoes. In contrast, Becalski et al. (2003) found that adding ground rosemary to olive oil resulted in a 25% reduction in acrylamide formation in fried potato slices.

Amrein et al. (2003) measured the ability of different amino acids to reduce acrylamide formation in gingerbread. The amino acids, glycine and L-glutamine, did not affect acrylamide levels, but addition of L-cysteine reduced the acrylamide content of gingerbread. Similarly, Biedermann et al. (2002b) reported that addition of cysteine to potato reduced acrylamide levels by 95%. Unfortunately, cysteine addition is not a practical method for affecting acrylamide formation since the amino acid imparts an unpleasant odor and flavor to food (Amrein et al., 2004). Other amino acids that have been found to reduce acrylamide formation in heated potato include glycine, alanine, lysine, glutamine and glutamic acid (Rydberg et al., 2003). However, it is not clear whether amino acid treatments can be used with any practicality to prevent or reduce acrylamide formation in cooked food.

Several papers discuss the effects of pH reducing treatments on acrylamide formation. Citric acid added at levels of 0.5 and 1.0 g/100 g to gingerbread reduced acrylamide formation by factors of 4 and 40, respectively (Amrein et al., 2004). These results support those by Jung et al. (2003) and Rydberg et al. (2003) who found that citric acid treatments prevented acrylamide formation in fried French fries, baked tortilla chips and an oven-heated potato homogenate. Rydberg et al. (2003) found major reductions (>90%) in acrylamide formation when ascorbic acid was added to homogenized potato and then microwave heated (3 min, 750 W). Acid treatments are effective at preventing acrylamide formation since they lowered pH into the range where acrylamide formation is minimized (<pH 5) (Jung et al., 2003).

Zyzak et al. (2003) reported that treating potato with asparaginase, an enzyme converts asparagine into aspartic acid, reduced acrylamide levels by 99%. There has been some interest in using the enzyme to reduce acrylamide formation in some commercial fried potato snack products (JIFAN, 2004; Zyzak et al., 2004).

Work-to-date suggests that, in general, additives have little to no effect on acrylamide formation. With treatments such as citric acid and asparaginase, where significant acrylamide reductions are possible, more work is needed to determine if they can be used for reducing acrylamide formation in home-prepared foods. In addition, more research is needed to determine if these treatments affect the acceptability (flavor, color, texture, etc.) of food.

3.4 Washing treatments and acid dips

Rinsing and soaking treatments have been effective at reducing acrylamide formation in French fries prepared from fresh-cut potatoes. We found that soaking potato slices in room temperature water for at least 15 min before frying resulted in 63% reduction in acrylamide (Jackson et al., 2004). Similarly, Grob et al. (2003) found that soaking potato pieces for 10 min in cold or warm water resulted in desirable flavor and texture when fried, yet had only half the acrylamide content of the comparable untreated slices. Soaking treatments reduce acrylamide formation by leaching out sugars and asparagine from the surface of the potato slice.

Using acid solutions rather than water to soak potatoes had enhanced ability to reduce acrylamide levels. We found that soaking the potatoes in 1:3 vinegar:water reduced the acrylamide forming potential by 75%. No greater effect was seen when the wash solution contained higher vinegar levels (1:1 vinegar:water). Jung et al. (2003) showed that dipping potato cuts in 1% and 2% citric acid solutions for one hour before frying inhibited acrylamide formation in French fries by 73.1% and 79.7%. Use of acid washes for preventing acrylamide formation likely is due to the drop in pH at the surface of the potato slice as well as to leaching losses of asparagine and reducing sugars.

4. CONCLUSIONS

Although current research indicates that some simple measures can be used by consumers to reduce acrylamide formation during home-preparation of foods, no method has been successful at totally preventing formation of acrylamide. One of the most practical and most efficacious methods for reducing acrylamide formation is to avoid frying, baking, broiling and grilling foods at excessively high temperatures and for long times. Since formation of acrylamide increases exponentially toward the end of the cooking process, an important factor for minimizing its formation is to determine the proper cooking endpoint. Research-to-date points to the need to control the cooking process to achieve the desired qualities, such as flavor and texture, yet prevent excessive browning. Degree of surface browning rather than cooking times and temperatures appears to be a good method of estimating the "degree of doneness" of some foods.

Several approaches have been identified for reducing acrylamide formation in home-prepared foods made from potato. They include proper storage of the raw potato tubers (i.e. >10°C) and soaking potato slices in water or acid solutions for at least 15 min before cooking (baking, frying or roasting). Acrylamide formation in bakery products can be minimized by using sodium hydrogen carbonate rather than ammonium hydrogen carbonate as a leavening agent and sucrose rather than sweetening agents high in reducing sugars. More work is needed to identify methods for consumers to reduce acrylamide formation in home-prepared foods. In addition, the consumer acceptability of foods cooked under conditions that prevent or reduce acrylamide levels needs to be determined.

REFERENCES

Amrein, T., Bachman, S., Noti, A., Biedermann, M., Barbosa, M., Biedermann-Brem, S., Grob, K., Keiser, A., Realini, P., Escher, F. and Amado, R., 2003, Potential of acrylamide formation, sugars, and free aparagine in potatoes: A comparison of cultivars and farming systems, *J. Agric. Food Chem.* **51**: 5556-5560.

Amrein, T.M., Schonbachler, B., Escher, F., and Amado, R., 2004, Acrylamide in gingerbread: Critical factors for formation and possible ways for reduction, *J. Agric. Food Chem.* **52**: 4282-4288

Bachman, M., Myers, J, and Bezuidenhout, B., 1992, Acrylamide monomer and peripheral neuropathy in chemical workers, *Am. J. Ind. Med.* **21**: 217-222.

Becalski, A., Lau, B., Lewis, D., and Seaman, S., 2003, Acrylamide in Foods: Occurrence, Sources, and Modeling., *J. Agr. Food Chem.* **51:** 802-808.

Becalski, A., Lau, B.P.-Y., Lewis, D., Seaman, S.W., Hayward, S., Sahagian, M., Remesh, M. and Leclerc, Y., 2004, Acrylamide in French fries: Influence of free amino acids and sugars, *J. Agric. Food Chem.* **52:** 3801-3806.

Biedermann, M., Biedermann-Brem, S., Noti, A., and Grob, K., 2002a, Methods for determining the potential of acrylamide formation and its elimination in raw materials for food preparation, such as potatoes, *Mitt. Lebensm. Hyg.* **93:**653-667.

Biedermann, M., Noti, A., Biedermann-Brem, S., Mozzarti, V., and Grob, K., 2002b, Experiments on acrylamide formation and possibilities to decrease the potential of acrylamide formation in potatoes, *Mitt. Lebensm. Hyg.* **93:** 668-687.

Biedermann-Brem, S., Noti, A., Grob, K., Imhof, D., Bazzocco, D., and Pfefferle, A., 2003, How much reducing sugar may potatoes contain to avoid excessive acrylamide formation during roasting and baking, *Eur. Food Res. Technol.* **217:** 369-373.

Callemen, C.J., Wu, Y., He, F., Tian, G., Bergmark, E., Zhang, S., Deng, H., Wang, Y., Crofton, K.M., Fennell, T.F., and Costa, L.G., 1994, Relationships between biomarkers of exposure and neurological effects in a group of workers exposed to acrylamide, *Toxicol. Appl. Pharmacol.* **126:** 361-371.

Dearfield, K. L., Douglas, G.R., Ehling, U.H., Moore, M.M., Sega, G.A., and Brusick, D. J., 1995, Acrylamide: a review of its genotoxicity and an assessment of heritable genetic risk. Mutat. Res. **330:** 71-99.

DiNovi, M. and Howard, D. 2004, The updated exposure assessment for acrylamide. Acrylamide in Food: Update - Scientific Issues, Uncertainties, and Research Strategies, (April 13, 2004); http://www.jifsan.umd.edu/presentations/acry2004/acry_2004_dinovihoward_files.

Food and Agricultural Organization/World Health Organization (FAO/WHO), 2002, FAO/WHO consultations on the health implications of acrylamide in foods. Summary report of a meeting held in Geneva, June 25-27, 2002

Friedman, M.A., Dulak, L.H., and Stedman, M.A., 1995, A lifetime oncogenicty study in rats with acrylamide, *Fundam. Appl. Toxicol.* **27:** 95-105.

Grob, K., Biederman, M., Biedermann-Brem, S., Noti, A., Imhof, D., Amrein, T., Pfefferle, A., and Bazzocco, D., 2003, French fries with less than 100 µg/kg acrylamide. A collaboration between cooks and analysts, *Eur Food Res Technol.* **217:** 185-194.

Health Canada, 2002, Acrylamide and Food. http://www.hc-sc.gc.ca/food-aliment/cs-ipc/chha-edpcs/e_acrylamide_and_food.html

IARC, 1994, Acrylamide. In *IARC Monographs on the Evaluation of Carcinogen Risk to Humans: Some Industrial Chemicals*; International Agency for Research on Cancer: Lyon, France, 1994; vol. 60, pp 389-433.

Jackson, L., Al-Taher, F., Jablonski, J., and Bowden, T., 2004, Unpublished data.

JIFSAN. Acrylamide in Food Workshop: Scientific Issues, Uncertainties, and Research Strategies, Rosemont, IL, (October 28-30, 2002); http://www.jifsan.umd.edu/acrylamide2002.htm

JIFSAN. Acrylamide in Food Workshop. Update: Scientific Issues, Uncertainties, and Research Strategies, Rosemont, IL, April, 13-15 2004 http://:vww.jifsan.umd.edu/acrylamide2004_anmt.htm

Johnson, K.A., Gorzinski, S.J., Bodner, K.M, Campbell, R.A., Wolf, C.H., Friedman, M.A., and Mast, R.W., 1986, Chronic toxicity and oncogenicity study on acrylamide incorporated in the drinking water of Fischer 344 rats. *Toxicol. Appl. Pharmacol.* **85:** 154-168.

Jung, M.Y., Choi, D.S. and Ju, J.W., 2003, A novel technique for limitation of acrylamide formation in fried and baked corn chips and in French fries. *J. Food Sci.* **68:** 1287-1290.

Konings, E.J.M., Baars, A.J., van Klaveren, J.D., Spanjer, M.C, Rensen, P.M., Hiemstra, M., van Kooij, J.A., and Peters, P.W.J., 2003, Acrylamide exposure from foods of the Dutch population and an assessment of the consequent risks, *Food and Chem. Toxicol.* **41:** 1569-1579.

Leung, K.S., Lin, A., Tsang, K, and Yeung, S.T.K., 2003, Acrylamide in Asian foods in Hong Kong. *Food Addit. Contam.* **20:** 1105-1113.

LoPachin, R.M., 2004, The changing view of acrylamide neurotoxicity, *Neurotoxicol.* **25:** 617-630.

Mottram, D.S., Wedzicha, B.L. and Dodson, A.T., 2002, Acrylamide is formed in the Maillard reaction, *Nature* **419:** 448-449.

NFCA. 2002, Risk assessment of acrylamide intake from foods with special emphasis on cancer risk. Report of the Scientific Committee of the Norwegian Food Control Authority. (June 6, 2002); http://snt.mattilsynet.no/nytt/tema/Akrylamid/acrylamide.pdf

Noti, A., Biedermann-Brenn, S., Biedermann, M., Grob, K., Albisser, P., and Realini, P. 2003, Storage of potatoes at low temperature should be avoided to prevent increased acrylamide formation during frying or roasting, *Mitt. Lebensm. Hyg.* **94:** 167-180.

Ono, H., Chuda, Y., Ohnishi-Kameyama, M., Yada, H., Ishizaka, M., Kobayashi, H., and Yoshida, M., 2003, Analysis of acrylamide by LC-MS/MS and GC-MS in processed Japanese foods, *Food Addit. Contam.* **20:** 215-220.

Roach, J.A.G., Andrzejewski, D., Gay, M.L., Northrup, D. and Musser, S.M., 2003, Rugged LC-MS/MS survey analysis for acrylamide in foods, *J. Agric. Food Chem.* **51:**7547-7554.

Rydberg, P., Eriksson, S., Tareke, E, Karlsson, P., Ehrenberg, L, and Tornqvist, M., 2003, Investigations of factors that influence the acrylamide content of heated foodstuffs, *J. Agric. Food Chem.* **51:** 7012-7018.

SNFA, 2002, Analytical methodology and survey results for acrylamide in foods, http://www.slv.se/engdefault.asp

Stadler, R.H., Blank, I., Varga, N., Robert, F., Hau, J., Guy, P. A., Robert, M. C., and Riediker, S., 2002, Acrylamide from Maillard reaction products, *Nature* **419:**449-450.

Surdyk, N., Rosen, J., Andersson, R., and Aman, P., 2004, Effects of asparagine, fructose and baking conditions on acrylamide content in yeast-leavened wheat bread, *J. Agric. Food Chem.* **52:** 2047-2051.

Svennsson, K., Abramsson, L., Becker, W., Glynn, A., Hellanäs, K.-E., Lind, Y., and Rosen, J., 2003, Dietary intake of acrylamide in Sweden, *Food Chem. Toxicol.* **41,** 1581-1586.

Tareke, E., Rydberg, P., Karlsson, P., Eriksson, S. and Tornqvist, M., 2002, Analysis of acrylamide, a carcinogen formed in heated foodstuffs, *J. Agric. Food Chem.* **50:** 4998-5006.

Taubert, D., Harlfinger, S., Henkes, L., Berkels, R. and Schomig, E., 2004, Influence of processing parameters on acrylamide formation during frying of potatoes, *J. Agric. Food Chem.* **52:** 2735-2739.

Towell, T.L., Shell, L., Inzana, K.D., Jortner, B.S., and Ehrich, M., 2000, Electrophysiological detection of the neurotoxic effects of acrylamide and 2,5-hexanedione on the rat sensory system, *Int. J. Toxicol.* **19:** 187-193.

US EPA, 1994, Chemical summary for acrylamide. Office of Pollution Prevention and Toxics. U.S. Environmental Protection Agency (September 1994); http://www.epa.gov/opptintr/chemfact/s_acryla.txt.

US FDA, 2004, United States Food and Drug Administration (FDA): Exploratory Data on Acrylamide in Foods, CFSAN/Office of Plant & Dairy Foods, (March 2004); http://www.cfsan.fda.gov/~dms/acrydata.html

WHO, 1996, *Guidelines for Drinking-Water Quality*, 2nd ed.; World Health Organization: Geneva, Switzerland, Vol.2, pp. 940-949, www.who.int/foodsafety/publications/chem./en/acrylamide_summary.pdf

Vattem, D.A. and Shetty, K., 2003, Acrylamide in food: a model for mechanism of formation and its reduction, *Innov Food Science Emerg Tech.* **4:**331-338.

Zyzak, D.V., Sanders, R.A., Stojanovic, M., Tallmadge, D.H., Eberhart, B.L., Ewald, D.K., Gruber, D.C., Morsch, T.R., Strothers, M.A., Rizzi, G.P. and Villagran, M.D., 2003, Acrylamide Formation Mechanism in Heated Foods, *J. Agr. Food Chem.* **51:**4782-4787.

Zyzak, D.V., Lin, P.Y.T., Sanders, R.A., Stojanovic, M., Gruber, D.C., Villagran, M.D. M.-S., Howie, J.K., and Schafermeyer, R.G., 2004, Method for reducing acrylamide in foods, foods having reduced levels of acrylamide, and article of commerce, U.S. Patent Application #20040101607, May 27, 2004, U.S. Patent Office.

AUTHOR INDEX

Adami, H.-O., 39
Albright, K.J, 343
Al-Taher, F., 447
Amado, R., 431
Amrein, T.M, 431
Andersen, M.E, 117
Baum, M, 77
Becalski, A, 271
Bertow, D., 77
Blank, I, 171
Burch, R., 387
Channell, G.A, 303
Chuda, Y, 405
Clarke, M, 387
Cook, D.J, 303
Devaud, S, 171
Dodson, A.T, 235, 255
Ehling, S, 223
Ehrenberg, L, 317
Eisenbrand, G, 77
Elmore, J.S, 235, 255
Eriksson, S, 285, 317
Escher, F, 431
Fauth, E, 77
Fennell, T.R, 109
Franke, K, 357
Friedman, M., 135
Friedman, M.A., 109
Fritzen, S, 77
Goldmann, T, 171
Granvogl, M., 205
Hamlet, C., 415
Hanley, A.B, 387
Hayashi, N, 405
Hengel, M, 223
Herrmann, A, 77
Hobayashi, H, 405
Hyunh-Ba, T, 171

Ing, B, 387
Jackson, L.S, 447
Jang, S., 329, 393
Kamendulis, L.M., 49
Karlsson, P, 285, 317
Klaunig, J.E, 49
Köhler, P, 205
Koutsidis, G, 235, 255
Lau, B.P.-Y, 271
Lewis, D, 271
Licea-Perez, H, 89, 97
Lindsay, R.C, 329, 343, 393
Liu, W, 343
Locas, C.P, 191
LoPachin, R.M., 21
Matissek, R, 293
Matsura-Endo, C, 405
Mayers, G.L., 89, 97
Mertes, P, 77
Mi, L, 97
Mori, M, 405
Mottram, D.S, 235, 255
Mucci, L.A, 39
Myers, T, 89, 97
O'Brien, J., 191
Offen, C, 387
Ohara-Takada, A, 405
Ohnishi-Kameyama, M, 405
Olin, S.S., 117
Ono, H, 405
Ospina, M, 89, 97
Pariza, M.W., 343
Park, Y, 343
Paulsson, B, 127
Petersen, B.J, 63
Pollien, P, 171
Rannug, A, 127
Raters, M., 293

Reimerdes, E.H., 357
Robert, F, 171
Roberts, M, 387
Rudolphi, M, 77
Rydberg, P, 317
Sadd, P, 415
Saucy, F, 171
Schieberle, P, 205
Schonbachler, B, 431
Scimeca, J, 117
Seaman, S.W, 271
Sell, M, 357
Shibamoto, K., 223
Silva, E.M, 371
Simon, P.W., 371
Spormann, T, 77
Stadler, R.H., 157, 171
Storkson, J.M, 343
Sun, W.F., 271
Tareke, E, 317
Taylor, A.J., 303
Törnqvist, M., 1, 127, 317
Tran, N., 63
Varga, N, 171
Vesper, H.W, 89, 97
Warholm, M, 127
Wedzicha, B.L, 235, 255
Wronowski, A, 191
Yada, H, 405
Yamaguchi, Y., 405
Yang, H, 343
Yaylayan, V.A, 191
Yoshida, M, 405
Zankl, H, 77

SUBJECT INDEX

Acetol, 206
Acidulants, 348
Acrolein, 128, 161, 172, 229
Acrylamide
  adducts, 1, 30
  analysis, 10, 97, 159, 174, 192, 205, 272, 288, 293
  and Maillard products, 135
  and nutritional status, 74
  binding, 286
  biological effects, 136
  biomarkers, 77, 89
  bromination, 276
  cancer risks, 15, 66
  carcinogenicity, 22, 39, 49
  chemistry, 287
  chronic exposure, 50
  communication of risk, 13
  crackers, 72
  derivatization, 208
  detoxification, 127
  dietary sources, 71
  distribution within sample, 359
  DNA reactions, 53
  dosimetry, 77
  effect on liver size, 22
  epidemiology, 41, 44
  exposure, 63, 89
  formation, 157, 165, 171, 191, 205, 210, 223, 256, 286, 298, 321, 334, 348, 394, 422, 435
  French fries, 163
  GC-MS, 274, 345
  genotoxicity, 23, 77
  historical perspective, 10
  history, 286
  human carcinogen, 39
  human metabolism, 113
  human studies, 5
  in breads, 72, 164, 455
  in breakfast cereals, 165
  in chocolate, 279
  in coffee, 14, 39, 166, 279
  in cookies, 72
  in French fries, 273, 358, 450
  in foods, 11, 41, 64, 70
  in gingerbread, 165
  in meat, 11
  in noodles, 408
  in peanut butter, 279
  in potato chips (crisps), 72, 89, 279, 297
  in potato products, 11, 162, 237, 256, 373, 395, 452
  in rice crackers, 408
  in roasted barley, 408
  in rusk, 210
  in rye products, 237, 256
  in snack foods, 72
  in tea, 411
  in toasted bread, 455
  in wheat products, 237, 256
  industrial applications, 21
  inhibition of formation, 163, 167
  intakes, 14, 22, 43, 67, 91
  interlaboratory tests, 297
  kinetics of formation, 235
  LC-MS/MS, 274, 294, 319
  lipids, 223
  mechanism of formation, 161, 171, 191, 205, 230, 236
  mechanism of nerve damage, 29

metabolism, 24, 98, 109, 117
minimization, 298, 358, 459
morphological effects, 22
neurotoxicity, 22
occupational exposure, 41
on-line monitoring, 303
oral consumption, 114
oxidation, 110
poisoning, 286
precursors, 22, 157, 171, 191, 205, 223, 255
rat studies, 10
reactivity, 1
recovery, 296
reduction, 157, 387, 396, 457
research needs, 16, 75, 167
risks, 1, 21, 63, 71, 77
safety, 135
sample preparation, 275, 345
sampling data, 69
sources, 90
stability in food, 160
suppression, 394
toxicity, 64, 77
variations in foods, 69
Acrylic acid, 161, 172, 229, 435
Adducts
albumin, 118
background, 84
Administered dose, 118
Adrenal gland damage, 54
Alanine, 231
Alkaline comet assay, 53
Alkylation, 30
Amadori products, 161, 179, 195
Amino acids, 191, 221, 231, 238, 257, 323, 373, 391, 409, 421
1-Aminobenzotriazole, 58
Amino-carbonyl reaction, 180
3-Aminoproionamide, 184, 207, 236
Ammonia, 161, 231, 435

Ammonium-based raising agents, 421
Ammonium bicarbonate, 433
Analysis
of acrylamide, *see Acrylamide analysis*
of amino acids, 257, 377
of sugars, 258, 377
Aneuploidy, 59
Antiallergenic Maillard products, 141
Antibiotic Maillard products, 139
Antimutagenic Maillard products, 139
Antioxidants, 137, 325
arginine-fructose, 138
bread, 138
garlic, 138
lysine-fructose, 138
lysine-glucose, 138
pyrrole derivatives, 138
APCI Mass Spectrometry, 303
Arabinose, 211
Ascorbyl palmitate, 321
Asparagine, 40, 158, 171, 179, 194, 231, 240, 262, 312, 323, 373, 391, 410, 435
Asparagine-triolein, 227
Aspartic acid, 197, 435
Ataxia, 23
Azomethine ylide, 180
β-Alanine acrylamide precursor, 191, 194

Bakery wares, 164, 417
Baking agents, 433
Beef fat, 227
Benzaldehyde, 179
Benzo[a]pyrenes, 74
Benzoyl peroxide, 321
Benzyl isocyanate, 128
Beta-elimination, 161

SUBJECT INDEX

Beverage consumption, 68
Bioavailability, 64
Biomarkers, 77, 89, 91
Biotransformation, 127
Biscuits, 41, 420
Blanching, 350, 395
Bladder cancer, 39
Blood
    acrylamide levels, 77, 113, 127
    arterial, 118
    cells, 100, 128
    detoxification, 127
    glycidamide levels, 127
    pools, 103
    proteins, 80, 100
    venous, 118
Brain synaptosomes, 29
Breads, 72, 164, 455
Breakfast cereals, 41, 165, 408
Browning, 436, 453
Browning controlled, 137
Browning inhibition, 148
Browning kinetics, 236
Browning products, 135, 148
Butanal, 177
Butane-2,3-dione, 177

Caco-2-cells, 138
Calcium ions, 335, 351, 398
Cake, 419
Cancer risks, 15, 42, 66
Canola oil, 227
Carbohydrate-rich foods, 11
Carbohydrate-rich foods, 171
Carbohydrates, 73
Carcinogenesis, 39, 49
Carcinogenic response, 120
Carcinogens in food, 3, 74
Carnosine acrylamide precursor, 200
Carnosine in meat, 200

Cattle, 9, 285
Cell transformations, 52, 56
Cellulosoe, 211
Central nervous system, 25
Cereals, 72, 164, 264
Chinese factory, 5
Chitosan, 395
1-Chloro-2,4-dinitrobenzene, 129
Chocolate, 279
Chromosomal aberrations, 59, 78
Chronic exposure, 94
Cigarette smoking, 43
Circulating concentrations, 121
Citric acid, 367, 391
Clastogenicity, 59, 78
Cocoa, 279
Cod liver oil, 227
Coffee, 41, 72, 279, 408
Coffee drinker, 106
Cold storage, 379, 409, 457
Color, 312, 363, 433, 453
Concentration curve, 119
Consumer-prepared foods, 447
Consumption of acrylamide, 22
Cookies, 72, 408
Cooking oil, 347, 450
Corn chips, 41
Corn oil, 227
Corn starch, 339
Crackers, 72, 408
Creatine in meat, 200
Crispbreads, 41, 279, 297
Cystamine, 217
Cysteine adducts, 30, 110
Cysteine as acrylamide precursor, 201, 231
Cysteine/glucose, 194
Cytochrome P-450 2E1, 58, 128
Cytokinesis micronucleus assay, 83

Decarboxylations, 180, 199

Decreased nerve transmission, 28
Deep frying, 331, 454
Deterministic method, 65
Detoxifying mechanism, 78, 127
Diacetyl, 182
Dicarbonyl compounds, 172, 236
Dietary calories, 67
Diets, 65
Diol epoxide, 128
Distal axonopathy, 25
DNA adducts, 2, 51
DNA damage, 22, 53, 81
DNA synthesis, 52, 55, 120
Dominant lethal mutations, 22
Dopamine receptor affinity, 60
Dose-rate models, 24
Double quantum transfer spectra, 114

Edman protein degradation, 3, 90
Edman fluorinated reagent, 102
Electrophiles, 1, 16, 30
β-Elimination, 162, 236
β-Enaminone, 176
Energy, 73
Enzymes, 127, 374
    asparagine synthetase, 374
    aspartate aminotransferase, 374
    epoxide hydrolase, 128
    glutamate synthase, 374
    glutamine synthetase, 374
    glutamylcysteine synthetase, 56, 123
    glutathione transferase, 128
    pectinmethylesterase, 396
    polymorphic, 128
Epidemiology, 15, 39, 41, 44
Erythrose, 211
Ethacrynic acid, 129
Ethylene, 5
Ethylene oxide 5, 129

Exposure to acrylamide, 9, 63, 94
    low level, 110
    multi-dose, 121
    multi-route, 121
    routes, 109
    workplace, 109

Fat, 73, 361
Ferric ions, 335
Fiber, 73
Fiberglass model systems, 331
Flour, 164
Forward gene mutation, 79
Food, 64
    frequency questionnaire, 43
    processing, 90
Formation of acrylamide, *see* Acrylamide formation
French fries, 41, 72, 163, 273, 358, 450
Fried feed, 11, 286
Fried potatoes, 41
Fried potato products, 334, 373, 395, 452
Fructose, 175, 211, 244, 265, 324, 379, 410, 435

Gait abnormalities, 23
Garlic antioxidants, 138
Gas chromatography, 229
Gel electrophoresis, 81
Genotoxic Maillard products, 142
Genotoxicity, 4, 77
Ginger, 421
Gingerbread, 165, 431, 457
Globin isolation, 100
Glucose, 177, 211, 244, 265, 330, 324, 379, 410, 435
Glutamine, 231, 323, 373
Glutathione, 56, 60, 112
    acrylamide adducts, 112
    detoxification, 127

glycidamide adducts, 112
  transferase, 127
Glutathione, 56, 60, 128
Glycidamide, 1, 15, 24, 51
  analysis, 97
  DNA adducts, 79
  formation, 111
  hemoglobin adducts, 78
  metabolism, 122
  reactions, 79
  synthesis, 129
  synthesis of adducts, 130
Glycine, 231, 323
Glycoconjugates, 172, 178
Glyoxal, 177
Glyoxylic acid, 177
Grouting agent, 7, 285
Gyrometrin, 74

Hallandsås tunnel, 286
Hemoglobin adducts, 5, 42, 90, 98, 119
  analysis, 90, 97, 99
  and acrylamide intake, 14
  background levels, 9, 94
  cysteine, 3, 99
  detection limit, 92
  Edman degradation, 99
  in blood, 130
  histidine, 3
  N-terminal valine, 3, 99
  species differences, 9
Herb oils, 390
Hindlimb foot splay, 23
Histidine, 231
HPLC, 101, 208, 217, 221
Home-cooked foods, 447
Human,,
  blood, 77, 79
  disease, 136
  enzymes, 127
  intoxication, 23
  pharmacokinetic models, 117, 121
  studies, 91, 129
Humidity, 308
Hydroxyacetone, 177, 212
2-Hydroxy-1-butanal, 173
Hydroxyketone, 173
3-Hydroxypropionamide, 175
Hydrogen peroxide, 321

Inhibition
  browning, 148
  neurotransmission, 22
Ionizing radiation, 3
Intakes of acrylamide, 68, 91
Internal dose, 119
Intestinal flora, 5
Ionic complexes, 338, 397

Kinetic model, 235, 256

Lactamide, 202
Lactic acid, 202, 367
Lactose, 211
Lard, 227
LC/MS/MS methods, 101, 159, 174
Linoleic acid, 228
Lipid oxidation, 334
Lipids, 227
Liver effects, 55
Lung fibroblasts, 79
Lowest observed adverse levels, 27
Lymphocytes, 80
Lysine, 231, 323, 391

Macronutrients, 72
Maillard products,
  acrylamide formation, 155, 167, 171
  allergic, 141

antiallergic, 141
antibiotic, 139
antimutagenic, 139
antioxidative, 137
biological effects, 135
carcinogenic, 146
casein-ketose, 144
chromosome-damaging, 145
cytotoxic, 145
DNA-damaging, 144
from vitamins, 146
honey-lysine, 138
β-lactoglobulin, 138
lysine-fructose, 144
lysine-glucose, 144
lysine-xylose, 144
metallo-complexes, 148
pyrazine radicals, 144
structures, 140, 173
synthesis, 173
Maillard reaction, 22, 136, 235, 313, 358, 372
Maltose, 258
Mass spectra, 200, 219
Mass spectrometry, 101
Mathematical model, 352, 415
Mechanism of nerve damage, 29
Mechanism of acrylamide formation, 161, 171, 179, 213, 216, 217, 230, 236
Membrane turnover, 22
Mercapturic acid, 113
Metabolism, 94, 98
   in humans, 109
   in rodents, 109
   mechanisms, 117, 123
   pharmacokinetics, 117
   research needs, 123
Metabolites, 111
   analysis, 115
   in humans, 113
   in rodents, 111

   species differences, 115
   urinary, 112
Methional, 229
Methionine, 231
Methylgloxal, 177
Michael reaction, 239
Micronuclei, 81
Micronutrients, 72
Mitigation of acrylamide formation, 167
Mitosis, 83
Model systems, 272, 305, 331, 347, 361, 435
Model studies, 207
Modelling acrylamide formation, 235, 352, 415
Moisture, 186, 211, 260
Monte Carlo modeling, 66
Morphological effects, 57
Mycotoxins, 78
Mutagenicity, 3, 77

N-acetylcysteine, 56, 401
N-acetyl-S-3-(3-amino-3-oxopropyl)cysteine, 110
N-β-alanyl-L-histidine, 192
N-ethylmaleimide, 30
N-(D-glucos-1-yl)-L-asparaginate, 178, 206
N-glycosides, 172
N-glycosylasparagine, 179
N-methylacrylamide, 191
N-methylolacrylamide, 6
N-nitrosoamines, 78
N, N-dimethylacrylamide, 201, 209
Nematode lifespan, 136
Neoplasia, 51
Ninhydrin reaction, 137
NMR spectroscopy, 110
   Maillard compounds, 173
   two-dimensional, 111

Nerve terminals, 22
Neurological deficit, 22
Neuropathy, 25
Neurotoxicity, 9, 21, 23
   animal models, 21
   cumulative, 23
   mechanism, 30, 79
   symptoms, 22
Nitrogen fertilization, 375,
Non-coffee drinker, 106
Non-linear threshold risk model, 119
Non-smoker, 106
No-observed-adverse effect, 6, 27
Noodles, 408
Nucleophilic reactivity, 30
Nucleotide polymorphism, 128
Nutrition, 67, 71
Nutritional, 84

Occupational exposure, 41, 83, 90
Oleic acid, 228
Olive oil, 227
Oxidative lipid degradation, 172
Oxazolidine-5-one, 182, 192
Oxidants, 325
Oxidative stress, 59, 6
Oxo-compounds, 212
2-Oxopropanal, 212
3-Oxopropanal, 177
3-Oxopropionamide, 176, 184
3-Oxopropionic acid, 212

PBPK model, 117
Peanut butter, 279
Pectin, 396
Peripheral nervous system, 7, 25
pH and acrylamide formation, 187, 212, 324, 347, 366
Pharmaco-kinetics, 9, 117
Phytate, 336, 350, 395
Phytohemagglutinin, 83

Pilot study, 89
Poisoned cattle, 7
Polyacrylamide gels, 6, 50
Potato chips (crisps), 41, 72, 89, 93, 279, 297
Potato products, 11, 162, 237, 256, 373, 392, 395, 452
Potato tubers, 374, 408
Potatoes, 307, 320, 332, 390
Preservatives, 349
Presynaptic effect, 28
Probabilistic approach, 66
Probable human carcinogen, 39, 50
Protein, 322
Protein adducts, 119
   accumulation, 73, 98
   analysis, 98
   and cell lifespan, 98
   formation, 98
   metabolism, 119
   steady-state, 98
   turnover rates, 98
Py-GC/MS, 192
Pyrolysis, 173, 199
Pyruvic acid, 194, 202, 212

Quality, 293, 361

Rat studies, 52
Raw material variability, 164
Reducing sugar, 255, 299, 353, 372, 436, 457
Renal cancer, 43
Research needs, 46
Reproductive toxicity, 22
Restaurant-prepared food, 63
Ribose, 211
Rice crackers, 408
Risk
   assessment, 3, 118
   cancer, 42

communication, 15
management, 15
model, 119
neurotoxicity, 23
reduction, 67
zones, 8
Roasted barley, 283, 408
Rodents, 6
Rye products, 237, 256, 307
Rusk, 210, 220

S-carboxyethylcysteine, 110
S-2-carboxy-2-hydoxyethylcysteine, 110
Safe levels, 67
Safety/uncertainty factors, 119
Sardine oil, 227
Schiff base, 161, 172
Serine as acrylamide precursor, 201
Simulation modeling, 124
Sister chromatid exchange, 59, 78
Skeletal muscle weakness, 23
Smoking exposure, 6, 43, 90, 106
Snack foods, 72
Sodium ascorbate, 321
Sodium chloride, 365, 402
Solid phase extraction, 101, 272
Soybean oil, 227
Species differences, 115
Spinal cord injury, 26
Squalene, 227
Stability of acrylamide in food, 160
Starch, 211
Steady-state conditions, 94
Stearic acid, 228
Strecker aldehyde, 175, 236
Strecker degradation, 172, 236
Styrene, 179
Sulfhydryl adducts, 30, 123
Sucrose, 211, 258, 377

Synaptic dysfunction, 26
Syrian hamster embryos, 59

Tautomerization, 177
Tea, 408
Toasted bread, 455
Temperature and acrylamide formation, 187, 212, 214
Terminalopathy, 26
Testicular effects, 55
Thioguanine, 80
Thiol-directed systems, 30
Throughput method, 106
Thyroid gland damage, 54
Thyroid hormone regulation, 5
Time-course of acrylamide formation, 215
Tissue dose, 118
Tissue sulfhydryls, 123
Tobacco smoke, 10, 94
Toxicity, 64, 78
Toxifying mechanisms, 78
Triolein, 228
Tryptophan-P-1, 74
Tunnel construction, 6

Valine adducts, 84
Vinyologous compounds, 183
Vitamin browning, 146
Vitamins, 73

Water and acrylamide formation, 185, 211
WHO water guideline, 8
Wheat products, 237, 256, 307
Yeast, 401
Ylides, 196